GELS Handbook

Fundamentals, Properties and Applications

GELS Handbook

Fundamentals, Properties and Applications

Volume 3: Application of Hydrogels in Drug Delivery and Biosensing

Lifeng Kang and **Sheereen Majd** *Volume Editors*

Editors

Utkan Demirci
Stanford

Ali Khademhosseini
Harvard

World Scientific

NEW JERSEY • LONDON • SINGAPORE • BEIJING • SHANGHAI • HONG KONG • TAIPEI • CHENNAI • TOKYO

Published by

World Scientific Publishing Co. Pte. Ltd.
5 Toh Tuck Link, Singapore 596224
USA office: 27 Warren Street, Suite 401-402, Hackensack, NJ 07601
UK office: 57 Shelton Street, Covent Garden, London WC2H 9HE

British Library Cataloguing-in-Publication Data
A catalogue record for this book is available from the British Library.

GELS HANDBOOK
Fundamentals, Properties and Applications
(In 3 Volumes)

Volume 1: Fundamentals of Hydrogels
Volume 2: Applications of Hydrogels in Regenerative Medicine
Volume 3: Applications of Hydrogels in Drug Delivery and Biosensing

ISBN 978-981-4656-10-8 (Set)
ISBN 978-981-4656-13-9 (Vol 1)
ISBN 978-981-4656-14-6 (Vol 2)
ISBN 978-981-4656-15-3 (Vol 3)

In-house Editors: Rhaimie Wahap/Dipasri Sardar

Typeset by Stallion Press
Email: enquire@stallionpress.com

Printed in Singapore

Contents

Chapter 1

Modeling Drug Release from Synthetic Hydrogels

Chien-Chi Lin* and Han Shih
Department of Biomedical Engineering,
Indiana University-Purdue University Indianapolis
Indianapolis, Indiana, IN 46202, USA
**lincc@iupui.edu*

Hydrogels are hydrophilic but insoluble polymer networks capable of imbibing large quantity of water (typically above 90% of the mass). The high water content of hydrogels renders this class of biomaterial particularly suited for delivering therapeutically relevant biomolecules, such as peptides and proteins. Depending on the macromer compositions and polymerization mechanisms, hydrogels can be categorized into chain-growth, step-growth, and mixed-mode networks. The delivery of drug payload from a highly permeable hydrogel is usually related to the network structure of the gel, diffusivity of the drug, the amount of drug in the device available for delivery, and potential drug–polymer interactions. The amount of releasable drug can be controlled by methods such as mesh size reduction, affinity sequestration, and reversible reactions, etc. In addition to experimental work, no drug delivery strategy is complete without mathematical model development. Accurate mathematical models not only reduce the labor intensive and time consuming experimental work, but also validate release mechanisms. With prior knowledge regarding hydrogel network structure and compositions, characteristics of the drugs, and potential interactions between drug and polymer, one can predict the rate and duration of the drug release profiles. Understanding the basic mechanisms by which drug releases from hydrogels will facilitate the design of a drug carrier that can deliver relevant dosage for desired applications. This chapter aims at describing the fundamental and new development of drug delivery mechanisms with a focus on using synthetic hydrogel matrices as the delivery carriers.

1. Introduction

1.1. *Synthetic hydrogels for drug delivery applications*

Controllable drug delivery has evolved from simple disease treatment to more recently in promoting tissue regeneration *in vitro* and *in vivo*.[1–5] The efficacy of the released drugs depends on many variables, including the amount of payloads, structure and stability of the drugs, mechanisms of release, and physicochemical properties of the matrices.[1,2] One must thoroughly consider every aspect related to the availability and stability of the drugs to be delivered in order to achieve desired release rate for reaching therapeutic satisfaction. Added to this complexity is the increasingly diverse body of drug molecules identified over the years. These drugs range from small synthetic molecules to peptides, proteins, and nucleic acids. Conventional drug carriers such as solid tablets, capsules, and micro/nano-scale particles are great candidates for delivering small synthetic drugs. These 'dry' carriers, however, are less ideal for delivering sensitive biological molecules such as peptides and proteins because the manufacturing processes often involve using organic solvents that may be too harsh for bio-macromolecules. The delivery of these sensitive biomolecules often rely on matrices capable of imbibing large amount of water because an aqueous environment can facilitate the preservation of structural and functional integrity of bio-macromolecules. One example carrier for delivering biomolecules is polymeric hydrogel.[1,6] Hydrogels are prepared from macromers soluble in aqueous solutions. Through an appropriate chemical or physical crosslinking process, these initially water-soluble polymer chains become an insoluble yet hydrophilic network that imbibes a large quantity (~90–99% in mass) of water. Hydrogels have been used extensively in a variety of drug delivery applications due to their highly tunable material properties. Macromers derived from natural or synthetic origins can be used to fabricate hydrogels. From a material standpoint, however, synthetic macromers offer many more advantages and will be the focus of this chapter. Synthetic polymers such as poly (hydroxyl ethylmethacrylate) (PHEMA),[7] poly(vinyl alcohol) (PVA),[8] poly(acrylic acid) (PAA),[9] poly (ethyleneglycol)(PEG),[10,11] poly(N-isopropylacrylamide) (PNIPAAM), and poly (propylene fumarate-co-ethylene glycol) (PPF-co-EG)[12] are some of the most commonly used synthetic materials for preparing hydrogels.

While hydrophilic macromers/polymers are increasingly developed for preparing hydrogels,[13] the methods by which these polymers crosslink into a hydrogel network fall into two categories: chemical reaction or physical interactions.[14] Nucleophilic reactions, redox- or radical-mediated polymerization, and orthogonal click reactions are considered chemical methods in which the bonds formed following crosslinking reactions are typically covalent and irreversible. On the other hand, hydrogelation or "sol-gel transition" can be induced by changing physical parameters in the pre-polymer solutions, such as pH, temperature, ionic strength, and hydrophobic interactions. Regardless of the method of gelation, the

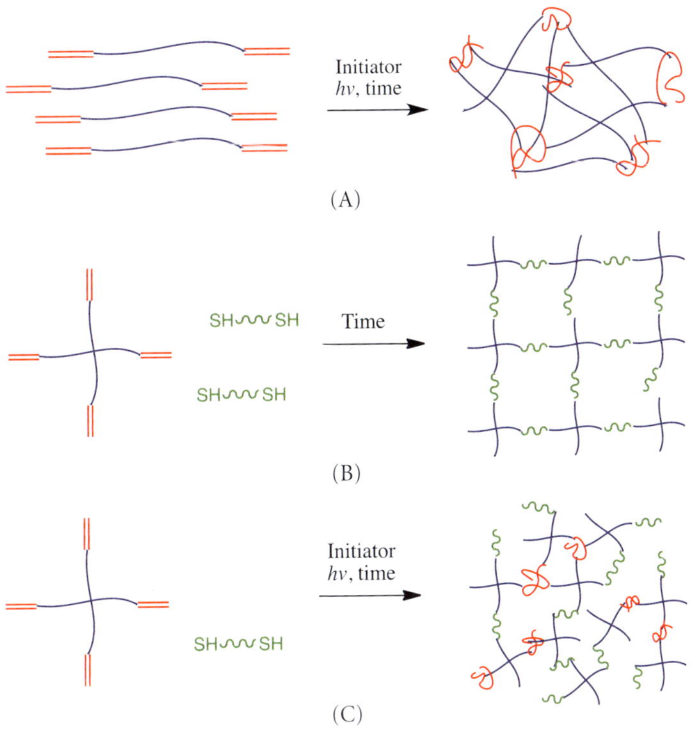

Fig. 1. Schematics of synthetic hydrogel network structures: (A) Chain-growth network containing heterogeneous crosslinks; (B) Step-growth network containing orthogonal crosslinks (e.g., thiolene click reaction without homopolymerization); (C) Mixed-mode network containing both orthogonal and heterogeneous crosslinks (e.g., thiol-vinyl crosslinking with homopolymerization).

network structures of all chemically crosslinked synthetic hydrogels belong to one of the following three groups (Fig. 1):

1. Chain-growth network resulted from random homopolymerization of macromers containing the same crosslinkable moiety (Fig. 1(A)).
2. Step-growth network formed from mutually reactive but different macromers/polymers (Fig. 1(B)).
3. Mixed-mode step-and-chain-growth network containing features from the above two network structures (Fig. 1 (C)).

1.2. *Mesh size of hydrogels*

Regardless of the network structures, drug molecules within a hydrogel must 'migrate' through the crosslinked polymer chains through a Brownian motion prior to their release into the surrounding medium. This process is referred to as diffusion, which is a well-studied phenomenon critical in many physical processes.

The diffusion of drugs from a hydrogel is highly affected by the degree of polymer network crosslinking, which affects hydrogel 'mesh size' (ξ).[1,15] Mesh size is a critical parameter in determining the permeability of the hydrogel. When the hydrodynamic radius (R_H) of the drug molecule is much smaller than the mesh size of a hydrogel, one can assume the diffusion of this drug from hydrogel will not be significantly retarded if there is no additional interaction existing between the drug and the polymer network. On the other hand, if the mesh size of a hydrogel network is on the same order of magnitude to R_H of the drug molecule, diffusion of the drug molecule will be significantly affected. If the mesh size is close to or smaller than the size of the molecule, the drug cannot be liberated to the medium unless significant network degradation has occurred.

Mesh size of a chemically crosslinked hydrogel can be described by the classic Flory-Rehner theory.[16] First step towards this is the understanding of the compositions and reactivity of the polymers used, as well as the mechanism of crosslinking in order to obtain accurate parameters necessary for acquiring the mesh size of the hydrogel. In one example, multi-arm PEG-acrylate (e.g., PEG-tetra-acrylate) and dithiol-containing linear linker (e.g., PEG-dithiol) can be used to form chemically crosslinked hydrogels via a nucleophilic reaction mechanism (i.e., Michael-type addition reaction) that yields a purely step-growth network structure. These two macromers, however, can also form a crosslinked hydrogels via a radical-mediated mixed-mode step-and-chain growth polymerization mechanism because acrylates can undergo homopolymerization under certain initiation conditions. Albeit formed from the same polymers, these two hydrogel network structures are drastically different (Figs. 1(B) and 1(C)).

To estimate the mesh size of a hydrogel, the first parameter to acquire is the polymer volume fraction in the swollen state ($v_{2,s}$).[1] $v_{2,s}$ is defined as a ratio of the polymer volume (V_p) to the swollen hydrogel volume (V_g). It is also a reciprocal of the volumetric swollen ratio (Q), which can be related to the densities of the solvent (ρ_1) and polymer (ρ_2) and the mass swollen ratio of the hydrogel (Q_m). Equation (1) correlates these parameters:

$$v_{2,S} = \frac{V_p}{V_g} = \frac{1}{Q} = \frac{\frac{1}{\rho_2}}{\frac{Q_m}{\rho_1} + \frac{1}{\rho_2}}. \tag{1}$$

In most cases, water is the solvent for a hydrogel. In addition, Q_m is a ratio of swollen hydrogel weight to dried weight and is more readily measurable compared to Q. By measuring Q_m for a given polymer formulation and at a given time, one can easily calculate $v_{2,s}$.

The number average molecular weight between crosslinks ($\overline{M_c}$) describes the average molecular weight between two adjacent crosslinks with each of these crosslinks joined by at least two polymer chains. $\overline{M_c}$ in a chain-growth di-vinyl crosslinked network is written as:

$$\frac{1}{\overline{M_c}} = \frac{2}{\overline{M_n}} - \frac{\left(\frac{\bar{v}}{V_1}\right)\left[\ln(1-v_{2,s}) + v_{2,s} + \chi_{12}\, v_{2,s}^{\ 2}\right]}{v_{2,s}^{\ 1/3} - \frac{v_{2,s}}{2}}, \tag{2}$$

Here, $\bar{v}$ is the specific volume of the polymer, V_1 is the molar volume of the solvent (18 cm^3/mole for water), and χ_{12} is the Flory–Huggins interaction parameter for polymer–water interaction.

If a hydrogel is prepared through a mechanism involving homopolymerization where the constituent polymer chains react with each other (i.e., chain-growth and mixed-mode polymerization), Eq. (2) can be used to estimate the average molecular weight of a linear polymer chain connected by two adjacent crosslinks. Each of these crosslinks may contain an undefined number of polymer chains. For a hydrogel crosslinked by a step-growth reaction where all polymer chains are linked by mutually reactive moieties, the calculation of $\overline{M_c}$ can be greatly simplified if sufficient prior knowledge regarding the structural information of this step-growth network is known.[17] A classical step-growth network is prepared by reacting two multifunctional and mutually reactive polymers (A and B) with molecular weight MW_A and MW_B, and functionality f_A and f_B. Note that to produce a stable step-growth network, the average functionality of the two polymers should be greater than two. After complete reaction, any linear polymer within the crosslinked network contains part of polymers A and B. If polymer A has a functionality equal to or greater than 3 (i.e., $f_B \geq 3$) and polymer B is linear (i.e., $f_B = 2$), the average molecular weight between crosslinks is expressed as:[17]

$$\overline{M_c} = 2\left(\frac{MW_A}{f_A} + \frac{MW_B}{f_B}\right). \tag{3}$$

If both polymer A and polymer B have functionalities that are both equal to and greater than 3 (i.e., $f_A, f_B \geq 3$), $\overline{M_c}$ of this network is simply the summation of the molecular weights of each 'arm' from the two polymers:

$$\overline{M_c} = \left(\frac{MW_A}{f_A} + \frac{MW_B}{f_B}\right). \tag{4}$$

Note that Eqs. (3) and (4) are valid only for step-growth network hydrogels under the assumption that polymer A and polymer B do not undergo homopolymerization.

In the case of a highly swollen gel (i.e., $Q > 10$), Equation (2) can be simplified to correlate gel swelling with the average molecular weight between crosslinks ($\overline{M_c}$):

$$Q = \left[\frac{\bar{v}\left(\frac{1}{2} - 2\chi_{12}\right)\overline{M_c}}{V_1} \right]^{3/5}. \tag{5}$$

$\overline{M_c}$ derived from Equation (2) is crucial in obtaining another critical parameter related to hydrogel mesh size: The root-mean-squared end-to-end distance of network chains between two crosslinks in the neutral state or $(\overline{r_0^2})^{1/2}$, which can be defined as:

$$(\overline{r_0^2})^{1/2} = l(C_n N)^{1/2} = l\left(C_n \frac{n\overline{M_c}}{M_r} \right)^{1/2}, \tag{6}$$

Here, C_n is the Flory characteristic ratio, l is the average bond length along the polymer backbone, N is the number of chemical bonds between two crosslinks, and n is the number of chemical bonds within a repeating unit (e.g., n equals to 2 for PVA and 3 for PEG). Finally, M_r is the molecular weight of the repeating units of the polymer chain.

After obtaining $v_{2,s}$ and $(\overline{r_0^2})^{1/2}$, hydrogel mesh size (ξ) can be readily determined as:[18]

$$\xi = v_{2,s}^{-1/3} (\overline{r_0^2})^{1/2} = Q^{1/3} (\overline{r_0^2})^{1/2}. \tag{7}$$

A typical synthetic hydrogel in its swollen state has a mesh size of a few tens of nanometer. This size range is much larger than most small-molecule drugs, thus it does not significantly hinder small molecular weight drug diffusion. However, the sizes of many therapeutically relevant biomolecules (e.g., proteins, growth factors, antibodies, etc.) are in the range of a few nanometers, making the use of hydrogels ideal (in addition to the hydrophilic nature of the network) for delivering these biomolecules. It is worth noting that mesh size also correlates to other physical properties of synthetic hydrogels, including mechanical strength and degradability. These factors can and will affect the rate of drug delivery from synthetic hydrogels.

2. Considerations of Using Hydrogels for Drug Delivery

2.1. *Design criteria for hydrogel-based drug delivery devices*

Material selection and network fabrication method will influence the rate and mode of drug release from the hydrogel matrices. Several design criteria are crucial

Table 1. Design criteria for hydrogel in drug delivery applications. Adapted from.[5]

Design criteria	Design variables
Transport properties — Molecule diffusion	• Molecular weight and size of protein • Molecular weight of polymer • Crosslinking density • Polymer–protein interactions • Hydrogel degradation rate • Additional functionalities
Physical properties — Gelling mechanisms/conditions — Structural properties — Biodegradability — Stimuli-responsiveness	• Polymer/crosslinker/initiator concentrations • Temperature, pH, ionic strength • Molecular weight of polymer • Mechanical strength • Concentration of degradable groups • Concentration of responsive groups
Biological properties — Biocompatibility	• Cytotoxicity of the hydrogel • Leachable molecules • Capsule formation

for drug delivery formulations and have to be evaluated prior to hydrogel fabrication and drug loading. These criteria are also important in the mathematical modeling for predicting drug release.[5] Table 1 lists these important criteria and variables for designing hydrogel-based drug carriers. Within the realm of transport properties, the most notable variable is the drug diffusion coefficient, which is affected by the molecular size of the drug and characteristics of the polymer network, such as the crosslinking density of hydrogel. If special functionalities, such as ionic groups, are introduced into the hydrogel networks, interactions between these functionalities and encapsulated drugs will also affect drug diffusivity.

The understanding of transport and physical properties of the hydrogel is especially crucial in modeling molecule release. Even if a hydrogel delivery formulation is designed with the appropriate physical and transport properties, it may still fail to perform its therapeutic role when implanted *in vivo* due to a localized inflammatory response. For example, the formation of a fibrous capsule surrounding the delivery device creates additional diffusion barriers that may reduce drug release rates or even prevent drug diffusion from the device. Furthermore, increased proteolytic activity in the implantation site may increase the rates of matrix and drug degradation. Proper material selection, fabrication process, and surface texture of the device are always critical in designing biocompatible hydrogel formulations for controlled release applications *in vivo*.

2.2. Methods of drug loading in hydrogel

2.2.1. *In situ loading — Direct drug entrapment*

In situ loading often refers to direct loading of drugs in hydrogel matrices during network crosslinking and polymerization. The simplest way of *in situ* loading is direct entrapment of drug in the crosslinked network. The drug to be loaded is simply mixed with the other precursor components (i.e., polymer, crosslinkers, initiators, etc.) prior to network crosslinking. After gelation is complete, the drug molecules are physically entrapped within the hydrogel network. This process is possible for many biologically relevant molecules because most of the hydrogels for controlled release and drug delivery applications are prepared in aqueous solutions. This method is direct and allows one to easily and precisely estimate the amount of payload in a hydrogel. *In situ* loading is a preferential method of drug loading when a large amount of drug is needed or when multiple agents are to be packaged within the same hydrogel matrix. Although entrapping drug during network crosslinking is easy and efficient, care must be taken if certain adverse reactions exist between reactive polymers and biological molecules.[19–21] Higher degree of polymer–drug interactions/reactions often reduces the amount of drug available for delivery.

2.2.2. *In situ loading — Particulate systems*

Another type of *in situ* loading is the encapsulation of drug-containing micro- or nano-particles. In this process, drugs are first emulsified or partitioned into polymeric particles, followed by entrapment of these drug-loaded polymeric particles or gel spheres into a bulk hydrogel matrix. Drug will only become available for delivery after degradation or erosion of the particles has occurred and the rate of drug delivery may be retarded by slow degradation of the particles. Because of the multistep processes involved in the delivery of drugs from these multi-phase systems, accurate modeling of drug release from these intelligently designed delivery devices is complicated. However, this approach allows for local and more sustained presentation of the drug to cells simultaneously encapsulated within the hydrogels. This approach also allows for the delivery of multiple drugs at various rates.[22]

2.2.3. *In situ loading — Drug immobilization*

One benefit of using chemically crosslinked hydrogels for drug delivery is that certain drugs can be immobilized within the hydrogel during network crosslinking.[23] The drug of interest can be directly linked to the network or conjugated to the network through a cleavable linker. The later approach is more attractive because the linker can be designed to become cleavable by biologically relevant stimuli (e.g., water, proteases, light). The immobilized drug will be liberated only in the presence of appropriate stimuli.[23–26] This approach allows the experimenter to create complex drug delivery profiles, such as extending the duration of the drug delivery or releasing multiple drugs at different rates due to the difference in linker cleavage rates. To immobilize drug in a hydrogel through a cleavable linker, the drug (synthetic molecules or proteins) must be genetically or chemically modified.[27–30] As a result, the released drug molecules will contain a short linker segment following cleavage. Therefore, one must first evaluate the reaction route for preparing the chemically modified drug in order not to disturb the biological function of the released drug.

2.2.4. *Post-loading*

Post-loading is the loading of drug after the completion of hydrogel network crosslinking. This process is typically achieved through incubating hydrogel in a buffer solution containing high concentration of drug.[31] The drug molecules partition into the hydrogel matrices due to the concentration difference between the gel and the surrounding buffer. Post-loading is a thermodynamically driven process and it stops when the chemical potential of the system (i.e., hydrogel) is in balance with the surrounding (i.e., buffer). The efficiency of drug loading depends on many factors. For example, if the mesh size of a hydrogel is smaller than or similar to the size of the drug, very limited amount of drug can be loaded in the hydrogel. Furthermore, if certain interactions exist between the drug and the polymer network, such as electrostatic or hydrophobic interaction, the loading can be improved.[32] Altering solvent conditions (e.g., pH, ionic strength, salt concentrations, etc.) in the buffer phase may increase the efficiency of post-loading.[33] Post loading is typically more time-consuming and required the use of a bath of solution with high drug concentration. This process, however, prevents the potential complications caused by adverse chemical reactions between the drug and the polymer during *in situ* loading.

2.2.5. *Implication of drug loading method in mechanism of drug release*

The method of drug loading in a hydrogel largely dictates how this drug releases into the surrounding medium.[6] For example, if a drug is loaded into the hydrogel through physical entrapment, the mechanism by which this drug releases is diffusion-mediated, given that the drug size is smaller than the mesh size of the gel. If, however, the size of the drug is larger than the mesh size of the gel, no release is possible unless some degradation in the hydrogel network has occurred. On the other hand, if a drug is tethered in the hydrogel through chemical conjugation, its eventual release to the surrounding medium is a collective result of a series of events including the introduction of the degradation trigger (e.g., protease), the cleavage kinetics of linker, and the diffusion of the drug. Once the drug has become free and available for delivery, the mechanism governing its release is once again depends on the relative sizes of the drug and the gel mesh size.

3. Mechanisms and Modeling of Drug Release from Hydrogels

Drug release from a hydrogel involves a sequence of events including water (or molecular triggers) diffusion into the matrix, polymer swelling, polymer degradation or linker cleavage, drug dissolution and diffusion out of the matrix, and other related phenomena. If any of these processes occur significantly slower than the others, it is the rate-limiting step of the entire release sequence. Identifying the rate-limiting steps in a drug delivery device will greatly simplify the mathematical modeling of the release behavior. For a highly swollen hydrogel-based drug delivery device, diffusional transport is almost always a critical governing step. Even in the cases where other governing mechanisms are introduced (e.g., degradation, polymer swelling, stimuli-responsive, affinity sequestration, etc.), diffusional mass transport still plays a major role in determining the drug release rate. Therefore, two parameters are especially important in modeling drug release from a hydrogel: The diffusivity of a drug within a hydrogel matrix and the mesh size of the hydrogel during drug release.[18]

In general, drug delivery from a hydrogel matrix can be categorized into the following mechanisms[5]:

1. Diffusion-controlled: In this mechanism, diffusion is the rate-limiting step. Drugs are typically loaded in a highly swollen hydrogel and the release rate is governed by the diffusion of drug molecules. During drug release, hydrogel mesh size will not change significantly (i.e., no degradation). Therefore, the diffusivity of the drug in the hydrogel will remain relatively constant.[18,34–37]
2. Swelling-controlled: In swelling-controlled drug delivery, drugs are entrapped in an initially dried polymer matrix (e.g., hydroxypropyl methylcellulose, HPMC). Drugs can only be released to the medium after significant swelling of the polymer matrix has occurred. During drug release, the polymer chains undergo a glassy-rubbery transition, which increases network mesh size significantly. Drug diffusivity in the polymer matrix is usually a function of time and mesh size of the hydrogel.[38–42]
3. Chemically-controlled: Chemically-controlled drug delivery devices are increasingly being developed for biomedical applications. Depending on the

hydrogel design strategy, drugs can be released by a combination of several mechanisms. A few examples include: pendant chain cleavage, gel surface erosion, gel bulk degradation, affinity sequestration, etc. A common feature among these diverse delivery mechanisms is the involvement of chemical reactions during drug release. These chemical reactions may (e.g., bulk degradation, stimuli-responsive swelling) or may not (e.g., pendant chain cleavage, gel surface erosion, or affinity sequestration) change hydrogel mesh size.[10,25,43–47]

3.1. *Diffusion controlled delivery systems*

Diffusion-controlled release is the most predominant mechanism that describes drug release from a swollen hydrogel matrix.[48] Fick's law of diffusion with either constant or variable diffusion coefficient can be used to describe the diffusion of molecules within the hydrogel. Prior to applying Fick's law of diffusion, one must first acquire the diffusivity of the drug. Free volume, hydrodynamic or obstruction-based theories provide methods by which drug diffusivity can be determined.[36] For highly swollen hydrogels, diffusivity at a constant temperature is often assumed to be a constant and independent of solute concentration. For non-porous hydrogels and porous gels with mesh size similar to release particles, the diffusion coefficients decreased as steric hindrance occurred due to the presence of polymer chains within crosslinked network.

Fick's law of diffusion can be used to describe drug transport within highly swollen hydrogels. Diffusion-controlled mechanism can be either reservoir or matrix systems. For a reservoir system (i.e., drug depot) where the drug is surrounded by a polymeric hydrogel membrane, Fick's first law of diffusion can be used to describe drug release through the membrane.[15,37]

$$J_A = -D\frac{dC_A}{dx}, \tag{8}$$

Here, J_A is the steady-state drug transfer rate across a unit area (i.e., flux) and is related to the mass average velocity of the system. D is the drug diffusion coefficient or diffusivity (cm^2/sec). C_A is the concentration of the drug. D may or may not be a function of drug concentration. Experimentally, D can be measured by nuclear magnetic resonance and fluorescence correlation spectroscopy.[49] If a hydrogel does not swell or degrade significantly during drug release, D is often assumed to be a constant. This assumption is often reasonable and greatly simplifies mathematical modeling of drug release. If, however, the properties of the hydrogel system changes as time, especially in gels with significant degradation or phase transition, the diffusivity of the encapsulated molecules will not be a constant.

Diffusion coefficient will also be affected by changes in other environmental conditions, such as temperature and potential drug–matrix interactions. In general, drug diffusion coefficients can be described using the following general form[1]:

$$\frac{D_g}{D_0} = f(r_s, v_{2,s}, \xi), \tag{9}$$

Here, D_g and D_0 are the drug diffusion coefficients in the swollen hydrogel network and in pure solvent, respectively. r_s is the radius of the drug to be delivered. $\upsilon_{2,s}$ is the polymer volume fraction and ξ is the mesh size of the hydrogel. Equation (9) accounts for structure of the hydrogel, polymer composition, water content, and the size of the molecules. For degradable hydrogel, D_g changes as the network degrades because both gel mesh size and polymer volume fraction change over time.

For a matrix system (or monolithic system) where the drug molecules are dispersed homogeneously throughout the polymeric matrix, Fick's second law of diffusion is used for modeling:

$$\frac{dC_A}{dt} = D\frac{d^2C_A}{dx^2}. \tag{10}$$

Here, D is assumed as a constant. Equation (10) is limited to the modeling of thin planar geometry matrix (i.e., slab) where the majority of drug molecules are released from the main surface of the slab and the release through the edges is negligible. A 'perfect sink' condition is also used to simplify the mathematical modeling. For other geometries, the mathematical modeling is more complex and required special considerations.

When diffusion is concentration-dependent, the following equation can be used:

$$\frac{\partial C_A}{\partial t} = \frac{\partial}{\partial x}\left(D(C_A)\frac{\partial C_A}{\partial x^2} \right). \tag{11}$$

If D is known, Equation (11) can be solved with the consideration of matrix geometry (e.g., slab, sphere, or cylinder) and with appropriate initial and boundary conditions. For a slab geometry where the drugs are assumed to release only from the surface of the slab (i.e., one-dimensional and neglecting edge effect), an exact analytical solution to the non-steady state diffusion equation (i.e., Eq. (11)) can be obtained using separation of variable technique. Subsequently, the cumulative amount of drug released as a function of time can be expressed as:

$$\frac{M_t}{M_\infty} = 1 - \frac{8}{\pi^2}\sum_{n=0}^{\infty}\frac{\exp\left[\frac{-(2n+1)^2\pi^2}{L^2}Dt\right]}{(2n+1)^2}, \tag{12}$$

where M_t and M_∞ are the cumulative amounts of drug released at any time t and at infinite time, respectively. D is the diffusivity of the drug within the system and L is the total thickness of the slab.

Similar analytical answers for drug released from spheres (with a radius of R) and cylinders (with a radius R and height H) can be expressed as Eqs. (13) and (14):[48]

$$\frac{M_t}{M_\infty} = 1 - \frac{6}{\pi^2}\sum_{n=0}^{\infty} \frac{\exp\left[\frac{-n^2\pi^2}{R^2}Dt\right]}{\pi^2}, \quad (13)$$

$$\frac{M_t}{M_\infty} = 1 - \frac{32}{\pi^2}\sum_{n=1}^{\infty}\frac{1}{q_n^2}\exp\left(-\frac{q_n^2}{R^2}Dt\right) \cdot \sum_{p=1}^{\infty}\frac{1}{(2p+1)^2}\exp\left(-\frac{(2p+1)^2\pi^2}{H^2}Dt\right), \quad (14)$$

Equations 12–14 can be expressed using infinite series of exponential functions. However, practically it is more common to use short time and late time approximation (Table 2).

Under the condition that D is known, Equations 12–20 are commonly used to predict diffusion of molecules, from nano-scale size drugs to micro-scale proteins

Table 2. Approximation of diffusion-controlled drug release from hydrogel matrices with different geometry.[48]

Geometry	Time	Approximation	Equation
Slab	$\frac{M_t}{M_\infty} \le 0.6$	$\frac{M_t}{M_\infty} = 4\left(\frac{Dt}{\pi L^2}\right)^{\frac{1}{2}}$	(15)
	$\frac{M_t}{M_\infty} \ge 0.4$	$\frac{M_t}{M_\infty} = 1 - \frac{8}{\pi^2}\exp\left(-\frac{\pi^2 Dt}{L^2}\right)^{\frac{1}{2}}$	(16)
Sphere	$\frac{M_t}{M_\infty} < 0.4$	$\frac{M_t}{M_\infty} = 6\left(\frac{Dt}{\pi R^2}\right)^{\frac{1}{2}} - \frac{3Dt}{R^2}$	(17)
	$\frac{M_t}{M_\infty} > 0.6$	$\frac{M_t}{M_\infty} = 1 - \frac{6}{\pi^2}\exp\left(-\frac{\pi^2 Dt}{R^2}\right)$	(18)
Cylinder	$\frac{M_t}{M_\infty} \le 0.4$	$\frac{M_t}{M_\infty} = 4\left(\frac{Dt}{\pi R^2}\right)^{\frac{1}{2}} - \frac{Dt}{R^2}$	(19)
	$\frac{M_t}{M_\infty} > 0.6$	$\frac{M_t}{M_\infty} = 1 - \frac{4}{2.405^2}\exp\left(-\frac{2.405^2 Dt}{R^2}\right)$	(20)

or other macromolecules, from hydrogels.[50,51] It is worth noting that these equations only describe diffusion of molecules from hydrogels without the consideration of potential drug–matrix interactions or matrix degradation.

3.2. *Swelling controlled delivery systems*

Swelling controlled drug delivery systems are also referred to as matrix systems. In a matrix system, drug molecules are initially entrapped within the dry polymer matrix in the form of crystals or precipitates. Upon contact with body fluid, water penetrates into the dry polymer matrix and causes swelling of the polymer chains. Increased water concentration within the swollen matrix leads to a glassy-rubbery transition in the polymer matrix, as well as drug dissolution. The prediction of such mechanism involves moving boundary conditions and a water concentration-dependent drug diffusivity. Whether the drug release mechanism is diffusion-controlled or swelling-controlled depends on the relative time scales for polymer relaxation (λ) and for drug diffusion (t). The Deborah number (D_e) is defined as[39,52]:

$$D_e = \frac{\lambda}{t} = \frac{\lambda D}{\delta(t)^2}, \tag{21}$$

Here, D is the diffusivity of the drug and $\delta(t)^2$ is the time-dependent thickness of the swollen phase (i.e., $t = \delta(t)^2/D$). In diffusion-controlled delivery systems where t is the rate-limiting step (i.e., $t \gg \lambda$ or $D_e \ll 1$), the release can be described by the Fick's law of diffusion as discussed in the previous section. On the contrary, if λ is the rate-limiting step (i.e., $t \ll \lambda$ or $D_e \gg 1$), the release is swelling-controlled and the rate of drug release depends on the state of polymer chain swelling. In both cases, the diffusion of the drug molecules in the swelling polymer can be described by the Fick's law of diffusion. When t approaches λ or $D_e \sim 1$, the release becomes non-Fickian.

The first mathematical model to describe swelling-controlled drug delivery from a matrix system is the Higuchi equation based on a planar geometry. The Higuchi equation is a pseudo-steady state equation that assumes the following: 1. initial drug concentration in the matrix is much higher than its solubility; 2. the diffusion occurs in one-dimension with no edge effect; 3. constant drug diffusivity; 4. negligible matrix swelling and dissolution; and 5. a perfect sink condition. The model is given by:

$$\frac{M_t}{A} = \sqrt{D(2C_0 - C_s)C_s t} \quad C_0 > C_s, \tag{22}$$

where M_t is the amount of drug released at time t, A is the surface area, D is the drug diffusivity, C_0 and C_s are the initial drug concentration in the matrix and drug solubility, respectively. Equation (22) can also be expressed by:

$$\frac{M_t}{M_\infty} = K\sqrt{t}, \tag{23}$$

where M_∞ is the total amount of drug released from the matrix, which should be equivalent to the amount of drug loading prior to release. K is a constant taking into account the design variables of the delivery system. It is worth noting that Equation (23) has the same form as Equation (15) for diffusion-controlled release in the early time.

The prediction of drug delivery from a swollen hydrogel can be achieved by an empirical time-dependent power law function developed by Peppas *et al.*[38]:

$$\frac{M_t}{M_\infty} = k * t^n, \tag{24}$$

Here, a structural/geometric constant (k) and a release exponent (n) were used for a particular system and release mechanism.[1,38,42] Table 3 lists the n values for delivery systems with different geometries.

It is indicative when n equals unity, the drug fractional release M_t/M_∞ of a purely swelling-controlled slab-based delivery system appears to be zero-order.[42,50] On the other hand, for a diffusion-based system, power law is easily applicable but not robust enough to describe the complicated release kinetics. For example, in diffusion-controlled slab release systems where $n = 0.5$, the power law is only valid for the first 60% of the release profile (Table 2).

Empirical models are useful in providing information about drug release mechanisms and revealing the variation of release profiles due to the changes in the chemical or physical properties of the delivery system. Equation (24) can be modified to take into accounts both drug diffusion and polymer relaxation:

$$\frac{M_t}{M_\infty} = k_1 t^m + k_2 t^{2m}, \tag{25}$$

Table 3. Release exponent values (n) in the time-dependent power law function.[42]

Matrix Geometry	Diffusion-controlled delivery system	Non-Fickian transport	Swelling-controlled delivery system
Slab	0.5	$0.5 < n < 1$	1
Sphere	0.43	$0.43 < n < 0.85$	0.85
Cylinder	0.45	$0.45 < n < 0.89$	0.89

where k_1, k_2, and m are constants. $k_1 t^m$ and $k_2 t^{2m}$ represent the diffusion and polymer relaxation contribution to the release mechanism, respectively.

Although Equation (25) considers both the effects of diffusion and polymer relaxation during swelling, it does not account for "moving-boundary" conditions in which the gel phase expands as water penetrates into the matrix. In Korsmeyer and Peppas model, a dimensionless swelling interface number, S_w, was introduced to correlate the moving boundary phenomenon[39,53–55]:

$$S_w = \frac{v\delta(t)}{D}, \tag{26}$$

Here, v is the velocity of hydrogel swelling front and D is the drug diffusion coefficient in the swollen phase. For a slab system ($S_w \ll 1$), a zero-order release profile is obtained because drug diffusion is much faster than the movement of glassy-rubbery interface.

Siepmann and Peppas developed a more rigorous model for prediction drug delivery from swelling-controlled systems. This sequential-layer model accounts for drug diffusion, polymer relaxation, and dissolution.[40–42,56] Using Fick's second law of diffusion with a concentration-dependent diffusion coefficient, this model predicts drug transport in both radial and vertical directions in a cylindrical geometry[56]:

$$\frac{\partial C_k}{\partial t} = \frac{\partial}{\partial r}\left(D_k \frac{\partial C_k}{\partial r}\right) + \frac{D_k}{r}\frac{\partial C_k}{\partial r} + \frac{\partial}{\partial z}\left(D_k \frac{\partial C_k}{\partial z}\right), \tag{27}$$

where C_k and D_k are the concentration and diffusivity of the diffusible species (water and drug), respectively. Concentration-dependent diffusivities derived by a "Fujita-like" free-volume model can be expressed as:[35]

$$D_1 = D_{1eq} * \exp\left[-\beta_1\left(1 - \frac{C_1}{C_{1eq}}\right)\right], \tag{28}$$

$$D_2 = D_{2eq} * \exp\left[-\beta_2\left(1 - \frac{C_1}{C_{1eq}}\right)\right], \tag{29}$$

where β_1 and β_2 are dimensionless constants and "*eq*" represents the equilibrium drug concentration at the water-matrix interface where polymer disentanglement occurs. Although Eqs. (27)–(29) can only be solved numerically because of the concentration-dependent diffusion coefficient, the numerical solutions agree well with experimental results.

3.3. *Chemical controlled delivery systems*

The rate of drug delivery from a hydrogel matrix may be affected by physical transformation or chemical reactions occurring within the delivery matrix. These reactions may be: 1. reversible or irreversible binding between the drug and the polymer matrix;[46,47,57,58] 2. cleavage of drug-polymer linkers via hydrolytic, enzymatic, or other modes of degradation;[23,25,26,30,59] 3. degradation or erosion of the polymer network;[10,17,43] and 4. change of drug diffusivity or solubility due to polymer phase transition induced by external stimuli (e.g., lights, thermal energy, electromagnetic forces, ultrasound).[60–65] Some of these transformations or reactions will occur automatically without the intervention from the experimenters. This section concerns such delivery mechanisms. Readers are directed to Section 4 for "on-demand drug delivery system", which discusses the delivery of a drug due to user-controlled external stimuli.

3.3.1. *Bulk degradation mediated release*

Hydrogels that can be degraded in physiologically relevant conditions are increasingly being developed for tissue engineering and regenerative medicine applications.[66–68] The major rationale of developing degradable hydrogels is that the synthetic material initially served as a bioactive structural support could be replaced by matrices produced by the regenerated tissues. Hydrogels can be designed under bulk degradation and/or surface erosion. In bulk degradation, hydrogel crosslinking density decreases homogeneously and monotonically as time throughout the entire gel. This degradation mechanism occurs because the time-scale for water (for hydrolytically labile hydrogels) or enzyme (for enzymatically labile hydrogels) diffusion into the matrix is much faster than the time scale for polymer chain cleavage. In another word, water or enzyme concentration in the hydrogel matrix is assumed to be constant and the cleavage reaction of the chemical bonds is the rate-limiting step in the degradation process.

For highly swollen and bulk-degrading hydrogels, the degradation follows pseudo-first order kinetics:[43,69]

$$\frac{dM_w}{dt} = -kM_w. \tag{30}$$

Here, M is the molecular weight of the degradable polymer chains, t is degradation time, k is the kinetic rate constant of the degradation. In many occasions, M can be replaced by the concentration of labile chemical bonds or the moduli of the hydrogels. Integrating Eq. (30) gives:

$$M_{w,t} = M_{w,0} * e^{(-kt)}. \tag{31}$$

Equation (31) has been successfully used to predict the decrease in hydrogel modulus or increase swelling ratio due to hydrolytic degradation of the labile linkages presence in the hydrogel network.[17,43,70] From Eq. (9), it is clear that drug diffusivity is a function of hydrogel mesh size, which increases as time during hydrogel bulk degradation. The exponential decrease in polymer molecular weight and gel crosslinking density (or exponential increase in hydrogel swelling ratio) described by Eq. (31) suggest that drug release will be governed by both drug diffusion and network degradation. Macroscopically, bulk degradation of hydrogels can be precisely predicted with a given kinetic rate constant and a starting gel modulus (or crosslinking density). The prediction of drug released from a bulk degrading hydrogel is far more complex than predicting the rate of gel degradation. Principally, drug diffusivity and hydrogel network mesh size are correlated through the average molecular weight between crosslinks ($\overline{M_c}$). The Flory-Rehner equation for highly swollen hydrogels Eq. (5) predicts a scaling relationship between $\overline{M_c}$ and hydrogel volumetric swelling ratio (Q):

$$Q \sim (\overline{M_c})^{3/5}. \tag{32}$$

A scaling relationship also exists between the network mesh size (ξ) and $\overline{M_c}$:

$$\xi \sim \left(\overline{M_c}\right)^{7/10}. \tag{33}$$

The diffusivity of a drug in a bulk degrading hydrogels with time-dependent mesh size can be predicted by Equation (34):[18]

$$\frac{D_g}{D_0} = \left(1 - \frac{r_s}{\xi}\right) \exp\left[-Y\left(\frac{v_{2,s}}{1 - v_{2,s}}\right)\right]. \tag{34}$$

Here, Y is defined as the ratio of the critical volume required for a translational movement of the encapsulated drug molecule and the average free volume per molecule of solvent. A good approximation for Y is unity. For highly swollen hydrogels ($Q > 10$) with degradable crosslinks, the diffusivity can be expressed as:[43,69]

$$1 - \frac{D_g}{D_0} = \frac{r_s}{\xi} \cong \exp\left(-\frac{7}{5} jk'_E t\right). \tag{35}$$

Here, the combined parameter jk'_E is the pseudo-first-order reaction rate constant for a hydrolytically degradable network (j represents the number of cleavable liker in a polymer chain). Due to the occurrence of network degradation, mesh size is time-dependent. D_g increases as degradation proceeds and approaches D_0. Once the diffusivity is determined, the release profiles can be predicted using Fick's second law of diffusion. Experimental studies have verified the theoretical predictions using this scaling law.[43]

In addition to the scaling model, microscopic level Monte Carlo simulations have been used to predict drug delivery from degradable network.[71] Monte Carlo simulation is appropriate for predicting network property changes. However, it does not provide information regarding drug release. Therefore, Fick's law of diffusion must be incorporated to link the network degradation to drug diffusion.

3.3.2. *Surface erosion-mediated release*

Controlled drug delivery can be achieved by entrapping drug in an erodible polymer. The major benefit of using erosion-controlled drug delivery is that both hydrophilic and hydrophobic drugs can be delivered from the polymer. Furthermore, since the polymer erodes overtime, no secondary surgery is required to remove the drug carrier upon the depletion of the drug. Surface eroding matrices are also advantageous because the delivery profile is more likely to be zero-order (i.e., constant delivery rate). Surface erosion typically occurs on polymers that are hydrophobic but contains hydrolytically labile chemical bonds. A classic example is polyanhydride.[72–76] Water penetration in these hydrophobic polymers is usually slow compared to the rate of hydrolytic bond cleavage. This results in 'layer-by-layer' erosion. In these polymers, the mechanical properties of the bulk material will not change until significant portion of the polymer is eroded away.

Due to the inherently high water content, hydrogels typically do not degrade in surface erosion mechanism. However, if enzymatic sensitive linkages are incorporated in the hydrogel network structure, surface erosion may also occur, given that the time scale for enzymatic cleavage is faster than its penetration into the gel. For example, Rice *et al.* developed enzymatically sensitive poly (ethylene glycol)-polycaprolactone (PCL-b–PEG–b-PLC) block copolymer hydrogels that were susceptible for lipase-mediated cleavage.[77] Surface erosion also occurs in step growth hydrogels formed by four-arm PEG-norbornene and bis-cysteine peptide linker sensitive to enzyme-mediated cleavage.[78,79] Surface-erosion in hydrogels can only occur if enzyme diffusion into hydrogels is limited or when the cleavage of enzyme-sensitive linkers is faster than the penetration of enzyme in hydrogel.

Models predicting drug delivery from surface eroding polymers can also be applied to enzymatically degradable, surface-eroding hydrogels. Hopfenberg developed a drug release model applying to surface-eroding polymers. In this prediction, a zero-order surface detachment of the drug is the rate-limiting step.[80]

$$\frac{M_t}{M_\infty} = 1 - \left(1 - \frac{k_a t}{C_0 a_0}\right) n, \tag{36}$$

where C_0 is the initial drug concentration, k_a is the radial degradation constant, and a_0 is the initial dimension (radius for a spherical or cylindrical geometry and half-thickness for a slab-like geometrical structure). In addition, n is a geometrical factor and number 1, 2, or 3 is assigned accordingly for a slab, cylinder or sphere, specifically. When a slab-shaped device is used (i.e., $n - 1$), Equation (36) predicts that the drug will be released in a zero-order profile. Later, it was revealed by Katasulu that Eq. (36) is not a good approximation for a cylindrical tablet ($n = 3$).[81] To improve upon Eq. (36), Katzhendler, Hoffman and coworkers developed a general mathematical model to study heterogeneous eroding network accounting for different radical and vertical erosion rate constants for a flat cylindrical tablet, where k_a and k_b are radical and vertical degradation rate constant, respectively:[82]

$$\frac{M_t}{M_\infty} = 1 - \left(1 - \frac{k_a t}{C_0 a_0}\right)^2 \left(1 - \frac{2k_b t}{C_0 b_0}\right). \tag{37}$$

Here, a_0 is the initial radius of the tablet and b_0 is the thickness of the tablet. A range of drug release rates can be obtained by varying the radius to thickness ratio. This model does not consider swelling of matrices or assumes that gel swelling happens before the occurrence of erosion or drug release. Based on these fundamental efforts, there are other models developed to predict molecule release via surface-erosion.[40,81,83]

3.3.3. *Bulk- and erosion-based*

Mathematical models describing drug delivery from degradable hydrogels or polymers often assume that the delivery profiles are affected by a single, dominate degradation mechanism: surface erosion or bulk degradation. In bulk degradation models, the time scale for hydration is usually assumed to occur rapidly compared to random cleavage of degradable linkages throughout the hydrogel matrix. In surface eroding systems, water penetration is considered limited, leading to confined degradation/erosion on the surface of the matrix. Bulk degradation and surface erosion mechanisms can be simultaneously integrated in the design of drug delivery devices. This creates complex delivery patterns that may be more suited for particular therapeutic applications. Rothstein *et al.* developed a unified model for both such surface- and bulk-eroding systems. This model combines diffusion-reaction equations, which accounts for polymer hydration kinetics, polymer dissolution, and pore formation.[44,84] This model, however, is less applicable to hydrogel systems because of highly swollen nature of the gel network.

3.3.4. *Pendant-chain systems*

To increase therapeutic efficacy of the drug delivery devices, drugs can be covalently tethered in the swollen hydrogel through a cleavable linker. This section deals with hydrolytically labile linkers while Section 4.2.1 covers enzymatic linker cleavage. The delivery of covalently tethered pro-drugs is governed by the rate of linker hydrolysis and not by the diffusion of the drug. The most commonly used linker chemistry is ester. For example, polylactide and other linkers containing ester bonds have been used to covalently link synthetic drugs within hydrogel network.[26,27] The cleavage rate of the ester linkage is proportional to the concentration of the ester, which assumes the form of Eqs. (30) and (31). This pseudo-first-order cleavage kinetics allows one to obtain the amount of freely diffusible drug. Coupling with diffusion equation further permits one to acquire the rate of drug delivery from the hydrogels. If the hydrogels can be also degraded hydrolytically, the modeling requires considerations of not only hydrogel degradation rate but also network structural information. For example, Metters *et al.* developed a statistical-co-kinetic model to predict the delivery of ester-tethered fluorescent probe from a hydrolytically degradable step-growth hydrogels.[85] The delivery assumed a bi-phasic release profile where a zero-order release profile was first obtained prior to gel dissolution. When the gels degraded to near inverse gelation, however, an almost instantaneous burst release was observed. Because of the idealized step-growth network structure, the slope of the zero-order release and the extent of latent burst release could both be controlled via adjusting functionality of the polymer or gel degradation kinetics.

3.3.5. *Reaction-diffusion controlled release system*

Hydrogels capable of reversibly sequestering and releasing therapeutics have tremendous potential in promoting tissue regeneration. The concept of reversible sequestering of proteins within the hydrogel matrices comes from the ability of extracellular matrix (ECM) to reversibly sequester/release growth factors critical in tissue morphogenesis. Researchers have developed these "affinity" hydrogels bearing reversible binding motifs to alter release rates of certain protein therapeutics. In a highly swollen affinity hydrogel, the rate of protein/drug delivery depends on the concentration of the freely diffusible drug, which is determined by the binding affinity between the drug and the immobilized ligand. If the binding events occur strongly (i.e., high association and low dissociation rate), the concentration of freely diffusible drug will become too low, leading to limited amount of drug release.

The basic release model of a drug from affinity hydrogels was developed by Crank.[34] Let us consider the delivery of a protein (P) from a hydrogel immobilized with appropriate protein-binding affinity ligands (L). The binding equilibrium exists between freely diffusible protein; network immobilized ligand; and the bound complex (PL). The equilibrium can be described as:

$$\mathrm{P} + L \xleftrightarrow{K_D} PL, \tag{38}$$

Here, K_D is a time-independent equilibrium dissociation constant, which can be expressed as:

$$K_D = \frac{k_{off}}{k_{on}} = \frac{[P][L]}{[PL]}, \tag{39}$$

where k_{off} and k_{on} are the dissociation and association reaction rate constant, respectively. K_D (with unit M) has an inverse relationship with the strength of the affinity binding. Another time-independent equilibrium constant related to the concentration ratio of bound and free protein is given by:

$$K_b = \frac{[PL]}{[P]}. \tag{40}$$

Combining Eqs. (39) and (40) gives:

$$K_D = \frac{[L]}{K_b}. \tag{41}$$

If one assumes a binding event that is faster than protein diffusion, the following equation can be obtained for the transport of reversibly bound protein within an affinity hydrogel:[86]

$$\frac{\partial[P]}{\partial t} = \frac{D}{K_{b+1}} \nabla^2[P], \tag{42}$$

Here, D is the diffusivity of the protein in the absence of affinity ligand. Therefore, the ratio of D to (K_b+1) represents the apparent diffusivity of the protein in this affinity hydrogel. When compared to Fick's second law of diffusion (Equation (10)), Equation (42) shows that the retardation of protein release from the affinity hydrogel is a result of rapid and reversible protein-ligand binding, which decreases the apparent protein diffusivity by a factor of (K_b + 1). When the binding affinity is high, most of the proteins are bound to the ligand, leading to a high K_b values, and drastically decreased apparent diffusivity.

Equation (42) represents a highly simplified scenario where the binding events of soluble protein to the immobilized ligand occur much faster than diffusion. Because of this assumption, this model is limited to describing systems with simple yet rapid binding mechanisms or with a high ratio of ligand to protein. A generalized reaction-diffusion model without assuming fast binding equilibrium is given by a set of three equations:[57]

$$\frac{\partial[P]}{\partial t} = D\frac{\partial^2[P]}{\partial x^2} - k_{on}[P][L] + k_{off}[PL], \tag{43}$$

$$\frac{\partial[L]}{\partial t} = -k_{on}[P][L] + k_{off}[PL], \tag{44}$$

$$\frac{\partial[PL]}{\partial t} = k_{on}[P][L] - k_{off}[PL]. \tag{45}$$

This model assumes one-dimensional release geometry; with known k_{on}, k_{off}, and D values; perfect sink condition; protein-ligand binding is in equilibrium prior to the onset of release. With these assumptions, the initial and boundary conditions are:

$$[P] = [P]_{eq} \quad \text{at } t = 0, \tag{46}$$

$$[L] = [L]_{eq} \quad \text{at } t = 0, \tag{47}$$

$$[PL] = [PL]_{eq} \quad \text{at } t = 0, \tag{48}$$

$$[P] = 0 \quad \text{at } x = \text{d}, \tag{49}$$

$$\frac{\partial[P]}{\partial t} = 0 \quad \text{at } x = 0. \tag{50}$$

where x is the spatial coordinate in the gel perpendicular to the gel surface and δ is the half-thickness of the swollen hydrogel slab. Equation (43) describes the change in free protein concentration in the affinity hydrogel due to both affinity binding and diffusion. When no affinity binding exists (i.e., $k_{on} = k_{off} = 0$), Eq. (43) can be reduced to the Fick's second law of diffusion. Solving (43) to (50) numerically gives the time-dependent distributions of $[P]$, $[L]$, and $[PL]$ within the hydrogel. These concentration distributions can then be integrated, based on the geometry of the hydrogel slab, to obtain cumulative protein release.

Lin and Metters utilized this reaction-diffusion model to predict the release of hexa-histidine-tagged green fluorescent protein (his-GFP) from an affinity PEG hydrogels immobilized with metal-chelating ligand.[57,58] While the model

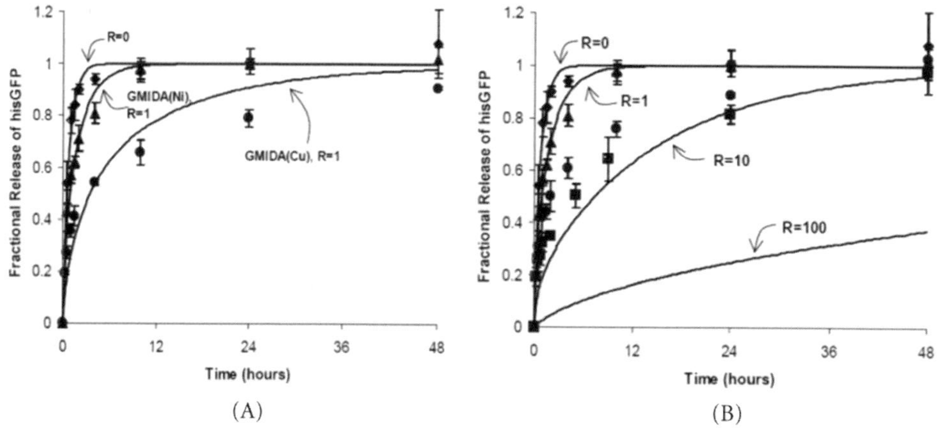

Fig. 2. *In vitro* his-GFP release from PEG-co-GMIDA hydrogels as functions of (A) Ligand affinity. No ligand (♦), R = 1, GMIDA(Ni) (▲), and R = 1, GMIDA(Cu) (■). (B) Ligand concentration with R = 0 (♦), 1 (▲), 10 (●), and 100 (■). Symbols are experimental data, and curves are theoretical predictions as indicated in the figure. Protein loading concentration:$1.82 * 10^{-4}$M. (Reprinted with permission from John Wiley & Sons, Inc.) [57]

prediction shows good agreement with ligand with different affinities were used (e.g., changing divalent metal ions to alter affinity binding constants, Fig. 3(A)), significant deviations were found when the same ligand was used at higher ligand concentrations (Fig. 3(B)). Specifically, the model over-predicts the amount of protein that can be released at high ligand concentrations. The authors attributed this phenomenon to decreasing hydrogel crosslinking efficiency (and hence higher diffusion of proteins) in the presence of free radical scavenging metal ions.

It is worth noting that Eqs. (43)–(45) are a set of equations describing reaction-diffusion occurring with one protein-ligand pair. Binding equilibrium equations can be derived and solved similarly if more than two binding components are involved.[46,47] This reaction-diffusion model Eqs. 43–45 could also predict the release of drugs from hydrogels containing reversible Diels-Alder bonds.[87]

4. Novel Developments in Hydrogel-based Drug Delivery Devices

4.1. *Bio-orthogonal click reactions*

Aforementioned nucleophilic thiol-acrylate (or maleimide) conjugation or radical-mediated thiol-ene reactions are considered bio-orthogonal reactions. Other bio-orthogonal chemistries useful in *in situ* cell labeling are also increasingly adapted for preparing cytocompatible hydrogels. For example, DeForest and Anseth developed PEG-conjugates that contain ring strain and electron-withdrawing di-fluorinated cyclooctyne. Without the use of metal catalyst, di-fluorinated cyclooctyne promotes the [3+2] dipolar cycloaddition with azides, which leads to gel network.[88–91] Alge and Anseth designed similar click hydrogels using tetrazine-norbornene click reaction.[92] In such case multi-arm PEG-tetrazine was used to react with bis-norbornene functionalized peptide for forming step-growth network. In another case, Maynard and colleagues developed oxime click chemistry to form hydrogels for supporting cell adhesion.[93] This was achieved by reacting multi-arm aminooxy-PEG with glutaraldehyde. Diels-Alder reaction has also been explored to form hydrogels that contain temperature-dependent reversible crosslinks.[87]

The uniqueness of these bio-orthogonal chemistries is that no initiator or catalyst is required to trigger gel crosslinking. This is a significant advancement since many of the initiators or catalysts available for hydrogel crosslinking are, at least to some degree, cytotoxic to the cells. A second common feature of these orthogonally-crosslinked hydrogels is that they have an idealized step-growth network structure (similar to structure shown in Fig. 1(B)). It is worth noting, however, that the time needed to reach complete gelation of the current 'click' hydrogels may be longer than conventional gelation methods that utilize initiators or catalyst. Because the gelation occurs as soon as the reacting components are mixed together, no spatial-temporal control in gelation kinetics is possible. These hydrogels also contain certain degrees of network non-ideality. One general rule to evaluate whether a network is truly 'ideal' is to look at the swelling ratios of the hydrogels. The equilibrium swelling ratio of a hydrogel with a truly ideal network structure should be independent of polymer concentrations in the precursor solution.[17,70]

4.2. *On-demand drug delivery*

Conventional drug delivery systems often rely on diffusion or degradation as delivery mechanisms. While these matrices can be easily prepared and drug release profiles from these matrices are highly predictable, these 'passive' drug delivery devices cannot meet the challenging needs in regenerative medicine and tissue engineering. It is known that tissue regeneration is a spatial-temporally process where different biophysical and biochemical cues occur and disappear in a highly regulated manner. In some cases, the delivery of a certain drug is not needed until at a later time or at a localized position during tissue regeneration. Hence, matrices capable of delivering drugs in a triggerable fashion will be beneficial for maximizing the efficacy of the drug or decreasing systemic drug toxicity.

In general, triggerable drug delivery can be categorized into two groups: cell-mediated and user-defined. In cell-mediated triggerable drug delivery, the drug of interest is usually tethered to the network through a protease-sensitive linker.[23,27,29,59,94] When a cell is nearby, proteases or enzymes secreted by the cell will diffuse and cause linker cleavage, which renders the otherwise immobilized drug freely diffusible. The major benefit of this approach is that the drug will only be released when a relevant protease or cell has migrated near the drug-loaded hydrogel. Cell-mediated drug delivery is particularly useful in promoting tissue regeneration. For example, Aimetti and Anseth developed an anti-inflammatory drug delivery system that responds to the presence of human neutrophil elastase (HNE),[94] a protease secreted by the first line of inflammatory cells, neutrophils, at the site of inflammation.

In user-defined triggerable drug delivery systems, a trigger is input by the user, which causes a physical or chemical response in the hydrogel matrices and subsequently results in drug release. Many types of triggers have been engineered into responsive hydrogel matrices, including near-infrared (NIR, ca. 650–900 nm),[61] visible or UV light (ca. 280–800 nm),[24] magnetic force,[60] ultrasound,[64] or other electrochemical processes. The application of NIR increases temperature in gold nanostructure (e.g., nanoparticles, nanorods, or nanoshell) embedded in temperature-responsive polymer hydrogels. Upon NIR irradiation, rising temperature causes the thermal-responsive gels to shrink or expand, thus altering drug release profiles. Magnetic force can trigger deformation of 'ferrogel' (i.e., *g* gel incorporated with superparamagnetic particles), which alters release kinetics of the incorporated drug. Visible or UV light induces cleavage of light sensitive moiety linked to both the hydrogel network and the drug. Upon light exposure, drug molecules are liberated due to the cleavage of the linker. Finally, ultrasound (longitudinal pressure wave with frequency >20 kHz) can be used to trigger degradation or heating of appropriate materials, which also leads to the release of the drug payload.

4.2.1. *Modeling cell-mediated drug delivery from hydrogel*

To predict drug release from such an enzyme responsive on-demand drug delivery system, a reaction-diffusion model was typically employed.[94] For an enzyme responsive delivery system, substrate cleavage is often governed by Michaelis-Menten kinetics. Since enzyme has to diffusion from the external into the hydrogel, its concentration within the hydrogel is usually heterogeneous. Fick's second law of diffusion is usually used to describe the concentration of the enzyme in the hydrogel. At the surface of the hydrogel, enzyme concentration is usually considered the same as that in the solution ($[E]_0$). On the other hand, a symmetry boundary condition was often used. With these boundary conditions, the concentration of enzyme in the hydrogels can be solved analytically from Equation (51) [94]:

$$[E](z,t)=[E]_0\left\{1-\frac{4}{\pi}\sum_{n=0}^{\infty}\frac{(-1)^n}{2n+1}\times\exp\left[-D_E(2n+1)^2\frac{\pi^2 t}{4h^2}\right]\cos\left[\frac{(2n+1)\pi z}{2h}\right]\right\}, \quad (51)$$

where h is the half-thickness of the gel. When the spatial profile of enzyme concentration is solved, the cleavage of the substrate can also be determined, assuming that it is governed by the Michaelis-Menten kinetics.

$$-\frac{\partial[S](z,t)}{\partial t}=\frac{\partial[P](z,t)}{\partial t}=[E](z,t)k_{cat}\frac{[S]}{K_m+[S]}. \tag{52}$$

Here, $[S]$ is the concentration of the substrate (i.e., cleavable linker), k_{cat} is the kinetic constant for formation of the product for a given enzyme-substrate pair, K_m is the Michaelis constant for the enzyme-substrate pair, and $[P]$ is the concentration of the cleaved product. After the product is produced, it has to diffusion out of the hydrogel and the process can also be modeled by the Fickian diffusion equation:

$$\frac{\partial[P](z,t)}{\partial t}=D_p\frac{\partial^2[P](z,t)}{\partial z^2}. \tag{53}$$

Here, D_p is the diffusivity of the cleaved product in the gel. When taking into account the hydrodynamic radius of the solute, the mesh size of the hydrogel, and the degree of swelling of the hydrogel, the diffusion of the product can be solved numerically. Aimetti *et al.* utilized this model to predict the release of pendant peptides that were covalently linked to the chain-growth PEG hydrogel network through human neutrophil elastase (HNE) sensitive linkers.[94] As shown in Fig. 3, the model agrees well with the release profile up to 80% release. The authors attributed the deviation between model predictions and experimental results in

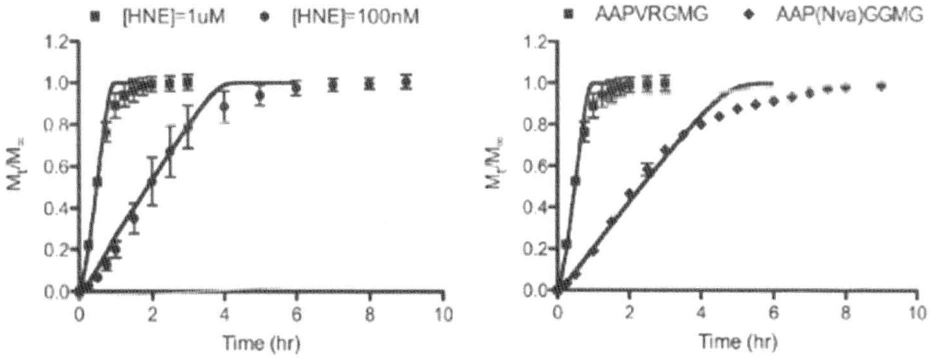

(A) (B)

Fig. 3. Model peptide release from enzyme (HNE) sensitive hydrogels. Effect of (A) varying HNE concentrations and (B) varying HNE substrates on release profiles. Solid lines indicate diffusion-reaction model results. Error bars indicate standard deviation (n = 3). (Reprinted with permission from Ref. 94. Copyright (2009) American Chemical Society.)

the later part of the release to the complexity of the system at later time points. As the substrate is being converted to product, free carboxylic acids accumulate in the hydrogel, forming a charged hydrogels that may affect the diffusivity of the drug.

4.2.2 *Modeling photo-responsive drug delivery from hydrogel*

Photochemistry has enjoyed a long and successful history in hydrogel fabrication. Numerous hydrogel systems, including chain-growth, step-growth, and mixed-mode step-chain growth polymerizations, are developed based on photochemistry. Photochemical reactions are also increasingly being utilized as mechanisms of user-controlled matrix degradation or drug delivery.[24,25,45,95] To achieve photo-responsive drug delivery, typically the drugs are modified with a crosslinkable moiety and a photolabile linker.[24] The crosslinkable moiety allows for covalent conjugation of the modified drug into the hydrogel matrix, whereas the photolabile linker permits light responsive cleavage and hence diffusion of the drug from the hydrogel. Various photosensitive macromers have been developed that allows for biased release of drugs by switching wavelengths.[45,95] Because the photolabile linkers available to date are mostly sensitive to UV light (e.g., 365 nm) and near-UV visible light (e.g., 405 nm), the majority of the photodegradable hydrogels are fabricated via non-photo-based gelation, such as redox-mediated[24,25,45,95] or Michael-type addition reactions.[96] A recent report has demonstrated the feasibility of using light with orthogonal wavelengths to prepare photodegradable hydrogel.[97] The concentration and the molar absorbability of the photodegradable (PD) linker are critical for the modeling of light-responsive drug delivery. The concentration of the photodegradable groups in the hydrogel as a function of time can be expressed as[96]:

$$[PD] = [PD]_0 * e^{(-kt)}, \tag{54}$$

where

$$k = \frac{\phi s}{N_A h v} I_0 = k_{eff} * I_0. \tag{55}$$

Here, Φ is quantum yield, ε is the molar absorptivity, N_A is Avogadro's number, h is the Planck constant, v is the frequency at the wavelength of irradiation, k_{eff} is the effective constant incorporating all constants except for I_0, which is the light intensity at the surface of the sample. The rate of cleavage/consumption of photodegradable groups is the derivative Eq. (25) with respect to time:[25]

$$R_{PD} = -\frac{d[PD]}{dt} = k_{eff} * I_0 e^{(k_{eff} \cdot I_0 t)}. \tag{56}$$

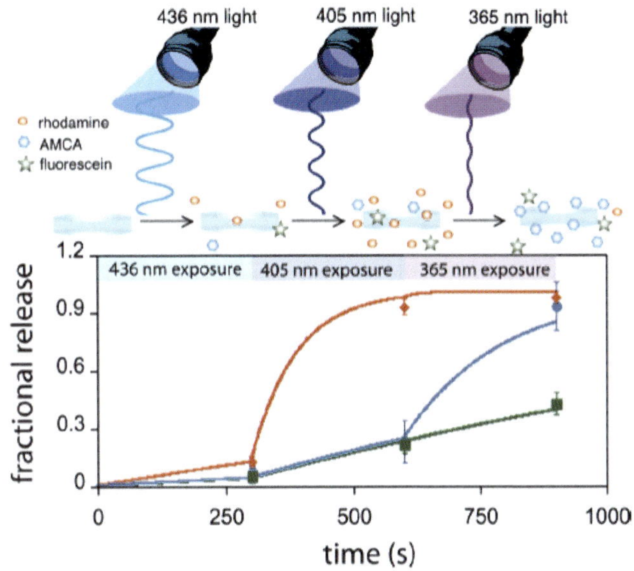

Fig. 4. Fractional release of rhodamine, AMCA, and fluorescein from a hydrogel as a function of light exposure. Solid lines depict predicted release, actual release shown as data points. (Reprinted with permission from Ref. 45. Copyright (2012) American Chemical Society.)

Since the cleavage/consumption of photodegradable linker means the accumulation of drug available for release, Eq. (56) is also the rate of drug release into the hydrogel. It is worth noting that, light is not a constant for an optically thick (thickness > 50 μm) gel sample because any light-absorbing molecules along the axis of incident irradiation will attenuate the incident light. To account for light attenuation as a function of species absorbance, an integrated form of Beer-Lambert's law can be used:

$$I(x,t) = I_0 * \exp\left[-\sum_{i=1}^{N} \varepsilon_i \int_0^x C_i(x',t)dx'\right]. \tag{57}$$

This model can be solved numerically by integration of the spatial derivative of light intensity and the temporal derivatives of the intact and degraded *PD* groups. The rate of conjugated drug release can be predicted after taking into account the rate constant of the photodegradable group, which roughly scales with the absorptivity.[24] As shown in Fig. 4, Kasko *et al.* used this model to accurately predict release of drug from hydrogels containing photosensitive linkers.[25,45]

5. Conclusions

Hydrogel-based controlled drug delivery systems are an important class of devices in disease treatment and for fundamental biomedical research. The design of hydrogel-based drug delivery systems have benefited from increasingly sophisticated mathematical modeling. A successful mathematical model can serve as a means to decrease the necessary experimental works that may be labor intensive and time consuming. It also helps researchers to validate the mechanism of drug release from the hydrogel matrices. In general, the modeling of drug delivery from hydrogels involves two steps: 1. predicting concentration of the drug available for delivery and 2. modeling diffusion of drug from hydrogels. Since hydrogels are highly swollen, the majority of the release mechanisms (except for purely surface erosion systems) are related to diffusion. Because diffusion is a process that will occur when drug molecules have become available in hydrogel matrices, efforts in the design of hydrogel-based drug delivery systems often involve tailoring the availability of the drugs. Conventional methods of controlling drug availability from the hydrogel rely on adjusting hydrogel mesh size, which controls the rate of drug diffusion. As more degradable macromers become available for hydrogel fabrication, the mechanisms and related mathematical modeling of drug delivery from hydrogel matrices also become increasingly complicated. Nonetheless, the ability to control the rate of hydrogel degradation (either by decreasing network crosslinking density or cleavage of drug-bound linkers) offers researchers an important tool of regulating drug diffusion, from which to affect cellular behavior or drug efficacy.

References

1. N. A. Peppas, P. Bures, W. Leobandung, and H. Ichikawa, *Eur. J. Pharm. Biopharm.*, **50**, (2000) 27–46.
2. N. A. Peppas, J. Z. Hilt, A. Khademhosseini, and R. Langer, *Adv. Mater,* **18**, (2006) 1345–1360.
3. K. Y. Lee and D. J. Mooney, *Chem. Rev.,* **101**, (2001) 1869–1879.
4. J. L. Drury and D. J. Mooney, *Biomaterials,* **24**, (2003) 4337–4351.
5. C. C. Lin and A. T. Metters, *Adv. Drug. Deliver. Rev.,* **58**, (2006) 1379–1408.
6. C. C. Lin and K. S. Anseth, *Pharm. Res.,* **26**, (2009) 631–643.
7. T. V. Chirila, *Biomaterials,* **22**, (2001) 3311–3317.
8. C. M. Hassan and N. A. Peppas, *Macromolecules,* **33**,(2000) 2472–2479.
9. L. C. Dong and A. S. Hoffman, *J. Control Release,* **15**,(1991) 141–152.
10. A. S. Sawhney, C. P. Pathak, and J. A. Hubbell, *Macromolecules,* **26**, (1993) 581–587.
11. N. A. Peppas, K. B. Keys, M. Torres-Lugo, and A. M. Lowman, *J. Control Release,* **62**, (1999) 81–87.
12. E. Behravesh, S. Jo, K. Zygourakis, and A. G. Mikos, *Biomacromolecules,* **3**, (2002) 374–381.
13. C. R. Nuttelman, M. A. Rice, A. E. Rydholm, C. N. Salinas, D. N. Shah, and K. S. Anseth, *Prog. Polym. Sci.,* **33**, (2008) 167–179.

14. W. E. Hennink and C. F. van Nostrum, *Adv. Drug. Deliv. Rev.*, **54**, (2002) 13–36.
15. N. A. Peppas, Y. Huang, M. Torres-Lugo, J. H. Ward, and J. Zhang, *Annu. Rev. Biomed. Eng*, **2**, (2000) 9–29.
16. F. J. Flory, *Principles of Polymer Chemistry*. Ithaca, NewYork: Cornell Univ. Press (1953).
17. A. Metters and J. Hubbell, *Biomacromolecules*, **6**, (2005) 290–301.
18. S. R. Lustig and N. A. Peppas, *J. Appl. Polym. Sci.*, **36**, (1988) 735–747.
19. C. C. Lin and A. T. Metters, *Pharm. Res.*, **23**, (2006) 614–622.
20. J. D. McCall, C. C. Lin, and K. S. Anseth, *Biomacromolecules*, **12**, (2011) 1051–1057.
21. J. D. McCall and K. S. Anseth, *Biomacromolecules*, **13**, (2012) 2410–2417.
22. T. A. Holland, Y. Tabata, and A. G. Mikos, *J. Control Release*, **101**, (2005) 111–125.
23. A. H. Zisch, U. Schenk, J. C. Schense, S. E. Sakiyama-Elbert, and J. A. Hubbell, *J. Control Release*, **72**, (2001) 101–113.
24. A. M. Kloxin, A. M. Kasko, C. N. Salinas, and K. S. Anseth, *Science*, **324**, (2009) 59–63.
25. D. R. Griffin, J. T. Patterson, and A. M. Kasko, *Biotechnol. Bioeng.*, **107**, (2010) 1012–1019.
26. D. S. W. Benoit, C. R. Nuttelman, S. D. Collins, and K. S. Anseth, *Biomaterials*, **27**, (2006) 6102–6110.
27. C. Yang, P. D. Mariner, J. N. Nahreini, and K. S. Anseth, *J. Control Release*, **162**, (2012) 612–618.
28. D. S. W. Benoit, A. R. Durney, and K. S. Anseth, *Biomaterials*, **28**, (2007) 66–77.
29. A. H. Zisch, M. P. Lutolf, M. Ehrbar, G. P. Raeber, S. C. Rizzi, N. Davies, H. Schmokel, D. Bezuidenhout, V. Djonov, P. Zilla, and J. A. Hubbell, *Faseb Journal*, **17**, (2003) 2260.
30. M. M. Martino, P. S. Briquez, A. Ranga, M. P. Lutolf, and J. A. Hubbell, *Proc. Natl. Acad. Sci. Am.*, **110**, (2013) 4563–4568.
31. S. W. Kim, Y. H. Bae, and T. Okano, *Pharm. Res.*, **9**, (1992) 283–290.
32. J. P. Schillemans, E. Verheyen, A. Barendregt, W. E. Hennink, and C. F. Van Nostrum, *J. Control Release*, **150**, (2011) 266–271.
33. S. H. Gehrke, L. H. Uhden, and J. F. McBride, *J. Control Release*, **55**, (1998) 21–33.
34. J. Crank, *The Mathematics of Diffusion*. 1975, Oxford: Oxford University.
35. M. Hedenqvist, A. Angelstok, L. Edsberg, P. T. Larsson, and U. W. Gedde, *Polymer*, **37**, (1996) 2887–2902.
36. B. Amsden, *Macromolecules*, **31**, (1998) 8382–8395.
37. A. M. Lowman and N. A. Peppas, *J. Biomater. Sci. Polym. Ed.*, **10**, (1999) 999–1009.
38. P. L. Ritger and N. A. Peppas, *J. Control Release*, **5**, (1987) 23–36.
39. C. S. Brazel and N. A. Peppas, *Biomaterials*, **20**, (1999) 721–732.
40. J. Siepmann, H. Kranz, R. Bodmeier, and N. A. Peppas, *Pharm. Res.*, **16**, (1999) 1748–1756.
41. J. Siepmann, K. Podual, M. Sriwongjanya, N. A. Peppas, and R. Bodmeier, *J. Pharm. Sci*, **88**, (1999) 65–72.
42. J. Siepmann and N. A. Peppas, *Adv. Drug. Deliv. Rev.*, **48**, (2001) 139–157.
43. K. S. Anseth, A. T. Metters, S. J. Bryant, P. J. Martens, J. H. Elisseeff, and C. N. Bowman, *J. Control Release*, **78**, (2002) 199–209.
44. S. N. Rothstein, W. J. Federspiel, and S. R. Little, *Biomaterials*, **30**, (2009) 1657–1664.

45. D. R. Griffin and A. M. Kasko, *Acs Macro Letters,* **1**, (2012) 1330–1334.
46. S. E. Sakiyama-Elbert and J. A. Hubbell, *J. Control Release,* **65**, (2000) 389–402.
47. S. E. Sakiyama-Elbert and J. A. Hubbell, *J. Control Release,* **69**, (2000) 149–158.
48. J. Siepmann and F. Siepmann, *J. Control Release,* **161**, (2012) 351–362.
49. G. Modesti, B. Zimmermann, M. Borsch, A. Herrmann, and K. Saalwachter, *Macromolecules,* **42**, (2009) 4681–4689.
50. L. Serra, J. Domenech, and N. A. Peppas, *Biomaterials,* **27**, (2006) 5440–5451.
51. S. Koutsopoulos, L. D. Unsworth, Y. Nagai, and S. Zhang, *Proceedings of the National Academy of Sciences of the United States of America,* **106**, (2009) 4623–4628.
52. R. Yoshida, K. Sakai, T. Okano, and Y. Sakurai, *Adv. Drug. Deliver. Rev.,* **11**, (1993) 85–108.
53. R. W. Korsmeyer and N. A. Peppas, *J. Membrane. Sci,* **9**, (1981) 211–227.
54. C. S. Brazel and N. A. Peppas, *Polymer,* **40**, (1999) 3383–3398.
55. C. S. Brazel and N. A. Peppas, European *J. Pharm. Biop.,* **49**, (2000) 47–58.
56. J. Siepmann and N. A. Peppas, *Pharm. Res.,* **17**, (2000) 1290–1298.
57. C. C. Lin and A. T. Metters, *J. Biomed. Mater Res. A,* **83A**, (2007) 954–964.
58. C. C. Lin and A. T. Metters, *Biomacromolecules,* **9**, (2008) 789–795.
59. M. Ehrbar, A. Metters, P. Zammaretti, J. A. Hubbell, and A. H. Zisch, *J. Control Release,* **101**, (2005) 93–109.
60. N. S. Satarkar and J. Z. Hilt, *J. Control Release,* **130**, (2008) 246–251.
61. D. V. Volodkin, A. G. Skirtach, and H. Mohwald, *Angew. Chem. Int. Ed.* **48**, (2009) 1807–1809.
62. N. S. Satarkar, D. Biswal, and J. Z. Hilt, *Soft Matter,* **6**, (2010) 2364–2371.
63. C. R. Thomas, D. P. Ferris, J. H. Lee, E. Choi, M. H. Cho, E. S. Kim, J. F. Stoddart, J. S. Shin, J. Cheon, and J. I. Zink, *J. Am. Chem. Soc.,* **132**, (2010) 10623–10625.
64. S. Bibi, E. Lattmann, A. R. Mohammed, and Y. Perrie, *J. Microencapsul,* **29**, (2012) 262–276.
65. J. Thevenot, H. Oliveira, O. Sandre, and S. Lecommandoux, *Chem. Soc. Rev.,* **42**, (2013) 7099–7116.
66. P. Martens, A. T. Metters, K. S. Anseth, and C. N. Bowman, *J. Phys. Chem. B,* **105**, (2001) 5131–5138.
67. A. T. Metters, C. N. Bowman, and K. S. Anseth, *J Phys Chem B,* **104**, (2000) 7043–7049.
68. P. Van de Wetering, A. T. Metters, R. G. Schoenmakers, and J. A. Hubbell, *J. Control Release,* **102**, (2005) 619–627.
69. M. N. Mason, A. T. Metters, C. N. Bowman, and K. S. Anseth, *Macromolecules,* **34**, (2001) 4630–4635.
70. H. Shih and C. C. Lin, *Biomacromolecules,* **13**, (2012) 2003–2012.
71. J. Siepmann, N. Faisant, and J. P. Benoit, *Pharm. Res.,* **19**, (2002) 1885–1893.
72. E. S. Park, M. Maniar, and J. C. Shah, *J. Control Release,* **52**, (1998) 179–189.
73. A. K. Burkoth, J. A. Burdick, and K. S. Anseth, *J. Biomed. Mater Res.,* **51**, (2000) 352–359.
74. A. Gopferich and J. Tessmar, *Adv. Drug. Deliv. Rev.,* **54**, (2002) 911–931.
75. M. J. Kipper, E. Shen, A. Determan, and B. Narasimhan, *Biomaterials,* **23**, (2002) 4405–4412.
76. D. J. Quick, K. K. Macdonald, and K. S. Anseth, *J. Control Release,* **97**, (2004) 333–343.

77. M. A. Rice, J. Sanchez-Adams, and K. S. Anseth, *Biomacromolecules,* **7**, (2006) 1968–1975.
78. C. C. Lin, A. Raza, and H. Shih, *Biomaterials,* **32**, (2011) 9685–9695.
79. A. A. Aimetti, A. J. Machen, and K. S. Anseth, *Biomaterials,* **30**, (2009) 6048–6054.
80. B. Hopfenberg H, Controlled Release from Erodible Slabs, Cylinders, and Spheres, in Controlled Release Polymeric Formulations. 33, *ACS Symposium Series.* Chapter 3. pp. 26–32 (1976).
81. H. Y. Karasulu, G. Ertan, and T. Kose, *Eur. J. Pharm. Biopharm.* **49**, (2000) 177–182.
82. I. Katzhendler, A. Hoffman, A. Goldberger, and M. Friedman, *J. Pharm. Sci.* **86**, (1997) 110-5.
83. K. Nam, J. Watanabe, and K. Ishihara, *Int. J. Pharm.* **275**, (2004) 259–269.
84. S. N. Rothstein, W. J. Federspiel, and S. R. Little, *J. Mater. Chem.,* **18**, (2008) 1873–1880.
85. J. W. DuBose, C. Cutshall, and A. T. Metters, *J. Biomed. Mater. Res. A,* **74A,** (2005) 104–116.
86. W. M. Saltzman, *Drug Delivery: Engineering Principles for Drug Therapy.* Oxford University (2001).
87. K. C. Koehler, K. S. Anseth, and C. N. Bowman, *Biomacromolecules,* **14**, (2013) 538–547.
88. C. A. DeForest and K. S. Anseth, *Nat. Chem.,* **3**, (2011) 925–931.
89. C. A. DeForest and K. S. Anseth, *Angew. Chem. Int. Ed.* **51**, (2012) 1816–1819.
90. C. A. DeForest, B. D. Polizzotti, and K. S. Anseth, *Nat. Mater.* **8**, (2009) 659–664.
91. C. A. DeForest, E. A. Sims, and K. S. Anseth, *Chem. Mater.,* **22**, (2010) 4783–4790.
92. D. L. Alge, M. A. Azagarsamy, D. F. Donohue, and K. S. Anseth, *Biomacromolecules,* **14**, (2013) 949–953.
93. G. N. Grover, J. Lam, T. H. Nguyen, T. Segura, and H. D. Maynard, *Biomacromolecules,* **13**, (2012) 3013–3017.
94. A. A. Aimetti, M. W. Tibbitt, and K. S. Anseth, *Biomacromolecules,* **10**, (2009) 1484–1489.
95. D. R. Griffin, J. L. Schlosser, S. F. Lam, T. H. Nguyen, H. D. Maynard, and A. M. Kasko, *Biomacromolecules,* **14**, (2013) 1199–1207.
96. M. W. Tibbitt, A. M. Kloxin, L. A. Sawicki, and K. S. Anseth, *Macromolecule,* **46,** (2013) 2785–2792.
97. C. S. Ki, H. Shih, and C. C. Lin, *Polymer,* **54**, (2013) 2115–2122.

Chapter 2

Natural Polysaccharide-Based Hydrogels for Controlled Localized Drug Delivery

Roman P. Blount, IV, and Narayan Bhattarai*

Department of Chemical, Biological and Bioengineering,
North Carolina A&T State University
1601 East Market street Greensboro, NC 27411, USA
**nbhattar@ncat.edu*

Hydrogels are liquid filled polymer networks that allow for improved drug encapsulation and delivery. In the field of drug delivery, hydrogels have a number of advantages for use in controlled localized drug delivery a major reason for those advantages is the ability of the hydrogel to undergo modifications that yield specified characteristics and traits. The polysaccharide polymer chosen for the fabrication of the hydrogel must be easily controlled for bioactivity and biocompatibility so that unfavorable reactions will not occur when exposed to target tissue sites. A large number of naturally occurring polysaccharides are frequently used as biocompatible hydrogels for controlled localized drug delivery applications. Hyaluronic acid, chitosan, and alginate are among those studied most frequently because they are available in sufficient quantities to be used, commercially and they possess the chemical and biological qualities needed for producing biocompatible hydrogels. These hydrogels which can come in a range of geometries and strengths, also have various drug loading capabilities, and controllable stimuli responsive characteristics. This review will observe the conventional natural polysaccharide hydrogel fabrication strategies, the system of drug loading and release, and the applications of hydrogels as microspheres, *in situ* gels, and porous 3D matrixes.

1. Introduction

For centuries drug delivery has been an important element in the area of healthcare. Drug delivery systems have been used to prevent the common cold, reduce the pain, and provide targeted therapy at specified release sites. Delivery time, absorption, excretion, etc. and other variables are important elements involved in effective targeted drug delivery systems. Medical professionals today are placing greater emphasis than ever before on the use of controlled localized drug delivery systems rather than standard drug delivery systems which provide more systemic therapies. This interest is a result of the need to deliver controllable drugs at predetermined rates, target specific cellular tissues, and increase the efficacy of the drug (i.e., stabilize against enzymatic degradation).[1,2] Accordingly, there has been a push for the development of therapeutic hydrogels made from polymeric materials that satisfy the targeted size range of micron to submicron and are multifaceted in the advantages that can be conferred on them.[1] The various size possibilities and the environmentally-sensitive qualities of hydrogels will lead to continued advances in the controlled localized drug delivery targeted to tissues.[2]

Hydrogels are hydrophilic polymer networks that can absorb large amounts of water or biological fluid based on the hydrophilic groups present.[3–8] The hydrogels form hydrated networks when of the polymer molecules are dispersed in water. The hydrogel hydration helps to body tissue compatibility by facilitating the diffusion of nutrients and cellular waste through the pores present in the gel.[9] Additionally, the porous nature of the hydrogels makes them ideal candidates for drug encapsulation and permits the release of the drug into the surrounding environment. Drug release depending on the hydrophobic or hydrophilic nature of the drug molecule loaded into the hydrogel. Unfortunately, the drug loading of hydrophobic molecules show limitations due to the high water content of most hydrogels, which leads to rapid drug release over the course of a few hours to a couple of days.[5] This rapid release proves to be unfavorable for prolonged hydrogels drug release models.

Crosslinking occurs as chemical bonds form between polymer chains (i.e., covalent, electrostatic interactions, etc.). The links between the polymer networks enables the hydrogel to absorb water and allows the polymer matrixes to form 3D structures (such as spheres, scaffolds, amorphous gels, and scaffolds. Hydrophobic or hydrophilic interactions, as well as other physical interactions can also lead to the formation of the hydrogels even though hydrogels are formed either through on chemical crosslinking or physical interactions, they can both be present simultaneously in a hydrogel matrix.[6,7] These network formation divisions will be discussed further in Sec. 3.[3] In addition, hydrogels can be further divided based on their classification of being homopolymers (e.g., alginate hydrogels) or copolymers (e.g., Chitosan-hyaluronic acid hydrogel derivatives), as well as on the charges of the monomeric units (i.e., neutral, anionic, and cationic).[10]

Hydrogels are important in a variety of applications and cell encapsulation, even so tissue engineering, delivery is drug uses of hydrogels. In that capacity, however, factors one of the primary. Present challenges that must be addressed such as material toxicity, immunogenicity, tissue accumulation, and low cellular uptake. Accordingly,[11] hydrogel, the ideal and, drug carrier models are those that possess biocompatibility, various biodegradability, and can be easily modified for controlled release and site-specific delivery. These characteristics which are present in natural polysaccharide hydrogels can be improved upon through although, the chemical modifications. The body of literature dedicated to specific polysaccharide preparation of hydrogels for delivery systems is very large. (i.e., alginate, hyaluronic acid, and chitosan), this review will summarize the advantages of natural polysaccharides in drug delivery, conventional polysaccharide hydrogel matrix formation, polysaccharide drug release, and the various applications of localized delivery systems (e.g., microspheres, *in situ* hydrogels, porous gels and scaffolds)).

2. Advantages of Natural Polysaccharide Hydrogels in Drug Delivery

Hydrogels can be fabricated from synthetic or natural polymers. However, natural polymers have advantages that synthetic polymers, lack. For instance, natural polymers are non-toxic, readily available, and less expensive than synthetic polymers.[12] Natural polysaccharides make natural polymers that are made up of monosaccharides are linked together through glycosidic bonds (Fig. 1(A)). These monosaccharide units in turn can lead to the branching or linearization of the polysaccharide.[11] Long lists of naturally occurring polysaccharides are frequently utilized in the preparation, characterization, and application of tunable, functional, and biocompatible material for a wide range of biomedical applications. Hyaluronic acid, chitosan, and alginate are among the natural polymers studied most frequently, for a variety of reasons. For instance, they are abundant in nature and commercially available, additionally, they offer desired chemical and biological advantages; and they provide diverse saccharide residues with varied functionalities for chemical modification and crosslinking to produce biocompatible hydrogels (Fig. 1(B)).[13,14]

Natural polymer sources, are readily available in abundance, from various sources such as from plants (e.g., alginate) animals (e.g., chitosan, hyaluronic acid, etc.), and cultures of microbial selected strains (e.g., dextran), as well. Moreover, because recombinant DNA techniques facilitate easy fabrication and production of the polysaccharide polymers natural polymers are preferable to synthetic polymers.[7,15] Additionally,

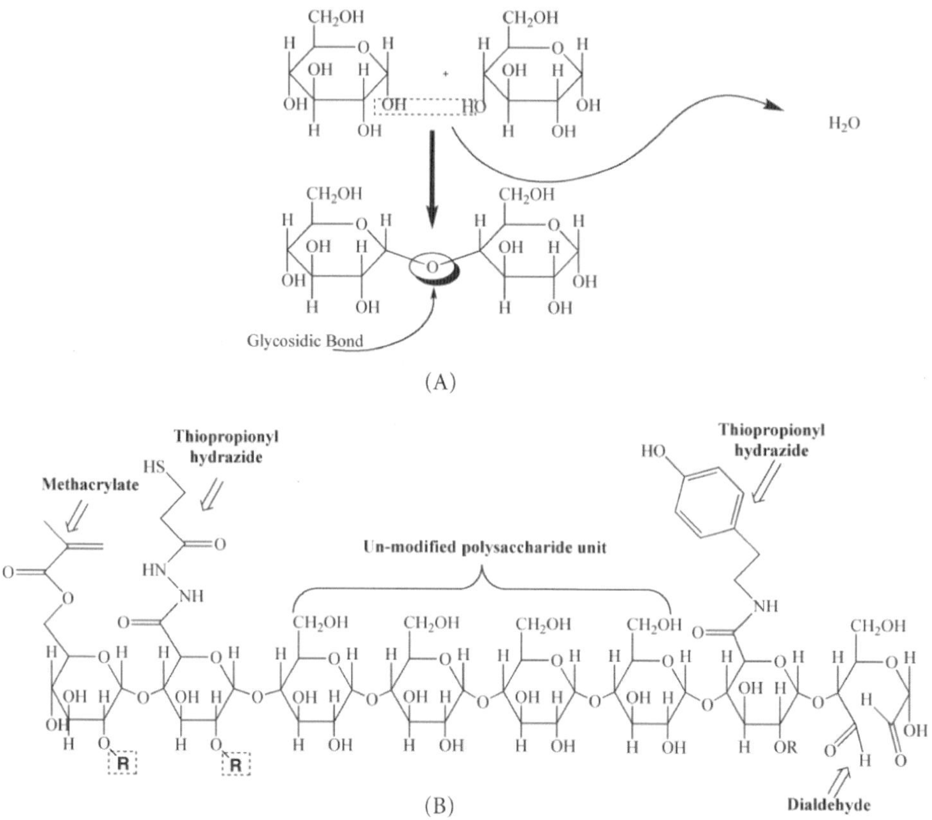

Fig. 1. (A) Structure of polysaccharide polymer formation highlighting the glycosidic bond of maltose and (B) polysaccharide with multiple functional groups attached (i.e., Methacrylate, Thiopropionyl hydrazide, Thiopropionyl hydrazid, and Dialdehyde).

Note: R-group = OH, H, COOH, NH, $COCH_3$, etc.

the controllable degradative properties, of natural polymers enable their drug loads to be delivered with low concentrations of drug release thus minimizing the likelihood of toxicity from drug overdose. The biocompatible capabilities of natural polysaccharides their ease of modification and their role in controlled localized drug delivery will be discussed later in this document.

2.1. *Biocompatibility and bioactivity*

High biocompatibility of polysaccharides is essential for the controlled delivery of drug molecules. A primary example of biocompatibility is seen in the low toxicity levels of natural polysaccharides. The advantages of biocompatibility are reflected in the ability to control the biodegradability or dissolution of the hydrogels via enzymatic, hydrolytic, or environmental (e.g., pH, temperature, etc.) pathways. Depending on the time scale and location of the drug delivery device however, degradation is not always desirable. As mentioned, earlier, natural polysaccharides can interact with various enzymes in the body to promote the degradation of the polysaccharides into their various monosaccharide components. This process is accomplished via the enzymatic breaking of glyosidic bonds. The resulting degradation provides more effective drug delivery by releasing any drug molecules that otherwise would not be able each the targeted area(s) to Beyond biocompatibility and degradation, bioactivity is a critical element in controlled drug delivery. Polymers possess an affinity for modification given the functional groups that their monomers contain (Fig. 1(B)). These modifying capabilities enable for the improvement of the hydrogel's drug delivery capabilities through altered biochemical properties, which in turn alters the bioactivity of the polysaccharide hydrogel.

Natural polysaccharides possess innate bioactivity as a result of their mucoadhesive, antimicrobial, and possessing anti-inflammatory characteristics.[11,15] Specifically, mucoadhesive hydrogels provide adhesive interactions between the polysaccharide hydrogel and the mucosal epithelia (Fig. 2). This interaction promotes the localized drug delivery to the target site. Various mechanisms play a role in effecting to mucoadhesive interactions. For example in the electrostatic interaction between mucin and positively charged polysaccharides, the reactive functional groups of the hydrogel hydrogen bond with the sugar moieties in the mucin. Mucoadhesion has been widely used as a means of achieving site-specific drug delivery through the incorporation of mucoadhesive hydrophilic polymers along with the active drug molecules.[16]

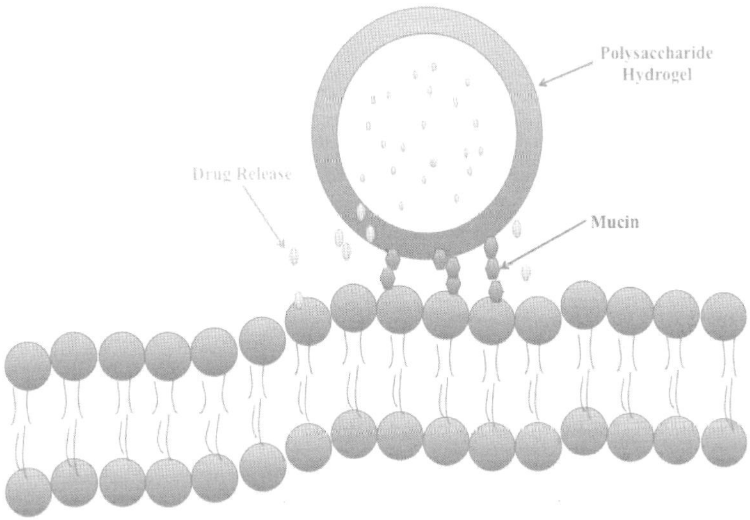

Fig. 2. Adhesive interaction between mucin and hydrogel providing localized drug delivery.

Chitosan and hyaluronic acid-based materials are very good mucoadhesives hydrogel, and can be made to bind to mucins through either electrostatic or hydrophobic interactions.[17,18] The amine and hydroxyl groups of the polysaccharide residues provide excellent mucoadhesive properties, and lead to prolonged residence time on the tissues where mucus is present. Several reactions with the thiolated chitosan and hyaluronic acid have been designed to improve mucoadhesive properties.[17–20] These properties make the gels useful for delivery of drugs locally where epithelia produce enough mucous such as vaginal, nasal, ocular, and oral tissues. The great advantage of the use of thiolated polysaccharides in gel formulations has been seen not only in the mucoadhesive, but also in the *in situ* gelling properties.[19,20]

3. Conventional Polysaccharide Hydrogel Preparations

Polysaccharide hydrogels have been prepared with a variety of different shapes, geometries, and formulations that include liquid gels, powders, beads, films, tablets, sponges, nanofibrils, textile fibers, and inorganic composites. In each preparation polysaccharide chains are either chemically crosslinked or physically associated to form the hydrogel. Chemical crosslinking is a result of covalent bonding among the polymer chains with crosslinker molecules or functionalized group within the polymer chains. In comparison, physical crosslinking makes no use of the formation of covalent bonds, but rather uses hydrophobic interactions, polyelectrolyte bonding, and host-guest complexing, among others, to form bonds. Our discussion below will focus on these two distinct hydrogel engineering approaches.

3.1 *Chemical crosslinking*

Chemical crosslinking involves the formation of covalent bonds between polysaccharides to produce hydrogel networks (Fig. 3). Formation of covalent bonds in chemically crosslinked hydrogel helps to precisely control the physical and chemical properties of hydrogels such as gel density, pore size, chemical functionalization, and degradation or dissolutions. These properties lead to consistent performance in their real word applications. The chemical crosslinking occurs via short chain reactive molecules, which are called crosslinkers. In addition, covalently crosslinked hydrogels are also obtained by attaching photo-reactive or enzyme-sensitive molecules onto the polysaccharide chains, followed by their subsequent exposure to UV or sensitive enzymes, respectively. The properties of crosslinked hydrogels depend mainly on their crosslinking density and the ratio of moles of crosslinker molecules to the moles of polymer repeating units.[4] In contrast to physically crosslinked hydrogel (see next section), covalent crosslinking prevents both dilution of the hydrogel matrix and diffusion of the polymer away from the site of implantation.[5]

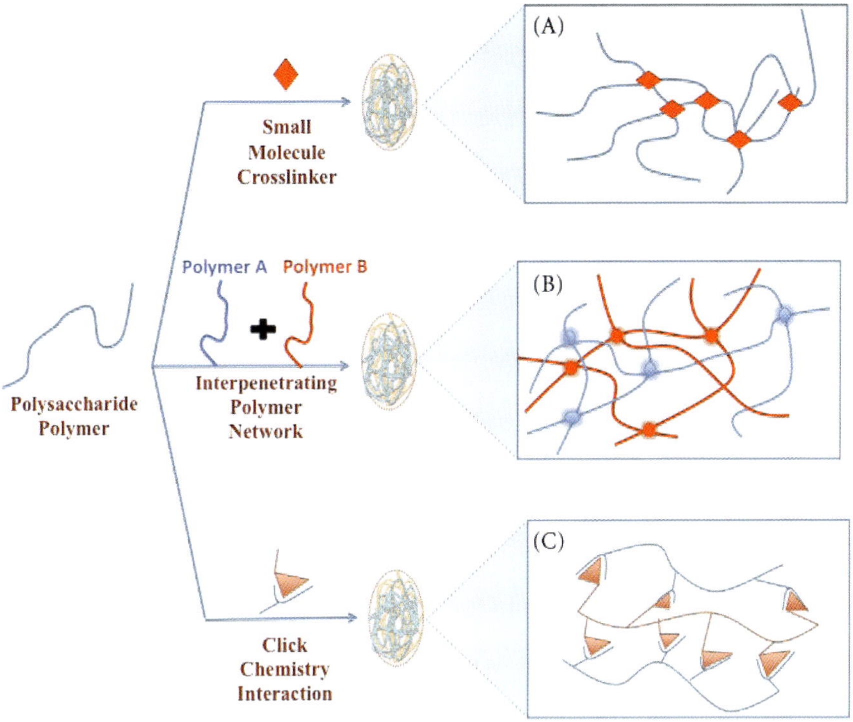

Fig. 3. Schematic of polysaccharide network formation via: (A) small molecule crosslinkers, (B) interpenetrating polymer interactions (IPN), and (C) click chemistry interactions, respectively.

In literature, many chemical reactions have been explored for synthesizing covalently-linking hydrogels. The most common crosslinking reactions include the: Reaction of an aldehyde and an amine to form a Schiff base; reaction of an aldehyde and hydrazide to form a hydrazine; and the Michael reaction of an acrylate and either a primary amine or a thiol to form a secondary amine or a sulfide.[5,21,22]

3.1.1. *Small molecule crosslinker reagents*

Small bi-functional molecules such as glutaraldehyde, diglycidyl ether, diisocyanate and diacrylate, etc. have been used to crosslink hydrogel networks.[5,23–26] The crosslinking that these molecules provide improves the mechanical and structural properties of hydrogels as the molecules create rigidity in the polymer networks. The improved rigidity is beneficial given the prevalence of low mechanical stiffness among natural polysaccharide hydrogels, the low stiffness can lead to rapid degradation of the polysaccharide polymers unless adequately fortified by high exposure to these crosslinking molecules.[5,15,27–30]

Covalently crosslinked hydrogels usually offer desirable physical properties, but the biocompatibilities of many crosslinkers are unknown. Since the fate of many of these molecules in the body has not been established. To prevent trace amounts of unreacted crosslinker agents *in vivo*, hydrogels must undergo vigorous purifications prior to implantation in the body. Moreover, crosslinking agents can react with the hydrogel payload, thus deactivating or limiting therapeutic release. Therefore, the search for safe and biocompatible covalent crosslinkers in the hydrogel research community is always of high priority.

The new crosslinking agent, genipin, is a chemical that is naturally derived from the gardenia and that has been shown to be one such biocompatible crosslinking agent.[31] Genipin has been reported to bind biological tissues[32] and biopolymers, such as chitosan and hyaluronic acid,[33] leading to covalent coupling.[32,33] Also, genipin it works as an effective crosslinking agent for polymers containing amino groups and it is much less cytotoxic than glutaraldehyde.[34] Chitosan memberanes crosslinked chitosan membranes with genipin exhibit a slower degradation rate than chitosan membranes crosslinked with their glutaraldehyde.[35] The drug release properties of both chitosan and chitosan-alginate hydrogels crosslinked *in situ* with genipin.[36,37] In addition, the crosslinking of hydrogels with enhanced physical properties of the hydrogels including a more compact microstructure, slower mass loss and higher compressive modulus.[33] Even though genipin shows good biocompatibility, it is still prons to interact negatively interact with encapsulated drugs, an unavoidable problem for gelation in the presence of a therapeutic.

3.1.2. *Click chemistry polymerization*

The internal structure of the hydrogels formed using small bi-functional molecules lacks chemical homogeneity.[38] However, "click chemistry has proven to be sufficient method of generating hydrogels with homogenous polymer networks click chemistry. Involves reactions that, create byproducts that can be removed without chromatography, and are high yielding wide in scope and simple to perform.[39] This process and makes use of the joining of polymeric subunits which allows for the crosslinking between the two components click chemistry has shown success in efficiency, and stability, and is easily applied to polysaccharides.[27] The large scope, high selectivity, mild reaction conditions, and nearly quantitative yields of these conversions perfectly fit the requirements of chemical, polymer-analogous polysaccharide modification by avoiding side reactions caused by the multifunctional and multi-chiral nature of these natural polymers.[40] Various reactions (Fig. 4) have been utilized in the formation of hydrogels including the Diels–Alder cycloaddition reaction, Cu (I)-catalyzed azide alkyne cycloaddition (CuAAC) reaction, and strain-promoted

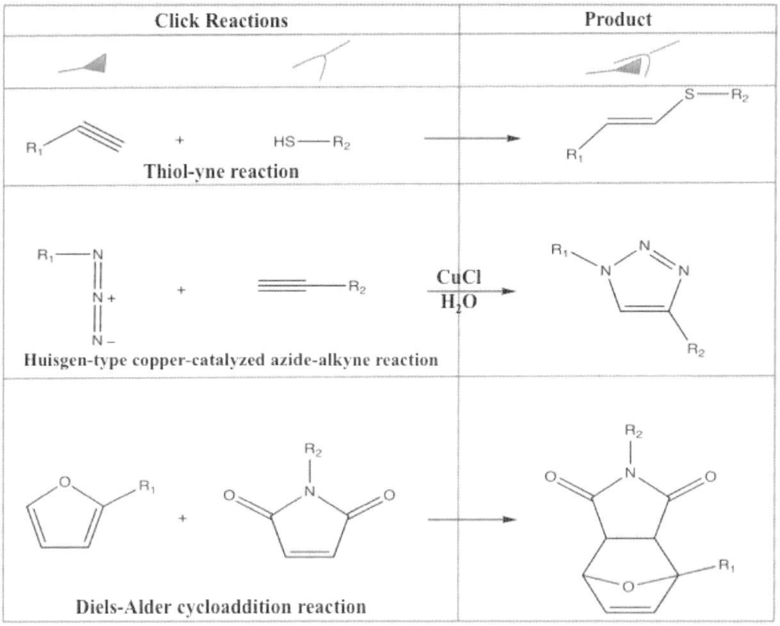

Fig. 4. Various common click reactions utilized in hydrogel fabrication and functionalization.

azide-alkyne cycloaddition (SPAAC) reaction. The thermoreversible nature of the Diels–Alder cycloaddition reaction has shown usefulness in localized delivery especially since the fabricated hydrogel's gelation time is strongly dependent on temperature, which makes this an attractive means to prepare stimuli-sensitive hydrogels.[41] Of the various forms of click chemistry reactions (Fig. 4) that can be utilized there is much interest in the use of Cu-free based click chemistry (e.g., SPAAC) as compared to the other reactions that require the use of toxic copper reactants, which are unfavorable for biomedical applications.[6,27,38] Studies have successfully shown that CU free click reactions can be used to create *in situ* hyaluronic acid hydrogels via modifications with azide and cyclooctyne groups.[42] The hyaluronic acid hydrogels that have been studied are noted for their applications as a means of preventing peritoneal adhesions, treating cartilage injury, improving tissue engineering, and providing drug delivery.[27,43,44]

3.1.3. *IPN interactions*

Advancements in the drug delivery field have led to the use of IPNs and semi-interpenetrating polymer networks. IPN fabrication provides a cheap option for fabricating hydrogels and strengthening their mechanical and physical properties.[45–49] These IPNs are composed of two or more networks that have been molecularly entwined through non-covalent chemical bonding and cannot be separated without breaking chemical bonds (i.e., formed in the presence of a crosslinker).[46,50,51] Semi-IPNs are the result of a polymer composed of one or more networks and one or more linear or branched polymer(s) characterized by the molecular scale penetration of at least one of the networks by some of the linear or branched polymers (i.e., formed in the absence of crosslinkers to create a network of embedded linear polymers encased in the original hydrogel).[46,50,51]

IPNs fabrication provides an IPN component meshing that can have interactions that can require multi-step preparations and purification processes, which can result in cytotoxic reactive groups and can limit their hydrogel applications when dealing with living cell or sensitive biological agents.[46] The Michael–Addition reaction, where the nucleophile reacts with the vinyl group on another polymer, has been used to prepare these polymer-polymer networks due to its rapid reaction time, ability to form different types of bonds, and relatively benign reactivity with biomolecules.[4,29] Several types of covalent linkages can be formed depending on the desired speed of crosslinking and selection of targeted reactive functional groups. These networks have allowed for injectable alginate–dextran methacrylate IPN systems to be developed for drug delivery as well as chitosan-gelatin IPN hydrogels crosslinked with genipin.[45,52]

3.2 *Physical interactions*

As noted in the previous section, hydrogels can be fabricated through the formation of covalent bonds. However, hydrogels can also be formed through non-covalent physical interaction strategies. These physical interactions enable for hydrogel formations that can be undone easily as opposed to existing covalently bonded hydrogels that are permanent in nature. Accordingly physically bonded hydrogels existing for a shorter period of time than covalently bonded hydrogels and serve as short-term drug release models rather than long-term one in aqueous *in vivo* environments.[4,53] These aforementioned non-covalent strategies include molecular entanglement and/or secondary forces (e.g., H-bonding, ionic or hydrophobic interactions).[3,54,55] Physical gels are non-homogeneous as a result of molecular entanglement or hydrophobically and ionically associated domains.[3] Further discussion of hydrophobic interactions, polyelectrolyte complexes, and host-guest complexes will be discussed in the following sections (Fig. 5).

3.2.1. *Hydrophobically assembled systems*

Hydrophobic interactions serve as a means of providing physically gelated matrixes. Polysaccharides can undergo chemical modification by the grafting of

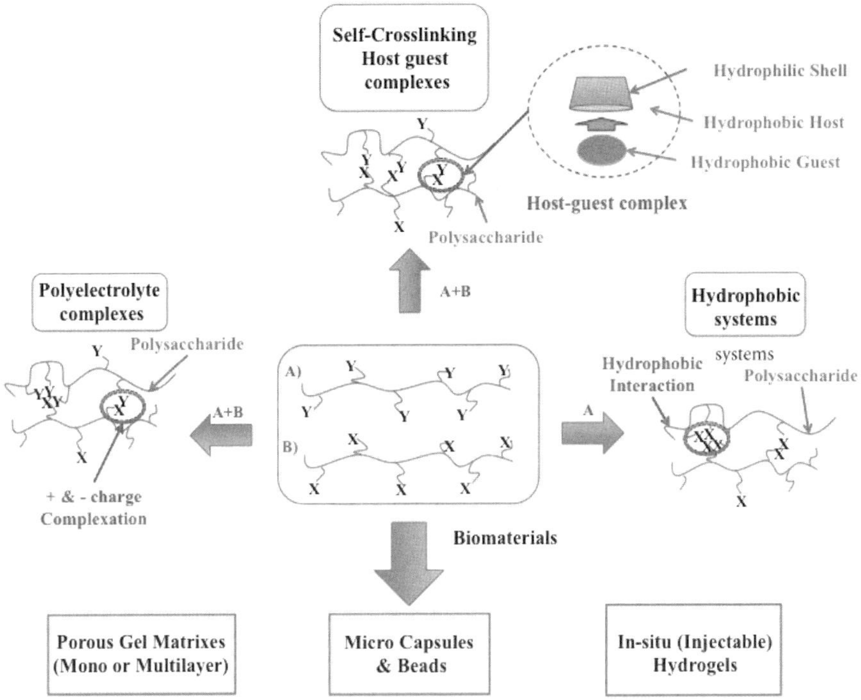

Fig. 5. Schematic of polysaccharide network formation via physical interactions.

hydrophobic groups.[56] Amphiphilic polysaccharide polymers with hydrophobic core domains and hydrophilic outer shells hydrophobically aggregate in aqueous environments given this hydrophobi interaction mechanism. The inter and intra-molecular hydrophobic interactions allowing for the self-association in aqueous solutions lead to different types of drug delivery systems such as micelles, nanoparticles, microspheres, liposomes and hydrogels.[56–60] The specific structures of the self-assembling polysaccharides are chosen based on the physiochemical properties of the loaded drug and the mode of administration.[56] These hydrophobic core domains can allow for the encapsulation of hydrophobic molecules. Amphiphilic polymers form self-assembled particles when introduced to aqueous environments as noted in the creation of amphiphilic PLGA-chitosan nanoparticles.[61] Amphiphilic alginate hydrogels have also been created through the covalent binding of hydrophobic alkyl chains. This method of self-assembly has in turn shown itself to be an efficient and versatile technique for the preparation of hydrogels.[62,63]

3.2.2. *Self-crosslinking host-guest complexes*

Hydrophobic interactions also play a role in host-guest complexation. Host-guest complexes create specific molecular recognition sites as the hydrophobic interior cavities have a high affinity for specific hydrophobic guest moieties, and lead to the ordered arrangement of molecules into defined structures.[5,64] Cyclodextrins are widely utilized hosts that are cyclic oligosaccharides synthesized from starch and have molecularly compatible cavities and negligible toxicity.[65] It has been noted that host-guest complexation can provide minimally invasive injectable hydrogels via the use of hyaluronic acid with a CD host and an adamantine guest.[64] Supramolecular assemblies with different viscosities were developed based on inclusion complex formation between a CD-grafted chitosan and various guest macromolecules such as adamantyl groups attached to chitosan.[66] Water-soluble CD-grafted chitosan derivatives and their inclusion complex with eugenol (EG) have been found to be a good candidates in terms of high EG entrapment, mucoadhesive property and anti-microbial activity for mucosal drug delivery.[67]

3.2.3. *Polyelectrolyte complexes (PECs)*

Some of the charged polysaccharides can be divided into positively charged and negatively charged electrolytes. Ionic interactions occur between systems of mixed charge (e.g., cations and anions). Ionic interactions with oppositely charged polyelectrolyte polysaccharides can lead to the creation of polyelectrolytes complexes. Polyelectrolytes are macro molecules that have a relatively large number of functional groups that are either innately charged or become charged under suitable conditions.[68,69] The primary interactions between two polyelectrolyte polymers are the reversible, but strong, electrostatic and dipole–dipole association, as well as hydrogen and hydrophobic bonds.[68,70] Chitosan is a positively charged polysaccharide and has been combined with other negatively charged natural polysaccharides (e.g., alginate, hyaluronic acid, etc.) to create chitosan-PEC membranes, hydrogels, and micro/nanoparticles for use in drug delivery for nutrients with delayed digestiability and controlled release for 3D scaffolding to mimic tumor microenvironment, and for bone tissue engineering.[68, 71–73]

Multivalent cations, such as calcium, magnesium, manganese, and aluminum are used to crosslink alginate. The vulnerable drugs and cell encapsulation.[3,74–76] The following are major disadvantages of these alginate/Ca^{++} systems; the gelation rate is hard to control; controlling the release kinetics of the encapsulated drugs is difficult; and in a process involving the loss of multivalent ions the beads degrade via into the surrounding medium.[77–79] In an attempt to precisely control the swelling properties and strength of alginate beads crossing with a second polymer has been adopted.[5] Although both complexes with polymer and metallic cations are based on ionic interaction, the polymer complexes possess additional macromolecular chain physical entanglement properties that are much stronger than hydrogen bonding van der Waals interactions and other secondary binding michanisns

D-glucono-δ-lactone, polyols, chitosan, polycationic polymers, and other secondary polymers have been used to obtain alginate/Ca^{++} beads with a more even structure.[77,82–84] This improvement is the resuet of alterations to the crosslinking process ionically crosslinked hydrogels including both microbeads and quaternized chitosan mixed with glycerophosphate have been widely used in drug and cell delivery applications.[5,15,28,29]

4. Drug Release

4.1. *Mechanisms of loading*

Major factors that are highly critical for successful drug release are: on the natural polysaccharide material chosen, the drug properties, of the drugs being used, and the drug loading method itself. There are three primary methods of drug loading: Entrapment, covalent linking, and diffusion.[4,62] Entrapment involves the formation of the hydrogel around large drug molecules to prevent the loss of drug function. Covalent linking binds the drug to the polysaccharide itself. This type of linking, in turn allows for a slower drug release model after gelation. It also reduces the potential for drug toxicity from high release concentrations. Diffusion makes use in diffusion, smaller drug molecules permeat the hydrogel network when the hydrogel is exposed to an aqueous solution containing the drug molecule. The diffusion and the entrapment process both lead to a rapid initial drug release. This rapid initial drug release also is highly dependent on the hydrogel's highly permeable nature. Beyond the primary drug loading models, drugs can be loaded via the functionalization of the hydrogel polymers with amphiphilic additives which will house the hydrophobic drugs internally when exposed to an aqueous environment. On the other hand the drugs can be loaded into small secondary release vehicles (e.g., microparticles, microgels, liposomes, and micelles) prior to hydrogel entrapment.

4.2. *Release triggers*

Drug release innately occurs through diffusive processes, however the use of stimuli-sensitive systems greatly increases the localized and target delivery of drug encapsulated hydrogels. The swelling and the physical breakup of the hydrogel cause the degradation of the hydrogels *in vivo*. These degradative processes, respectively, occur as the pore sizes are enlarged. The enlarged pores lead to a quicker drug release and the hydrogel disintegrates to allow additional drug release. This disintegration occurs primarily as the hydrogel crosslinks break apart in the aqueous environment.

The polymers of stimuli–sensitive systems change properties in response to exposure to biological targets (receptors, enzymes, physiological fluids, etc.) where

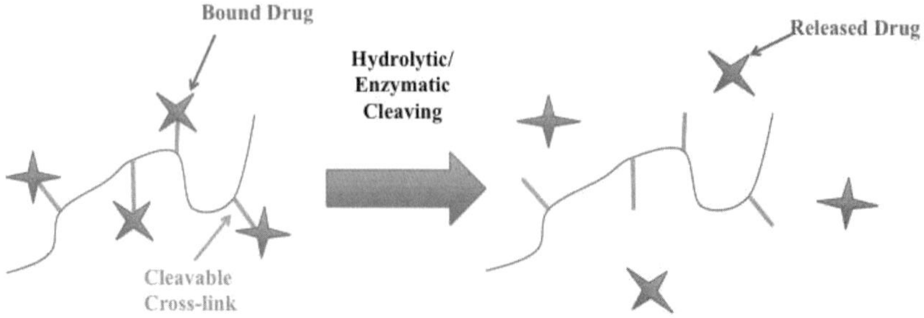

Fig. 6. Hydrolytic and enzymatic cleavage of crosslinker molecules to provide drug release.

molecular interactions lead to microscopic responses (i.e., solution to gel transitions, swelling, collapsing, etc.).[85,86] Stimuli that involve hydrolytic or enzyme-catalyzed reactions are useful, in releasing encapsulated drugs given drug inactivity *ex-vivo* and drug activity *in-vivo*. The inactivity and activity can be controlled by the use of crosslinks that can undergo hydrolytic or enzymatic cleavage (Fig. 6).

Hydrogel sensitivities provide controllable delivery in situations involving low pH, body-like temperatures, and electrical currents. Stimuli, found in environments that possess low pH and high temperatures in the body, which call for pH-sensitive and/or temperature-sensitive hydrogels for localized drug release.[2] The functionalization of polysaccharide polymers with weak acidic groups or weak basic groups can result in pH-sensitive hydrogels. This functionalization contributes to controlled drug release by causing the hydrogel to swell or de-swell in various pH ranges. Additionally, the development of pH sensitive-stealth coating (i.e., steric barrier that protects from opsonization and cellular interactions) has led to the establishment of electrostatic interactions with cell membranes in acidic pH.[61] This pH sensitivity, in turn, prevents the premature leakage of drugs from hydrogels in non-target sites and maximizes drug release at the target site.[11] Research on reverse thermo-sensitive gels has led to polymers that are injectable in the liquid state, but solidify in *in vivo* conditions.[2,87] Additionally, electrically-sensitive gels have been developed, and change the hydrogel permeability under electric fields (i.e., for neural, applications cardial, applications etc.).[88]

5. Localized Drug Delivery Systems

Through modification the physical-chemical properties and fabrication method as given in the previous sections, natural polysaccharides can produce exceptional localized hydrogel release systems. Natural polysaccharides are similar in characteristics to the extracellular matrix, ECM, they have ideal bio characteristics, and they have excellent cellular interactions.[89] These characteristics enable natural polysaccharides to be utilized in localized target sites in the form of microspheres, *in situ* hydrogels, and/or porous 3D matrixes. Accordingly, the localized delivery of polysaccharide drug systems is a key element in addressing tissue repair, cancer therapy, GI tract abnormalities, transdermal applications, and other medical processes and procedures. In addition to the incorporation of various drug molecules into the hydrogel delivery systems, polysaccharide also possess the ability to carry growth factors and whole cells for improved localized delivery for in therapeutic wound repair and disease treatment.

5.1. *Microspheres and nanospheres*

Microspheres, which are also referred to as microbeads and microcapsules, can possess either hollow or solid cores, both of which allow for internalization of therapeutic molecules and/or whole cells. Microspheres have been used in various biomedical engineering fields for their encapsulation and controlled release capabilities. These qualities have conferred led to many advances in tissue regeneration. For example, aliganate coupled with fibrin microbeads encapsulating human umbilical cord mesenchymal stem cells has been shown to be useful in proliferation, osteodifferentiation, and bone mineralization for bone tissue engineering.[90,91] In another instance human mesenchymal stem cells encapsulated in RGD modified alginate microspheres showed successful local controlled delivery of both cells and growth factors capable of facilitating myocardial infarction repairs.[92,93] Similar success has been achieved through sequential delivery of IGF-1 and HGF from affinity binding alginate-sulfate.[93] Beyond the use in tissue repair, microspheres are instrumental in many other situations. For example, chitosan-PLGA conjugates have been loaded with anti-tumor molecules to improve their chemotherapeutics and pH specific capsaicin-chitosan microspheres have shown potential for preventing obesity by increasing the rate of fat metabolism.[94,95]

The application of hydrogel microspheres, as discussed previously, is also done effective on the nanoscale in order to further providing improved targeting and drug release. The sub-cellular and sub-micron sizes allow for deep tissue penetration through fine capillaries, crossing of the fenestration present in the epithelial lining (e.g., liver), and efficient cellular uptake.[1] The are a promising drug delivery system due to high drug loading potential, sustained drug/gene effect in target tissue, elevated stability from enzymatic degradation, (i.e., within the cells themselves and increased oral and brain bioavailability of low molecular drugs and biomacromolecules.[62] Unlike nanoparticles which have a deuse polymeric are nonospheres possess an inner nanospace capable of drug encapsulation.[96] Functional modification of nanospheres is an important element in enhancing targeted delivery as this strategy increases the stability of the sphere, thus leading to prolonged localized circulation with sustained release of drugs not limited to cellular barriers.[29,62] Prolonged local circulation and reduced immune clearance are primary design criteria in tumor targeting, since the majority of administered nanocarriers depend on the leakiness of the vasculature surrounding tumors to reach targeted tumors.[11]

5.2. *In situ hydrogels*

In situ hydrogel are formed at the site of injection either through self-assembly or through photo polymerization. Self-assembling hydrogels are formed spontaneously after exposure to various physiological triggers such as pH or temperature while photo polymerization occurs through non-spontaneous irradiation *in situ* and is limited by depth of penetration.[97,98] These systems can be positioned in complex shapes according to the required dimensions of the injection site and the mechanical interlocking with surrounding tissues to strengthen the tissue hydrogel interface.[99] In addition, *in situ* forming hydrogels have proven to be attractive for cell delivery given their on-site introduction by a minimally invasive procedure that can be adapted for the specific tissue or defect to be repaired.[98] An example of *in situ* hydrogel delivery is seen in the loading of T31 human tracheal scar fibroblasts into hyaluronan thiolate derivative hydrogels for potential *in situ* clinical use.[99]

In situ hydrogel systems can also make use of thermally induced gelation when the in situ hydrogels is near physiological temperatures. This mode of induced gelation is an attractive route for delivery systems since that the method of crosslinking does not require harsh physical or chemical conditions. In the process the drugs or cells are mixed with hydrogel in the low temperature "sol" state, before being injected into a targeted tissue site, typically the subcutaneous layer of mammals, before gel at body temperature.[100] This process is also known as a sol-to-gel transition since natural polysaccharides do not possessing innate thermal responsiveness, various strategies, including the matrix fabrication methods mention earlier have been used to confer this physio-chemical property such as chemical modifications, formation of IPNs with other polymers possessing thermo-responsiveness, and complexation with hydroxylated salts.[98] A well-known example of thermogelling systems is Pluronic, where the sol-to-gel transition occurs by micellar packing through unimer-to-micelle transition as the temperature increases.[100,101] Copolymers of alginate and Pluronic F127 have been used to fabricate thermogels to deliver the antidepressant selegiline and provide novel selegiline delivery paths in addition to pre-existing systems such as Emsam.[102]Pluronic derivatives have also been used in tumor suppression as noted through conjugation with chitosan and alginate to encapsulate curcumin.[103]

5.3. *Porous 3D matrixes*

Scaffolds which can be designed to mimic the ECM are porous, three-dimensional, structures that can serve as structural support for tissues and provide an environment for cell seeding. Alginate, chitosan and hyaluronic acid possess glycosaminoglycans (GAGs), which are a major components of the native ECM.[104–106] The biochemical characteristics of GAGS allow for cellular proliferation by guiding cell colonization.[107] Aside from GAGS and other chemical cues, physical properties such as stiffness, hydrophilicity porosity, pore size and void fraction can influence cell alignment, orientation, multicellular organization, cell spreading, and cell attachment.[107–112] Applications of scaffolds in cartilage and cardiac regeneration make use of the cell organizing properties of scaffolds. Silk fibroin (SF) and hyaluronic acid (HA) blends have been prepared for use in both cartilage repair and cardiac repair patches which ahve been prepared from micro particles of SF and HA and crosslinked by genipin. Both the limits in the ability of damaged cardiac tissue — to repair itself after infarction and advances in cardiac donors make the shortage of cardiac tissue repair imperative. Even so, advances in engineering cardiac tissues by using alginate scaffolds revealed intense neovascularization with implantation onto rat infarcted myocardium.[109,115]

Additional scaffold applications have focused on the use of collagen-chitosan composites leading to the increased levels of platelet derived growth factor and nerve growth factor with elevated levels of antioxidants and decreased levels of peroxidation for wound repair.[116] When made into sponges and seeded with human fibroblasts and keratinocytes and grown *in vitro* before being transplanted onto the back of rats these collagen chitosan composites have helped to provide new insights into nerve regeneration for deep and extensive wounds.[117] Although, porous 3D matrixes are frequently used in wound repair bone regeneration is, one of the major applications for the matrines. Bone is a composite of collagenous fibrous matrix enriched with carbonated apatite.[118] As such, natural polysaccharide composites have been created to mimic the structure of bone tissue. For example, nano-hydroxyapatite/chitosan-pectin composites have been created to possess ideal porosity, water absorption, and degradative behaviors for mouse pre-osteoblasts (MC 3T3-E1) attachment and proliferation.[119] Chitosan-alginate scaffolds combined with mesenchymal stem cells have also been pursued for their enhanced bone regeneration capabilities.[120]

6. Perspectives and Conclusions

This review of using natural polysaccharide hydrogels for controlled drug release discusses the wide application of these hydrogel in drug delivery. Polysaccharide chains are highly biocompatible and easily modified through the addition of functional groups. The modification capabilities for controlled degradation at the targeted site and the conferring of additional biological properties for drug release, these biological properties serve to affect the cellular interactions of the polysaccharide hydrogels. Hydrogel physiochemical properties can be modified in the formation of the hydrogel itself, through chemical interactions, physical interactions, or a mixture of the two. These interactions as well as many other provide a myriad of applications within the field of drug release. As such, novel developments in hydrogel applications are allowing for the continued advancement of natural polysaccharide hydrogels for controlled localized drug delivery. For example microgels and nanogels have led to increased the localized delivery of drug molecules. Additionally research not only has shown the potential for amphiphilic nanogels to serve as artificial molecular, but also has led to breakthroughs in protein or peptide delivery.[96] Moreover, to improve dermal repair, *in situ* hydrogels loaded with T31 human tracheal scar fibroblasts have been place into hyaluronan thiolate derivative hydrogels. The development of polysaccharide hydrogels for localized delivery has become an important element in the push to develop ideal drug carrier models. These innovations will result in highly improved, more effective, and customized controlled drug systems that will provide greater advances not only in drug targeting, but also in drug delivery therapies.

References

1. J. Panyama and V. Labhasetwar, Biodegradable nanoparticles for drug and gene delivery to cells and tissue, *Adv. Drug Deliv. Rev.*, **64**, 61–71(2012).
2. Y. Qiu and K. Park, Environment-sensitive hydrogels for drug delivery, *Adv Drug Deliv. Rev.*, **64**, 49–60 (2012).
3. A. S. Hoffman, Hydrogels for biomedical applications, *Adv Drug Deliv Rev.*, **64**, 18–23 (2012) .
4. N. Bhattarai, J. Gunn, and M. Zhang, Chitosan-based hydrogels for controlled, localized drug delivery, *Adv Drug Deliv. Rev.*, **62**, 83–99 (2010).
5. T. R. Hoare and D. S. Kohane, Hydrogels in drug delivery: Progress and challenges, *Polymer*, **49**, 1993–2007 (2008).
6. C. Chang and L. Zhang, Cellulose-based hydrogels: Present status and application prospects, *Carbohydrate Polymers*, **84**, 40–53 (2011).
7. T. Coviello, P. Matricardi, C. Marianecci, and F. Alhaique, Polysaccharide hydrogels for modified release formulations, *J. Control Release*, **119**, 5–24 (2007).
8. N. Saha, A. Saarai, N. Roy, T. Kitano, and P. Saha, Polymeric biomaterial based hydrogels for biomedical applications, *J Biomat. Nanobiotechnol.*, **2**, 85–90 (2011).
9. V. Guarino, A. Gloria, M. G. Raucci, and L. Ambrosio, Hydrogel-based platforms for the regeneration of osteochondral tisue and intervertebral disc, *Polymers*, **4**, 1590–1612 (2012).

10. S. V. Vlierberghe, P. Dubruel, and E. Schacht, Biopolymer-Based Hydrogels as scaffolds for tissue engineering applications: A review, *Biomacromolecules*, **12**, 1387–1408 (2011).
11. N. Zhang, P. R. Wardell, and R. A. Bader, Polysaccharde-based micelles for drug delivery, *Pharmaceutics*, **5**, 329–352 (2013).
12. K. Reddy, G. K. Mohan, S. Satla, and S. Gaikwad, Natural polysaccharides: Versatile excipients for controlled drug delivery systems, *AJPS*, **6**, 275–286 (2011).
13. B. Balakrishnan and R. Banerjee, Biopolymer-based hydrogels for cartilage tissue engineering, *Chem. Rev.*, **111**, 4453–4474 (2011).
14. S. Fiejdasz, K. Szczubizlka, J. Lewandowska-Lancucka, A. M. Osyczka, and M. Nowakowska, Biopolymer-based hydrogels as injectable materials for tissue repair scaffolds, *Biomed. Mater.*, **8**, 035013 (2013).
15. Z. Liu, Y. Jiao, Y. Wang, C. Zhou, and Z. Zhang, Polysaccharides-based nanoparticles as drug delivery systems, *Adv. Drug Deliv. Rev.*, **60**, 1650–1662 (2008).
16. G. P. Andrews, T. P. Laverty, and D. S. Jones, Mucoadhesive polymeric platforms for controlled drug delivery, *Eur. J. Pharm. Biopharm.*, **71**, 505–518 (2009).
17. M. N. Kumar, R. A. Muzzarelli, C. Muzzarelli, H. Sashiwa, and A. J. Domb, Chitosan chemistry and pharmaceutical perspectives, *Chem. Rev.*, **104**, 6017–6084 (2004).
18. J. Ding, R. He, G. Zhou, C. Tang, and C. Yin, Multilayered mucoadhesive hydrogel films based on thiolated hyaluronic acid and polyvinylalcohol for insulin delivery, *Acta Biomater.*, **8**, 3643–3651 (2012).
19. A. Bernkop-Schnurch : A new generation of mucoadhesive polymers, *Adv. Drug. Deliv. Rev.*, **57**, 1569–1582 (2005).
20. A. Bernkop-Schnurch, M. Hornof, and T. Zoidl, Thiolated polymers-thiomers: Synthesis and *in vitro* evaluation of chitosan-2-iminothiolane conjugates, *Int. J. Pharm.*, **260**, 229–237 (2003).
21. N. Bhattarai, J. Gunn, and M. Zhang, Chitosan-based hydrogels for controlled, localized drug delivery, *Adv. Drug Deliv. Rev.*, **62**, 83–99 (2010).
22. X. Xu, A. K. Jha, D. A. Harrington, M. C. Farach-Carson, and X. Q. Jia, Hyaluronic acid-based hydrogels: From a natural polysaccharide to complex networks, *Soft Matter*, **8**, 3280–3294 (2012).
23. J. Berger, M. Reist, J. M. Mayer, O. Felt, N. A. Peppas, and R. Gurny, Structure and interactions in covalently and ionically cross-linked chitosan hydrogels for biomedical applications, *Eur. J. Pharm. Biopharm.*, **57**, 19–34 (2004).
24. W. E. Hennink and C. F. van Nostrum, Novel cross-linking methods to design hydrogels, *Adv. Drug. Deliv. Rev.*, **54** 13–36 (2002).
25. N. Bhattarai and M. Q. Zhang, Controlled synthesis and structural stability of alginate-based nanofibers, *Nanotechnol.*, **18**, 455601 (2007).
26. J. A. Burdick and G. D. Prestwich, Hyaluronic acid hydrogels for biomedical applications, *Adv. Mater.*, **23** H41–H56 (2011).
27. F. Yu, X. Cao, L. Zeng, Q. Zhang, and X. Chen, An interpenetrating HA/G/CS biomimic hydrogel via Diels–Alder click chemistry for cartilage tissue engineering, *Carbohydrate Polymers*, **97** 188–195 (2013).

28. W.-H. Tan and S. Takeuchi, Monodisperse Alginate Hydrogel microbeads for cell encapsulation, *Advanced Materials*, **19**, 2696–2701 (2007).
29. J. K. Oh, R. Drumright, D. J. Siegwart, and K. Matyjaszewski, The development of microgels/nanogels for drug delivery applications, *Prog. Polym. Sci.*, **33**, 448–477 (2008).
30. K. Y. Lee, K. H. Bouhadir, and D. J. Mooney, Controlled degradation of hydrogels using multi-functional cross-linking molecules, *Biomaterials*, **25**, 2461–2466 (2004).
31. J. Jin, M. Song, and D. J. Hourston, Novel chitosan-based films cross-linked by genipin with improved physical properties, *Biomacromolecules*, **5**, 162–168 (2004).
32. H. W. Sung, I. L. Liang, C. N. Chen, R. N. Huang, and H. F. Liang, Stability of a biological tissue fixed with a naturally occurring cross-linking agent (genipin), *J. Biomed. Mater. Res.*, **55**, (2001) 538–546.
33. H. P. Tan, H. Li, J. P. Rubin, and K. G. Marra, Controlled gelation and degradation rates of injectable hyaluronic acid-based hydrogels through a double crosslinking strategy, *J. Tissue Eng. Regen M*, **5**, 790–797 (2011).
34. H. W. Sung, R. N. Huang, L. L. H. Huang, and C. C. Tsai, *In vitro* evaluation of cytotoxicity of a naturally occurring cross-linking reagent for biological tissue fixation, *J. Biomat Sci-Polym. E*, **10**, 63–78 (1999).
35. F. L. Mi, Y. C. Tan, H. C. Liang, R. N. Huang, and H. W. Sung, *In vitro* evaluation of a chitosan membrane cross-linked with genipin, *J. Biomat. Sci-Polym. E*, **12** 835–850 (2001).
36. N. Bhattarai, H. R. Ramay, J. Gunn, F. A. Matsen, and M. Q. Zhang, PEG-grafted chitosan as an injectable thermosensitive hydrogel for sustained protein release, *J. Control Release*, **103**, 609–624 (2005).
37. S. C. Chen, Y. C. Wu, F. L. Mi, Y. H. Lin, L. C. Yu, and H. W. Sung, A novel pH-sensitive hydrogel composed of N,O-carboxymethyl chitosan and alginate cross-linked by genipin for protein drug delivery, *J. Control Release*, **96**, 285–300 (2004).
38. S. Yigit, R. Sanyal, and A. Sanyal, Fabrication and Functionalization of Hydrogels through "Click" Chemistry, *Chem. Asian J.*, **6**, 2648–2659 (2011).
39. K. B. Sharpless and R. Manetsch, *In situ* click chemistry: A powerful means for lead discovery, *Expert. Opin. Drug Dis*, 1 525–538 (2006).
40. T. Liebert, C. Hänsch, and T. Heinze, Click chemistry with polysaccharides, *Macromolecular Rapid Communications*, **27**, 208–213 (2006).
41. E. Lallana, F. Fernandez-Trillo, A. Sousa-Herves, R. Riguera, and E. Fernandez-Megia, Click chemistry with polymers, dendrimers, and hydrogels for drug delivery, *Pharm. Res.*, **29**, 902–921(2012).
42. A. Takahashi, Y. Suzuki, T. Suhara, K. Omichi, A. Shimizu, K. Hasegawa, N. Kokudo, S. Ohta, and T. Ito, *In Situ* cross-linkable hydrogel of hyaluronan produced via copper-free click chemistry, *Biomacromolecules*, **14**, 3581–3588 (2013).
43. Y. Yeo, T. Ito, E. Bellas, C. B. Highley, R. Marini, and D. S. Kohane, *In Situ* cross-linkable hyaluronan hydrogels containing polymeric nanoparticles for preventing postsurgical adhesions, *Annals of Surgery*, **245**, 819–824 (2007).
44. G. Huerta-Angeles, M. Němcová, E. Příkopová, D. Šmejkalová, M. Pravda, L. Kučera, and V. Velebný, Reductive alkylation of hyaluronic acid for the synthesis of biocompatible hydrogels by click chemistry, *Carbohydrate Polymers*, **90**, 1704–1711 (2012).

45. L. Cui, J. Jia, Y. Guo, Y. Liu, and P. Zhu, Preparation and characterization of IPN hydrogels composed of chitosan and gelatin cross linked by genipin, *Carbohydrate Polymers*, **99**, 31–38 (2014).
46. S. Choudhary, J. C. White, W. L. Stoppel, S. C. Roberts, and S. R. Bhatia, Gelation behavior of polysaccharide-based interpenetrating polymer network (IPN) hydrogels, *Rheologica. Acta*, **50**, 39–52 (2011).
47. L. L. J. Bastide, Large-scale heterogeneities in randomly cross-linked networks, *Macromolecules*, **21**, 2647–2649 (1988).
48. M. A. Haque, T. Kurokawa, and J. P. Gong, Super tough double network hydrogels and their application as biomaterials, *Polymer*, **53**, 1805–1822 (2012).
49. J. P. Gong, Why are double network hydrogels so tough? , *Soft Matter*, **6**, 2583–2590 (2010).
50. P. Matricardi, C. D. Meo, T. Coviello, W. E. Hennink, and F. Alhaique, Interpenetrating Polymer Networks polysaccharide hydrogels for drug delivery and tissue engineering, *Adv. Drug Deliv. Rev.*, **65**, 1172–1187 (2013).
51. A. D. Jenkins, P. Kratochvíl, R. F. T. Stepto, and U. W. Suter, Glossary of basic terms in polymer science (IUPAC Recommendations 1996), *Pure and Applied Chemistry*, **68**, 2281–2311 (1996).
52. P. Matricardi, M. Pontoriero, T. Coviello, M. A. Casadei, and F. Alhaique, *In Situ* cross-linkable novel alginate-dextran methacrylate IPN hydrogels for biomedical applications: Mechanical characterization and drug delivery properties, *Biomaterials*, **9**, 2014–2020 (2008).
53. N. A. Peppas, Hydrogels in medicine and pharmacy, *Br. Polym J.*, **21**, 184 (1989).
54. D. Campoccia, P. Doherty, M. Radice, P. Brun, G. Abatangelo, and D. F. Williams, Semisynthetic resorbable materials from hyaluronan esterification, *Biomaterials*, **19**, 2101–2127 (1998).
55. D. M. M. G. D. Prestwich, J. F. Marecak, K. P. Vercruysse, and M. R. Ziebell, Controlled chemical modification of hyaluronic acid, *J Control Release.*, **53**, 93–103 (1998).
56. L. N. Hassani, F. Hendra, and K. Bouchemal, Auto-associative amphiphilic polysaccharides as drug delivery systems, *Drug Discovery Today*, **17**, 608–614 (2012).
57. F.-L. Mia, C.-K. Peng, Mei-Feng Huan, S.-H. Lo, and C.-C. Yang, Preparation and characterization of N-acetylchitosan, N-propionylchitosan and N-butyrylchitosan microspheres for controlled release of 6-mercaptourine, *Carbohydrate Polymers*, **60**, 219–227 (2005).
58. W. Wang, A. M. McConaghy, L. Tetley, and I. F. Uchegbu, Controls on polymer molecular weight may be used to control the size of palmitoyl glycol chitosan polymeric vesicles, *Langmuir*, **17**, 631–636 (2001).
59. G. Liang, Z. Jia-Bi, X. Fei, and N. Bin, Preparation, characterization and pharmacokinetics of N-palmitoyl chitosan anchored docetaxel liposomes, *J. Pharm Pharmacol.*, **59**, 661–667 (2007).
60. C. Dufes, A. G. Schätzlein, L. Tetley, A. I. Gray, D. G. Watson, J. -C. Olivier, W. Couet, and I. F. Uchegbu, Niosomes and polymeric chitosan based vesicles bearing transferrin and glucose ligands for drug targeting, *Pharm. Res.*, **17**, 1250–1258 (2000).

61. Z. Amoozgar, J. Park, Q. Lin, and Y. Yeo, Low molecular-weight chitosan as a pH-sensitive stealth coating for tumor-specific drug delivery, *Mol. Pharm.*, 1262–1270 (2012).
62. A. V. Kabanov and S. V. Vinogradov, Nanogels as Pharmaceutical Carriers, in: Multifunctional Pharmaceutical Nanocarriers, 2008, pp. 67–80.
63. J. Shi, Z. Zhang, W. Qi, and S. Cao, Hydrophobically modified biomineralized polysaccharide alginate membrane for sustained smart drug delivery, *Int. J. Bio. Macromolecules*, **50**, 747–753 (2012).
64. C. B. Rodell, A. L. Kaminski, and J. A. Burdick, Rational design of network properties in guest–host assembled and shear-thinning Hyaluronic acid hydrogels, *Biomacromolecules*, **14**, 4125–4134 (2013).
65 G. Chen and M. Jiang, Cyclodextrin-based inclusion complexation bridging supramolecularn chemistry and macromolecular self-assembly, *Chem. Soc. Rev.*, **40**, 2254–2266 (2011).
66. R. Auzely-Velty and M. Rinaudo, New supramolecular assemblies of a cyclodextrin-grafted chitosan through specific complexation, *Macromolecules*, **35**, 7955–7962 (2002).
67. W. Sajomsang, O. Nuchuchua, P. Gonil, S. Saesoo, I. Sramala, A. Soottitantawat, S. Puttipipatkhachorn, and U. R. Ruktanonchai, Water-soluble beta-cyclodextrin grafted with chitosan and its inclusion complex as a mucoadhesive eugenol carrier, *Carbohydrate Polymers*, **89**, 623–631 (2012).
68. Y. Luo and Q. Wang, Recent development of chitosan-based polyelectrolyte complexes with natural polysaccharides for drug delivery, *Int. J. Bio. Macromolecules.*, In press (2013).
69. J. Du, J. Dai, J. L. Liu, and T. Dankovich, Novel pH-sensitive polyelectrolyte carboxymethyl Konjac glucomannan-chitosan beads as drug carriers, *React. Funct .Polym.*, **66**, 1055–1061 (2006).
70. A. V. Il'ina and V. P. Varlamov, Chitosan-based polyelectrolyte complexes: A review, Applied Biochemistry and Microbiology, **41**, 5–11 (2005).
71. X. Meng, F. Tian, J. Yang, C.-N. He, N. Xing, and F. Li, Chitosan and alginate polyelectrolyte complex membranes and their properties for wound dressing application, Journal of Materials Science: *Mater. Med.*, **21**, 1751–1759 (2010).
72. J. Berger, M. Reist, J. M. Mayer, O. Felt, N. A. Peppas, and R. Gurny, Structure and interactions in covalently and ionically crosslinked chitosan hydrogels for biomedical applications, *Eur. J. Pharm. Biopharm.*, **57**, 19–34 (2004).
73. S. Boddohi, N. Moore, P. A. Johnson, and M. J. Kipper, Polysaccharide-based polyelectrolyte complex nanoparticles from chitosan, heparin, and hyaluronan, *Biomacromolecules*, **10**, 1402–1409 (2009).
74. S. Van Vlierberghe, P. Dubruel, and E. Schacht, Biopolymer-based hydrogels as scaffolds for tissue engineering applications: A review, *Biomacromolecules*, **12**, 1387–1408 (2011).
75. F. Lim, and A. M. Sun, Microencapsulated islets as bioartificial endocrine pancreas, *Science*, **210**, 908–910 (1980).

76. A. S. Hoffman, Hydrogels for biomedical applications, *Adv. Drug. Deliver. Rev.*, **54**, 3–12 (2002).
77. C. K. Kuo and P. X. Ma, Ionically cross-linked alginate hydrogels as scaffolds for tissue engineering: Part 1. Structure, gelation rate and mechanical properties, *Biomaterials*, **22**, 511–521 (2001).
78. K. Y. Lee and D. J. Mooney, Hydrogels for tissue engineering, *Chem. Rev.*, **101**, 1869–1879 (2001).
79. L. S. Liu, J. Kost, F. Yan, and R. C. Spiro, Hydrogels from biopolymer hybrid for biomedical, food, and functional food applications, *Polymers Basel*, **4**, 997–1011 (2012).
80. K. Y. Lee, J. A. Rowley, P. Eiselt, E. M. Moy, K. H. Bouhadir, and D. J. Mooney, Controlling mechanical and swelling properties of alginate hydrogels independently by cross-linker type and cross-linking density, *Macromolecules*, **33**, 4291–4294 (2000).
81. K. H. Bouhadir, D. S. Hausman, D. J. Mooney, Synthesis of cross-linked poly(aldehyde guluronate) hydrogels, *Polymer*, **40**, 3575–3584 (1999).
82. A. D. Baldwin and K. L. Kiick, Polysaccharide-modified synthetic polymeric biomaterials, *Biopolymers*, **94**, 128–140(2010).
83. S. A. Abbah, W. W. Lu, D. Chan, K. M. Cheung, W. G. Liu, F. Zhao, Z. Y. Li, J. C. Leong, and K. D. Luk, Osteogenic behavior of alginate encapsulated bone marrow stromal cells: an *in vitro* study, *J. Mater. Sci. Mater. Med.*, **19**, 2113–2119 (2008).
84. N. Bhattarai, J. Gunn, and M. Q. Zhang, Chitosan-based hydrogels for controlled, localized drug delivery, *Adv. Drug Deliver. Rev.*, **62**, 83–99 (2010).
85. R. V. Ulijn, N. Bibi, V. Jayawarna, P. D. Thornton, S. J. Todd, R. J. Mart, A. M. Smith, and J. E. Gough, Bioresponsive hydrogels, *Materials Today*, **10**, 40–48 (2007).
86. J. Jagur-Grodzinski, Polymeric gels and hydrogels for biomedical and pharmaceutical applications, *Polym. Adv. Tech.*, **21**, 24–47 (2010).
87. B. Jeonga, S. W. Kim, and Y. H. Bae, Thermosensitive sol–gel reversible hydrogels, *Adv. Drug Deliv. Rev.*, **64**, 154–162 (2012).
88. J. Kosta, R. Langer, Responsive polymeric delivery systems, *Adv. Drug Deliv. Rev*, **64**, 327–341 (2012).
89. S. Huang and X. Fu, Naturally derived materials-based cell and drug delivery systems in skin regeneration, *J. Control Release*, **142**, 149–159 (2012).
90. H. Zhou and H. H. K. Xu, The fast release of stem cells from alginate-fibrin microbeads in injectable scaffolds for bone tissue engineering, *Biomaterials*, **32**, 7503–7513 (2011).
91. L. Zhao, M. Tang, M. D. Weir, M. S. Detamore, and H. H. K. Xu., Osteogenic media and rhBMP-2-induced differentiation of umbilical cord mesenchymal stem cells encapsulated in alginate microbeads and integrated in an injectable calcium phosphate-chitosan fibrous Scaffold, *Tissue Eng. A*, **17**, 969–979 (2011).
92. J. Yu, K. T. Du, Q. Fang, Y. Gu, S. S. Mihardja, R. E. Sievers, J. C. Wu, and R. J. Lee, The use of human mesenchymal stem cells encapsulated in RGD modified alginate microspheres in the repair of myocardial infarction in the rat, *Biomaterials*, **31**, 7012–7020 (2010).

93. E. Ruvinov, J. Leor, and S. Cohen, The promotion of myocardial repair by the sequential delivery of IGF-1 and HGF from an injectable alginate biomaterial in a model of acute myocardial infarction, *Biomaterials*, **32**, 565–578 (2011).
94. S. S. Chakravarthi and D. H. Robinson, Enhanced cellular association of paclitaxel delivered in chitosan-PLGA particles, *Int. J. Pharma.*, **409**, 111–120 (2011).
95. J. Chen, G.-D. Huang, S.-R. Tan, J. Guo, and Z.-Q. Su, The preparation of capsaicin-chitosan microspheres (CCMS) enteric coated tablets, *Int. J. Mol. Sci.*, **14**, 24305–24319 (2013).
96. Y. Sasaki and K. Akiyoshi, Nanogel engineering for new nanobiomaterials: From chaperoning engineering to biomedical applications, *Chem. Rec.*, **10**, 366–376 (2010).
97. S. R. V. Tomme, G. Storm and W. E. Hennink, *In situ* gelling hydrogels for pharmaceutical and biomedical applications, *Int. J. Pharm.*, **355**, 1–18 (2008).
98. T. Delair, *In situ* forming polysaccharide-based 3D-hydrogels for cell delivery in regenerative medicine, *Carbohydrate Polymers*, **87**, 1013–1019(2012).
99. X. Z. Shu, Y. Liu, F. S. Palumbo, Y. Luo, and G. D. Prestwich, *In situ* cross-linkable hyaluronan hydrogels for tissue engineering, *Biomaterials*, **25**, 1339–1348 (2004).
100. D. Y. Ko, U. P. Shinde, B. Yeon, and B. Jeong, Recent progress of *in situ* formed gels for biomedical applications, *Prog. Polym. Sci.*, **38**, 672–701 (2013).
101. A. Sosnik and D. Cohn, Reverse thermo-responsive poly (ethylene oxide) and poly (propylene oxide) multiblock copolymers, *Biomaterials*, **26**, 349–357 (2005).
102. C.-C. Chen, C.-L. Fang, S. A. Al-Suwayeh, Y.-L. Leu, and J.-Y. Fang, Transdermal delivery of selegiline from alginate–Pluronic composite thermogels, *Int. J Pharm.*, **415**, 119–128 (2011).
103. R. K. Das, N. Kasoju, and U. Bora, Encapsulation of curcumin in alginate-chitosan-pluronic composite nanoparticles for delivery to cancer cells, Nanomedicine: *NBM*, **6**, 153–160 (2010).
104. M. P. Lutolf and J. A. Hubbell, Synthetic biomaterials as instructive extracellular microenvironments for morphogenesis in tissue engineering, *Nat. Biotechnol.*, **23**, 47–55 (2005).
105. M. Rinaudo, Main properties and current applications of some polysaccharides as biomaterials`, *Polym. Int.*, **57**, 397–430 (2008).
106. S. J. Florczyk, G. Liu, F. M. Kievit, A. M. Lewis, J. D. Wu, and M. Zhang, 3D Porous chitosan-alginate scaffolds: New matrix for studying prostate cancer cell-lymphocyte interactions *In Vitro*, *Adv. Healthc. Mater.*, **1**, 590–599 (2013).
107. B. J. Lawrence and S. V. Madihally, Cell colonization in degradable 3D porous matrices, *Cell. Adh. Migr.*, **2**, 9–16 (2008).
108. A. P. Balgude, X. Yu, A. Szymanski, and R. V. Bellamkonda, Agarose gel stiffness determines rate of DRG neurite extension in 3D cultures, *Biomaterials*, **22**, 1077–1064 (2001).
109. A. Dar, M. Shachar, J. Leor, and S. Cohen, Optimization of cardiac cell seeding and distribution in 3D porous alginate scaffolds, *Biotechnol. Bioeng.*, **80**, 305–312 (2002).

110. A. M. Rajnicek, S. Britland, and C. D. McCaig, Contact guidance of CNS neurites on grooved quartz: Influence of groove dimensions, neuronal age and cell type, *J. Cell. Sci.*, **110**, 2905–2913 (1997).
111. C. S. Ranucci and P. V. Moghe, Substrate microtopography can enhance cell adhesive and migratory responsiveness to matrix ligand density, *Journal of Biomedical Materials Research Part B*: *Appl. Biomater.*, **54**, 149–161 (2001).
112. J. Zeltinger, J. K. Sherwood, D. A. Graham, R. Mueller, and L. G. Griffith, Effect of pore size and void fraction on cellular adhesion, proliferation, and matrix deposition, *Tissue Eng.*, **7**, 557–572 (2001).
113. C. Foss, E. Merzari, and C. Migliaresi, A. Motta, Silk fibroin/hyaluronic acid 3D matrices for cartilage tissue engineering, *Biomacromolecules*, **14**, 38–47 (2012).
114. M.-C. Yang, N.-H. Chi, N.-K. Chou, Yi-You Huanga, T.-W. Chung, Y.-L. Chang, H.-C. Liu, M.-J. Shieh, and S.-S. Wang, The influence of rat mesenchymal stem cell CD44 surface markers on cell growth, fibronectin expression, and cardiomyogenic differentiation on silk fibroin — Hyaluronic acid cardiac patches, *Biomaterials*, **31**, 854–862 (2010).
115. J. Leor, S. Aboulafia-Etzion, A. Dar, L. Shapiro, I. M. Barbash, A. Battler, Y. Granot, and S. Cohen, Bioengineered cardiac grafts to repair the infarcted myocardium and attenuate heart failure, *Circulation*, **102**, 56–61 (2000).
116. M. N. Rangasamy Judith, Chellan Rose, and Asit Baran Mandal, Biopolymer gel matrix as acellular scaffold for enhanced dermal tissue regeneration, *Biologicals,* 231–239 (2012).
117. I. P. Marie Gingras, and François Berthod, Nerve regeneration in a collagen–chitosan tissue-engineered skin transplanted on nude mice, *Biomaterials*, 1652–1661 (2003).
118. K. Gkioni, Sander C. G. Leeuwenburgh, T. E. L. Douglas, A. G. Mikos, and J. A. Jansen, Mineralization of hydrogels for bone regeneration, *Tissue Eng. B: Rev*, **16**, 577–585 (2010).
119. J. Lia, Hong Sunb, Da Suna, Yuli Yaob, and C.a.c.i. Fanglian Yaoa, E-mail the corresponding author, E-mail the corresponding author, K. Yao, Biomimetic multicomponent polysaccharide/nano-hydroxyapatite composites for bone tissue engineering, *Carbohydrate Polymers*, **85**, 885–894 (2011).
120. S. J. Florczyk, M. Leung, Z. Li, J. I. Huang, R. A. Hopper, and M. Zhang, Evaluation of three-dimensional porous chitosan alginate scaffolds in rat calvarial defects for bone regeneration applications, *J. Biomed. Mater. Res. A*, **101**, 2974–2983 (2013).

Chapter 3

In Situ Crosslinked Hydrogels for Drug Delivery

Jin Woo Bae and Ki Dong Park*
Department of Molecular Science Technology, Ajou University
5 Woncheon, Yeongtong, Suwon 443-749, ROK
**kdp@ajou.ac.kr*

In situ crosslinked hydrogels are an injectable-type of swollen hydrophilic matrices, particularly characterized by rapid gelation. Recently, these hydrogels hold great promise as drug delivery carriers due to their various advantages. They can facilitate not only encapsulation of drugs under mild conditions, but also *in situ* modulation of drug release kinetics by varying crosslinking reaction conditions. This chapter outlines various *in situ* crosslinking methods that are most commonly used, and briefly introduces hydrogel-based drug release mechanisms.

1. Introduction

Over many decades, polymeric systems for controlled drug delivery have been extensively exploited due to various advantages in safety, efficacy, and patient compliance.[1] Among numerous polymer-based drug delivery platforms, *in situ* crosslinked hydrogels have shown great promise as a versatile injectable therapeutic vehicle. This system facilitates formulation of various types of drugs in a pre-gel state, and then can be implanted into/onto the human body using a minimally invasive procedure to rapidly form drug-encapsulated hydrogels via a sol-gel transition. *In situ* crosslinked hydrogels are particularly advantageous for encapsulating biologically active and hydrophilic drugs such as proteins and genes, because these biomacromolecules are capable of being encapsulated without loss of their pharmacological activity under mild conditions.

In situ crosslinked hydrogels can be classified into two major categories based on the crosslinking methods: chemically-driven hydrogels and physicochemical association-driven hydrogels.[2] Chemical crosslinking methods allow the formation of covalent bonds between polymer chains, whereas physicochemical methods result in non-covalent association of polymers. Although both methods can confer mechanical stability and structural integrity suitable for drug delivery to hydrogel matrices, chemical crosslinking is still considered better for retaining mechanical properties of hydrogels. A variety of crosslinking methods have been proposed for the preparation of *in situ* crosslinked hydrogels: redox-polymerixation, photopolymerization, Michael addition, click reaction, enzymatic reaction, thermogelation, ion-induce gelation, inclusion complex formation, stereocomplex formation, and complementary binding.[3] Although all these methods can be applied for hydrogel-based drug delivery, the method needs to be capable of not only easily introducing distinctive molecular features to trigger *in situ* crosslinking, but also simply modulating key parameters to obtain desired hydrogel properties. This chapter provides an overview of the most commonly employed *in situ* crosslinking methods for hydrogel preparation, and drug release mechanisms in hydrogels.

2. *In Situ* Crosslinking Strategies for Hydrogel Formation

2.1. *Photopolymerization*

Photopolymerization is a light-based technique that uses to initiate and propagate a polymerization reaction. This technique has been extensively investigated to fabricate protective coatings, filling materials and films.[4] It offers several advantages including spatial and temporal control over polymerization, rapid curing rate, and minimal heat production, compared to conventional polymerization techniques. In particular, *in situ* photopolymerization enables to crosslink aqueous precursors in the presence of light-sensitive photoinitiators, thereby forming hydrogels.

To create *in situ* crosslinked hydrogels by photopolymerization, at least two reactive components are required: photoinitiator and photopolymerizable material. A photoinitiator creates free radicals upon exposure to specific wavelengths of light, and then initiates polymerization of liquid monomers or macromers through a photolytic mechanism. The rate of photopolymerization is dependent on several parameters such as photoinitiator efficiency, photoinitiator concentration, and light intensity.[5] However, free radical species created by a photoinitiator can damage the surrounding tissue due to oxidative damage to cell membranes, nucleic acids, and proteins.[6,7] In addition, radical polymerization *in vivo* accompanies a temperature increase. Considering these issues, Williams and his colleagues have evaluated the cytotoxicity of three commonly used photoinitiators (Irgacure 2959, 184, and 651) against six cell lines including bovine chondrocytes, goat bone marrow-derived mesenchymal stem cells, human bone marrow-derived mesenchymal stem cells, rabbit corneal epithelial cells, human fetal osteoblasts and human embryonic germ cells.[8] Since they reported that Irgacure 2959 was well tolerated by many cell types, it has been widely employed to produce *in situ* photo-crosslinked hydrogels. Another component is a photopolymerizable material that typically possesses double bonds. In most cases, water-soluble macromolecular precursors rather than monomers have been used due to their good biocompatibility.[4] They include poly(ethylene glycol) (PEG) acrylate derivatives, PEG methacrylate derivatives, oligo(PEG fumarate), nitrocinnmated chitosan and methacrylated hyaluronic acid[9–13] (Fig. 1).

For *in situ* photo-crosslinking, a photoinitiator is dissolved in an aqueous solution containing photopolymerizable macromers. Then, the precursor solution immediately forms a hydrogel upon UV light exposure. Incorporation of drugs into a hydrogel can be simply achieved using a drug–precursor mixture. As visible or UV light is transmitted through skin layers, a subcutaneously injected

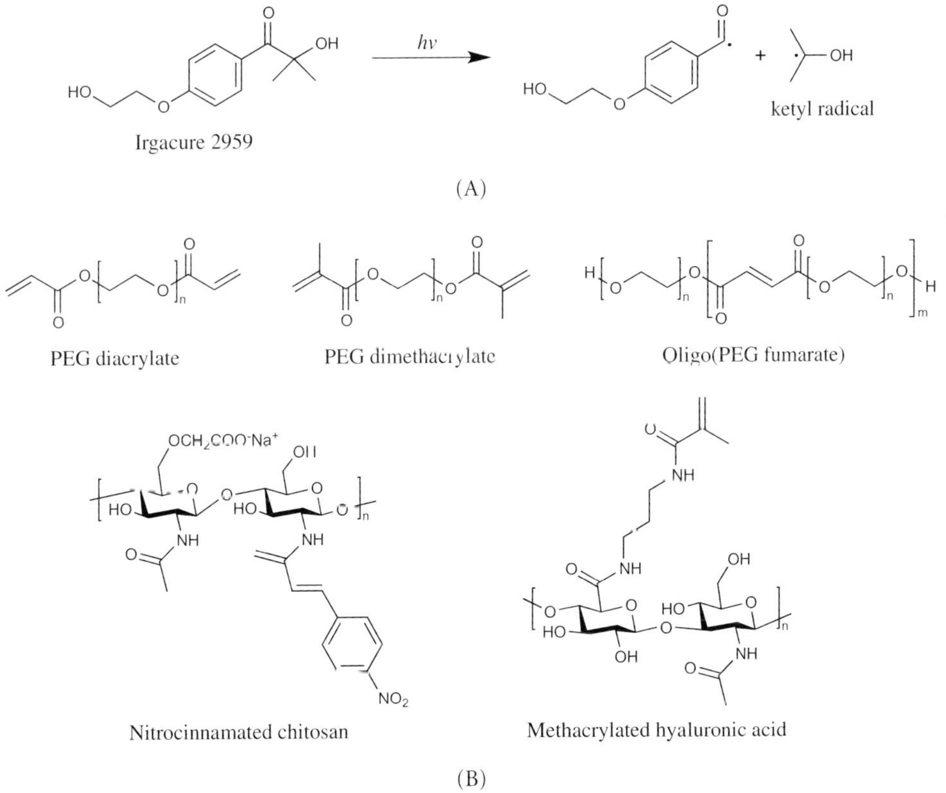

Fig. 1. Irgacure 2959 initiated photopolymerization (A)[14] and various photopolymerizable macromers (B).

hydrogel precursor solution can be converted to a hydrogel upon transdermal irradiation.[15]

2.2. *Michael addition*

Michael addition is a facile conjugation reaction between nucleophiles and electron poor olefins, typically referred to the base catalyzed addition of a nucleophile (Michael donor) to an activated α,β-unsaturated compound (Michael acceptor).[16] The reaction proceeds rapidly and efficiently at room temperature after mixing two aqueous solutions comprising Michael donors and acceptors. For the preparation of hydrogels using the Michael addition, water-soluble macromers need to be functionalized separately with either Michael donor or acceptor. Thiols and amines are commonly used as Michael donors, whereas there are numerous types of Michael acceptors characterized by olefins with an electron withdrawing and resonance-stabilizing group. Acrylates, methacrylates, maleimides and vinyl sulfones have been mostly used as Michael acceptors to prepare hydrogels (Fig. 2). Thiols are more reactive toward Michael acceptors than amines.[17] The Michael addition reaction rate increases with increasing pH, because deprotonation of thiols at high pH results in increased production of thiolate anions that are the active species in Michael

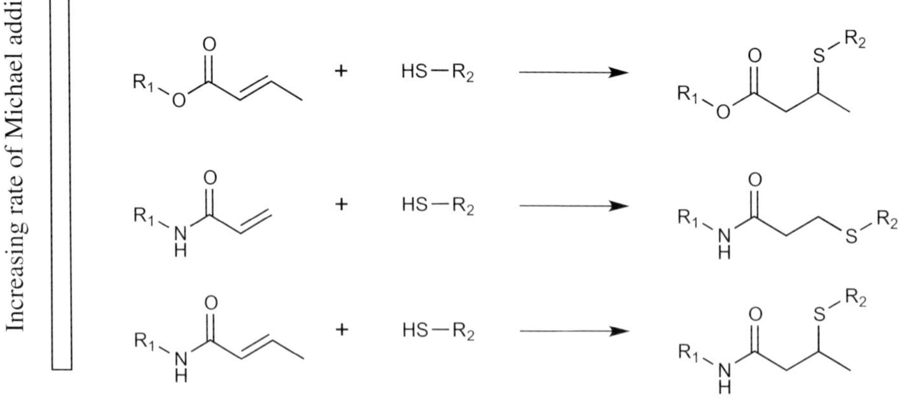

Fig. 2. Several examples of Michael addition reactions and comparison of their reaction rates. The Michael acceptors possessing double bonds react with thiolated compounds, thereby forming thioether bonds.

additions. In addition, they are found in proteins with cysteine residues that can be tethered to hydrogel precursors. As a result, their use in *in situ* preparation of hydrogels via the Michael addition has been widely reported.

Prestwitch's group has first reported *in situ* crosslinkable hydrogels based on the Michael addition between thiolated hyaluronic acids (thiolated HA) and α,β-unsaturated PEG derivatives.[18] PEG-diacrylate (PEGDA), PEG-dimethacrylate (PEGDM), PEG-diacrylamide (PEGDAA), and PEG-dimethacrylamide (PEGDMA) were used to investigate the relative reaction rates with cysteine as a model compound. They found that PEGDA exhibited the highest rate of reaction with cysteine, obtaining over 90% of yield within 2 min. PEGDM were less reactive with cysteine than PEGDA owing to the decreased electrophilicity of the double bond and increased steric hindrance. Other PEG derivatives had much lower reactivity than their corresponding esters. They also described that a higher pKa value of thiolated HA leads to a faster conjugate addition. By varying the mixing ratio of thiolated HA and PEGDA, the gelation time could be controlled from 5 to 31 min. Vinylsulfone as another Michael acceptor has shown much faster gelation time than acrylate.[19] However, derivatization of water-soluble macromers with vinylsulfone appears relatively complicated and needs an extensive purification to remove unreacted reagents.

2.3. *Horseradish peroxidase (HRP)-catalyzed reaction*

An enzyme-catalyzed crosslinking has received much attention lately as an emerging and promising approach to produce *in situ* crosslinked hydrogels. Several kinds of enzymes that can trigger covalent binding between molecules have been reported, including transglutaminase, HRP, tyrosinase, phosphopantetheinyl transferase, lysyl oxidase and plasma amine oxidase.[3,20] All these enzymatic systems are highly selective and operated under mild reaction conditions in respect of pH, temperature, and pressure. In particular, a HRP-catalyzed crosslinking approach has been extensively studied due to several advantages over other enzymatic systems.

HRP is a stable heme-containing protein that has the ability to catalyze the oxidation of phenols in the presence of hydrogen peroxide (H_2O_2).[21] The catalytic cycle of HRP for phenol–phenol conjugation is illustrated in Fig. 3.[22] First, the catalytic cycle is initiated by an interaction between H_2O_2 and HRP, which generates Compound I with a cation radical. Then, compound I is converted to compound II by using phenol as a reducing substrate. In this step, the produced phenoxy radical is consumed for phenol–phenol conjugation. An additional one-electron reduction step allows HRP to return to its original resting state, which produces the second phenoxyl radical available for further conjugation. Therefore, both HRP and H_2O_2 affect significantly various properties of hydrogels formed by phenol–phenol conjugation, which includes gelation time, stiffness and degradation rate.

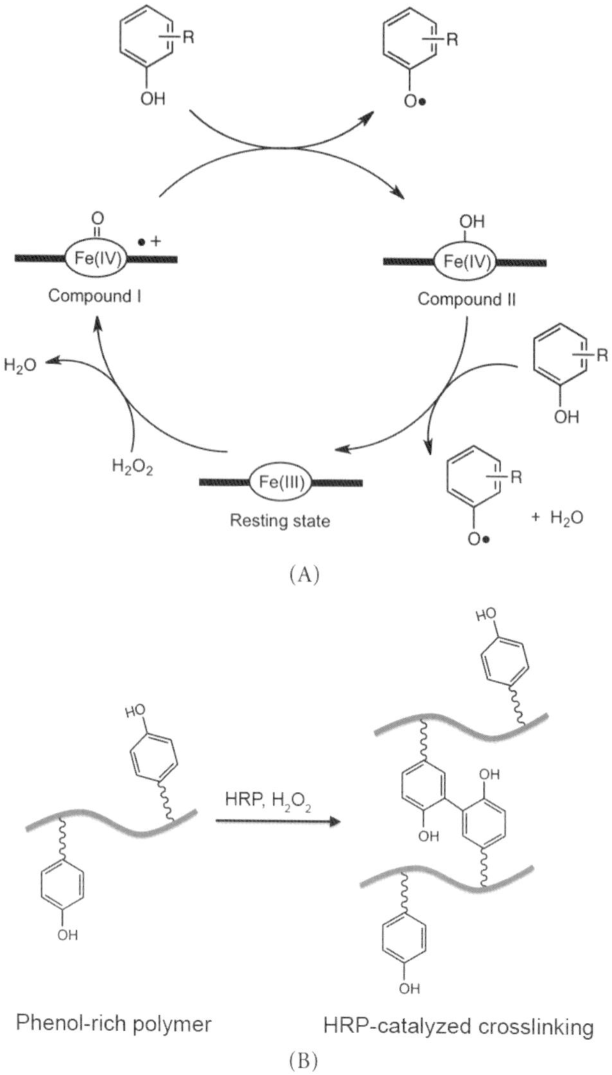

Fig. 3. The catalytic cycle of HRP with a phenolic compound as a reducing substrate (A) and HRP-catalyzed *in situ* crosslinking reaction between phenol-rich polymers in the presence of H_2O_2 (B). Adapted from Ref. 22.

The relationship between HRP/H_2O_2 and hydrogel properties has been well established by recent studies.[23–27] In the HRP-catalyzed crosslinking reaction, an increase in the HRP concentration leads to the decreased gelation time. It also leads to increased stiffness and degradation rate due to the increasing crosslinking density. As for H_2O_2, the gelation time increases with decreasing H_2O_2 concentration while increasing H_2O_2 has a positive impact on both hydrogel stiffness and degradation rate. However, an excessive use of H_2O_2 needs to be avoided, because it leads to H_2O_2-induced cell death.[25] The HRP-catalyzed crosslinking method is beneficial for facile *in situ* conjugation of phenolic compounds during gelation.[28] If a drug molecule has a phenol group in its intact or modified structure, it can be covalently linked to hydrogel matrices. Despite various advantageous features of HRP-catalyzed *in situ* crosslinking, there are still few obstacles to resolve for clinical use of HRP-catalyzed *in situ* forming hydrogels. As HRP is not clinically available at present, HRP entrapment inside hydrogels during gelation may have a potential safety issue.[29] Therefore, it is necessary to develop alternative agents that can mimic the catalytic activity of HRP, or a new methodology of HRP-catalyzed crosslinking.

2.4. *Thermoresponsive sol–gel transition*

Many polymers with balanced hydrophilicity and hydrophobicity undergo sudden changes in solubility in response to temperature changes. Using the thermoresponsive sol–gel transition phenomena, aqueous solutions of thermoresponsive polymers can form their hydrogels upon injection into the body. In the sol state at a low temperature, aqueous thermoresponsive polymer solutions flow freely, allowing drug molecules to disperse sufficiently. Injection of the aqueous mixture of polymer-drug into the body produces a drug-encapsulated hydrogel due to a temperature shift to 37°C. Thermoresponsive polymers typically possess two distinguishable domains that are either hydrophilic or hydrophobic, as shown in Fig. 4. With increasing

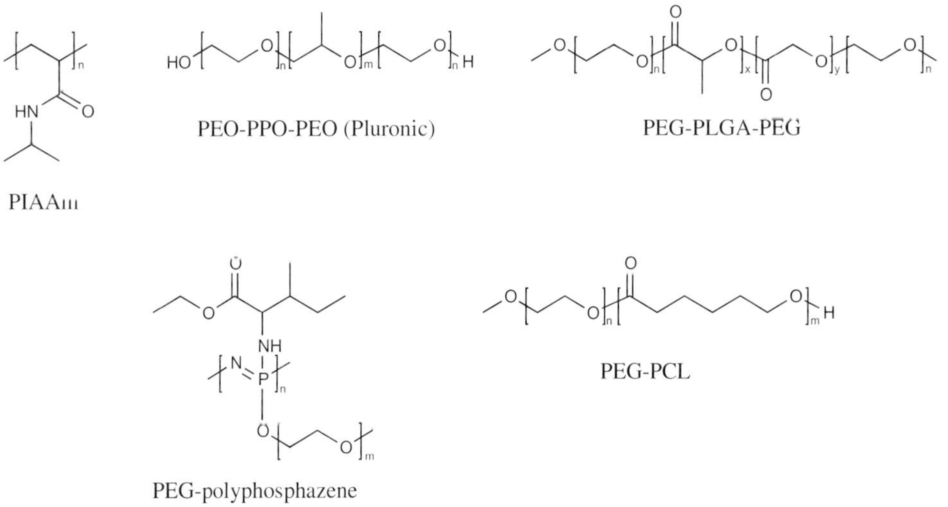

Fig. 4. Examples of thermoresponsive polymers. Thermoresponsive polymers are water-soluble and possess the amphiphilic nature resulted from a balanced hydrophilicity and hydrophobicity in their structures.

temperature, the hydrophobic domains of polymers physically associate with each other in a non-covalent manner to form a hydrogel.

Thermoresponsive hydrogels for drug delivery have been extensively reviewed in the literature.[30–33] Poly(*N*-isopropyl acrylamide) (PIPAAm) is a representative thermoresponsive polymer that undergoes a discrete phase transition in water around the lower critical solution temperature (LCST) of 32°C. To modulate physicochemical properties of PIPAAm hydrogels, hydrophobic or hydrophilic vinyl monomers can be copolymerized with IPAAm to obtain PIPAAm copolymers.[34] A change in the ratio of PIPAAm in block copolymers results in different thermal, mechanical and swelling properties that affect the drug release behavior. However, non-biodegradability of PIPAAm limits its *in vivo* drug delivery applications. For that reason, the vast majority of drug delivery systems use thermoresponsive block copolymers possessing good biodegradability and biocompatibility. One example of thermoresponsive block copolymers is Pluronics, which are commercially available with molecular weights ranging between 1,100 and 14,600. Pluronics, also called Poloxamers, are triblock poly(ethylene oxide)–poly(propylene oxide)–poly(ethylene oxide) (PEO–PPO–PEO) copolymers that exhibit LCST above the body temperature. There are a series of Pluronics varying in lengths of PEO and PPO, which form hydrogels by micellar packing as the temperature increases. However, they are quickly eroded in a few days due to the inherently poor mechanical properties. To improve the mechanical properties of Pluronic-based hydrogels, additional chemical crosslinking or hybridization with other materials can be employed.[35,36] Besides, various types of thermoresponsive polymers that use PEG as a hydrophilic block have been reported, including PEG-polyesters and PEG-polyphosphazenes. Thermoresponsive hydrogels for drug delivery are particularly advantageous from the clinical point of view, because they only need a small change in temperature to trigger temperature-induced gelation. However, this system generally exhibits low mechanical properties, implying its lack of ability for long-term drug release. In some cases, they expel water from hydrogels, which may cause an elevated local concentration of drugs. Therefore, these aspects need to be considered in designing thermoresponsive hydrogels suitable for drug delivery.

2.5. *Ion-induced complexation*

Multivalent ionic species can induce gelation by bridging molecules with opposite charges. Alginate is a well-known example of an ionically crosslinkable polymer.[37] Alginate is a linear polysaccharide that contains varying amounts of 1,4-linked β-D-mannuronic acid (M) and α-L-glucuronic acid (G) residues. It is currently used as the sodium salt of alginic acid (sodium alginate). Addition of divalent cations such as calcium into an aqueous alginate solution allows egg-box-like intermolecular bridging between two G blocks of adjacent alginate chains, leading to the formation of a gel network. Consequently, the properties of alginate hydrogels including stiffness and gelation time can be controlled by varying the stoichiometry of alginates and chelating cations.[38] However, *in vivo* degradation of alginate is poorly regulated due to the lack of alginase production in mammals.[39] In addition, alginate is relatively inert because mammalian cells do not express receptors for alginate.[40] Thus, chemical modification of alginate is required to control biodegradation and improve biocompatibility.

3. Drug Release from Hydrogels

Hydrogels are three-dimensional (3D) hydrophilic polymer networks that contain a large amount of water. Based on this unique nature, water diffusion and exchange occur through the hydrogel mesh (defined as the spacing between polymer chains) when drug-encapsulated hydrogels are implanted. The mesh size of a hydrogel network appears crucial for determining important properties of hydrogels including mechanical strength, degradation, and molecular diffusivity.[42,43] It can be typically manipulated by degree of crosslinking, chemical structure of the network components, and external stimuli such as pH, temperature, and ionic strength.[41] If the mesh size in a hydrogel is sufficiently larger than drug molecules, the drug molecules can easily diffuse out from swollen hydrogel matrices (Fig. 5). On the contrary, the diffusion of large drug molecules such as peptide, proteins and genes can be retarded. This is a simplified mesh size-dependent relationship with drug diffusion. However, there are many parameters affecting drug release kinetics in practical hydrogel systems. The parameters are related to material selection, crosslinking chemistry and hydrogel properties. To accurately predict drug release as a function of time, drug release mechanisms must be also considered in detail.

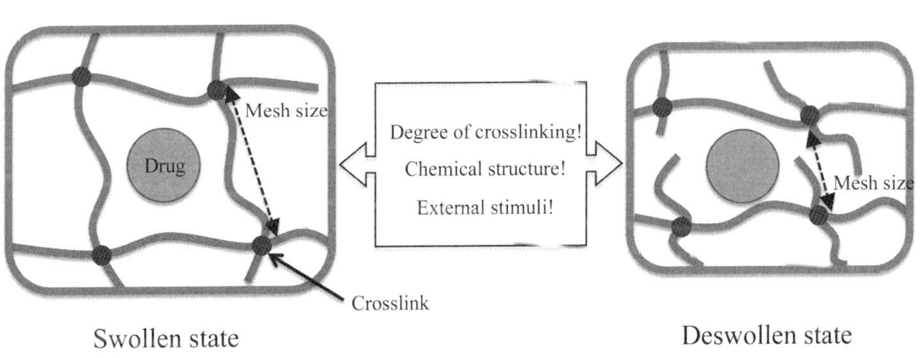

Fig. 5. A schematic illustration of mesh size in hydrogels at swollen or deswollen states. Adapted from Ref. 41.

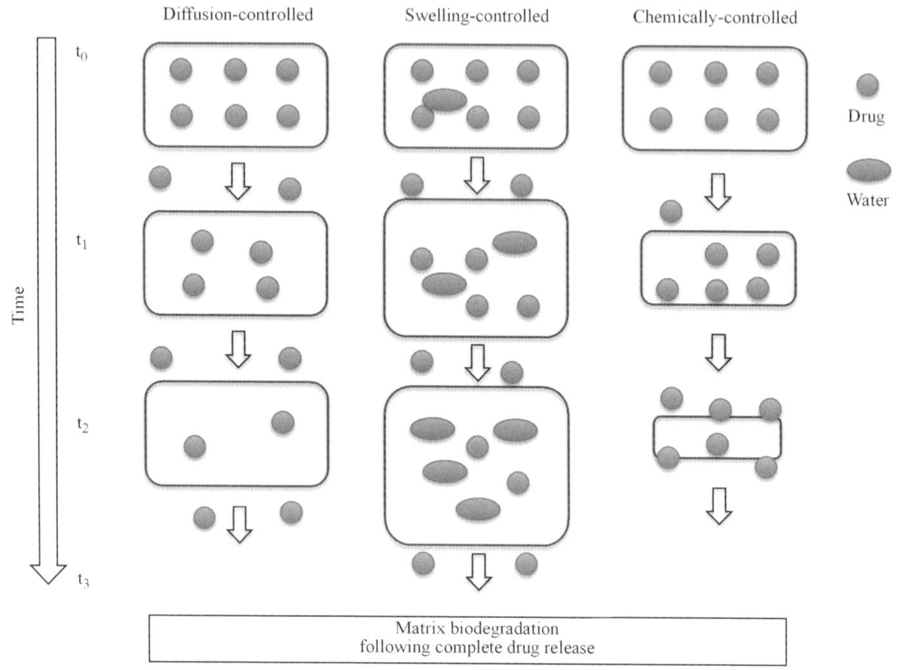

Fig. 6. A schematic illustration of three common mechanisms for controlled drug release from hydrogel matrices. Adapted from Ref. 44.

When hydrogel-based drug formulations are delivered into the body via several routes including oral, transdermal, nasal, and parenteral administrations, kinetics of drug release from hydrogel matrices can be controlled by three mechanisms based on the rate-limiting step, as illustrated in Fig. 6:[41,44] diffusion-controlled, swelling-controlled, and chemically-controlled. In diffusion-controlled systems, encapsulated drugs diffuse out through a hydrogel network that degrades eventually after complete release of the drugs. The diffusional behavior of encapsulated drugs commonly appears in most hydrogels. Another mechanism is swelling-controlled drug release, which is associated with hydrogel swelling that is the macroscopic evidence of glass-to-rubber transition of polymer. Below the glass transition temperature, T_g, of a polymer, the mobility of polymer chains is extremely low. In contrast, the polymer chains are highly mobile in the rubbery state above T_g, which allows much easier transport of water and drugs. Accordingly, the rate of hydrogel swelling determines the rate of drug release. Chemically controlled delivery can be further classified according to the type of chemical reaction (i.e., matrix biodegradation and bond cleavage of polymer–drug conjugates) occurring during drug release. Since many researchers prefer to use biodegradable hydrogels, this mechanism receives increasing attention in recent years. In particular, it gains tremendous importance for predicting drug release because current used hydrogels are designed to be stimuli-responsive for triggered drug release.

There have been many attempts to model the aforementioned mechanisms mathematically.[41] For example, Fick's law of diffusion can be used to describe diffusion-controlled drug release. This modeling is based on either highly swollen reservoir or matrix systems where drug is surrounded by a hydrogel membrane or uniformly dispersed throughout a matrix. For each system, a drug diffusion coefficient is empirically obtainable and is assumed to be constant to simplify the modeling. Consequently, the drug release profile can be simulatively predicted, which is comparable with that obtained by experiments. However, the equation to predict the drug diffusivity is not always available because the model becomes complicated in general by other mechanisms. Other two mechanisms have a similar limitation on their mathematical simulations, although several studies have shown good fits to experimental data. As for *in situ* crosslinked hydrogels, a simulation-based prediction of drug release is particularly difficult because crosslinking reactions occur within a few seconds or minutes, leading to irregular hydrogel geometries and unidentified polymer–drug interactions. In addition, the general trend for designing *in situ* crosslinked hydrogels is to combine multiple drugs plus stimuli-responsive properties (pH, temperature, redox potential, electric and magnetic fields) for improving therapeutic efficacy, which makes them more complex. Despite great advances in developing mathematical simulations, thus, there are still many challenges to accurately predict drug release in complex hydrogel-based drug delivery systems.

4. Conclusions and Outlook

In situ crosslinked hydrogels have become more attractive with increasing demands for the development of easy-to-use and versatile drug delivery platforms. One of the most excellent features in this hydrogel system is that many kinds of drugs can be easily formulated and delivered to the body via a minimally invasive procedure. *In situ* modulation of drug release kinetics is also achievable by altering experimental parameters corresponding to a crosslinking method to be employed. For long-term drug release, drug molecules can be covalently linked to gel-forming polymer backbones. Introduction of stimuli-responsive linkages to polymer–drug conjugates enables triggered drug release, which aims to improve therapeutic efficacy. *In situ* crosslinked hydrogels have also received a great deal of attention for various biomedical applications including tissue engineering. Cells of interest can be suspended in hydrogel precursor solutions and then injected at the desired site to regenerate and replace tissue. To promote tissue regeneration, simultaneous delivery of signaling molecules such as growth factors and cytokines has been performed using *in situ* crosslinked hydrogels as a therapeutic implant and vehicle. Therefore, we expect that *in situ* crosslinked hydrogels will serve as a versatile and injectable drug delivery platform for a broad range of biomedical applications.

Acknowledgments

This work was supported by the Basic Science Research Program through the National Research Foundation of Korea (NRF) funded by the Ministry of Science, ICT & Future Planning (Grant No. 2012-0006687).

References

1. K. E. Uhrich, S. M. Cannizzaro, R. S. Langer, and K. M. Shakesheff, *Chem. Rev.*, **99**, 3181 (1999).
2. W. E. Hennink and C. F. van Nostrum, *Adv. Drug Deliv. Rev.*, **54**, 13 (2002).
3. D. Y. Ko, U. P. Shinde, B. Yeon, and B. Jeong, *Prog. Polym. Sci.*, **38**, 672 (2013).
4. K. T. Nguyen and J. L. West, *Biomaterials*, **23**, 4307 (2002).
5. J. V. Crivello, *Nuclear Instruments and Methods in Physics Research Section B: Beam Interactions with Materials and Atoms*, **151**, 8 (1999).
6. T. Atsumi, J. Murata, I. Kamiyanagi, S. Fujisawa, and T. Ueha, *Arch. Oral Biol.*, **43**, 73 (1998).
7. J. Moan, K. Berg, E. Kvam, A. Western, Z. Malik, A. Ruck, and H. Schneckenburger, *Ciba Found. Symp.*, **146**, 95 (1989).
8. C. G. Williams, A. N. Malik, T. K. Kim, P. N. Manson, and J. H. Elisseeff, *Biomaterials*, **26**, 1211 (2005).
9. C. C. Lin and K. S. Anseth, *Pharm. Res.*, **26**, 631 (2009).
10. J. L. West and J. A. Hubbell, *React. Polym.*, **25**, 139 (1995).
11. R. Hu, Y.-Y. Chen, and L.-M. Zhang, *Int. J. Pharm.*, **393**, 97 (2010).
12. S. Jo, H. Shin, A. K. Shung, J. P. Fisher, and A. G. Mikos, *Macromolecules*, **34**, 2839 (2001).
13. M. R. Kim and T. G. Park, *J. Control. Release*, **80**, 69 (2002).
14. J. C. Scaiano, K. G. Stamplecoskie, and G. L. Hallett-Tapley, *Chem. Commun.*, **48**, 4798 (2012).
15. J. Elisseeff, K. Anseth, D. Sims, W. McIntosh, M. Randolph, and R. Langer, *Proc. Natl. Acad. Sci. U.S.A.*, **96**, 3104 (1999).
16. B. D. Mather, K. Viswanathan, K. M. Miller, and T. E. Long, *Prog. Polym. Sci.*, **31**, 487 (2006).
17. B. N. Naidu, M. E. Sorenson, T. P. Connolly, and Y. Ueda, *J. Org. Chem.*, **68**, 10098 (2003).
18. X. Zheng Shu, Y. Liu, F. S. Palumbo, Y. Luo, and G. D. Prestwich, *Biomaterials*, **25**, 1339 (2004).

19. C. Hiemstra, L. J. van der Aa, Z. Zhong, P. J. Dijkstra, and J. Feijen, *Macromolecules*, **40**, 1165 (2007).
20. L. S. Teixeira, J. Feijen, C. A. van Blitterswijk, P. J. Dijkstra, and M. Karperien, *Biomaterials*, **33**, 1281 (2012).
21. N. C. Veitch, *Phytochemistry*, **65**, 249 (2004).
22. S. Kobayashi, H. Uyama, and S. Kimura, *Chem. Rev.*, **101**, 3793 (2001).
23. M. Kurisawa, J. E. Chung, Y. Y. Yang, S. J. Gao, and H. Uyama, *Chem. Commun.*, 4312 (2005).
24. F. Lee, J. E. Chung, and M. Kurisawa, *J. Control. Release*, **134**, 186 (2009).
25. K. M. Park, Y. M. Shin, Y. K. Joung, H. Shin, and K. D. Park, *Biomacromolecules*, **11**, 706 (2010).
26. N. Q. Tran, Y. K. Joung, E. Lih, K. M. Park, and K. D. Park, *Biomacromolecules*, **11**, 617 (2010).
27. K. M. Park, K. S. Ko, Y. K. Joung, H. Shin, and K. D. Park, *J. Mater. Chem.*, **21**, 13180 (2011).
28. K. M. Park, Y. Lee, J. Y. Son, J. W. Bae, and K. D. Park, *Bioconjug. Chem.*, **23**, 2042 (2012).
29. S. Sakai, K. Moriyama, K. Taguchi, and K. Kawakami, *Biomacromolecules*, **11**, 2179 (2010).
30. A. Hatefi and B. Amsden, *J. Control. Release*, **80**, 9 (2002).
31. B. Jeong, S. W. Kim, and Y. H. Bae, *Adv. Drug Deliv. Rev.*, **54**, 37 (2002).
32. E. Ruel-Gariépy and J.-C. Leroux, *Eur. J. Pharm. Biopharm.*, **58**, 409 (2004).
33. D. Chitkara, A. Shikanov, N. Kumar, and A. J. Domb, *Macromol. Biosci.*, **6**, 977 (2006).
34. J. C. Garbern, A. S. Hoffman, and P. S. Stayton, *Biomacromolecules*, **11**, 1833 (2010).
35. K. W. Chun, J. B. Lee, S. H. Kim, and T. G. Park, *Biomaterials*, **26**, 3319 (2005).
36. K. M. Park, S. Y. Lee, Y. K. Joung, J. S. Na, M. C. Lee, and K. D. Park, *Acta Biomater.*, **5**, 1956 (2009).
37. H. H. Tønnesen and J. Karlsen, *Drug Devel. Ind. Pharm.*, **28**, 621 (2002).
38. C. K. Kuo and P. X. Ma, *Biomaterials*, **22**, 511 (2001).
39. T. Boontheekul, H.-J. Kong, and D. J. Mooney, *Biomaterials*, **26**, 2455 (2005).
40. A. D. Augst, H. J. Kong, and D. J. Mooney, *Macromol. Biosci.*, **6**, 623 (2006).
41. C.-C. Lin and A. T. Metters, *Adv. Drug Deliv. Rev.*, **58**, 1379 (2006).
42. B. Amsden, *Macromolecules*, **31**, 8382 (1998).
43. M. N. Mason, A. T. Metters, C. N. Bowman, and K. S. Anseth, *Macromolecules*, **34**, 4630 (2001).
44. A. K. Bajpai, S. K. Shukla, S. Bhanu, and S. Kankane, *Prog. Polym. Sci.*, **33**, 1088 (2008).

Chapter 4

Supramolecular Hydrogels for Drug Delivery

Prashant Sawant
Chief Scientist, Intraceuticals Pvt Ltd, Melbourne, Australia
pdsawant@yahoo.com

Recent growing interest in supramolecular hydrogels based drug delivery is evident from growing number of publications in the literature. Most of the supramolecular gels exhibit physically crosslinked three-dimensional inter-connecting networks of micronano fibers. The development of microstructure dictates the physico-chemical and drug delivery properties of supramolecular gels.

The present chapter is an overview of recent advances in supramolecular hydrogels and their application in the drug delivery. This chapter is divided into three parts: (i) types of supramolecular hydrogelators and mechanism of formation of microstructures, (ii) development of unique supramolecular hydrogels for drug delivery properties, and (iii) the development of a few pharmaceutical products.

This overview suggests that the field of supramolecular hydrogels for the drug delivery is still in its early stages and there is a lot of scope for continuous development to overcome product development and regulatory barriers.

1. Introduction

Supramolecular chemistry is the chemistry of the intermolecular bond, covering the structures and functions of the entities formed by the association of two or more chemical species.[1] Supramolecules or supramolecular assemblies are desecrate and forms self-assembled, physically three-dimensional interconnecting network structures formed by non-covalent interaction. The self-assembly the step is after formation of molecules from atoms via covalent interactions. Two main components of supramolecular assemblies are the host and the guest.[1] The host molecule is realized through atoms by covalent interaction whereas the guest molecule interacts with host by non-covalent forces and ultimately forms supramolecules or supramolecular assemblies. Most of the supramolecules (except biological supramolecules such as DNA) are low molecular weight. Therefore, most of the supramolecular hydrogel[2] are formed by interactions of low molecular weight gelator (LMGA) molecules (host) with water molecules (guest).

Literature suggests that LMGA molecules can be self-assembled[3] or self-organized[4] to form inter-connecting three dimensional network micro-nano structures that can entrap the water molecules in the interstitial spaces. Besides drug delivery vehicles, these types of gels have found potential applications in nanotechnology,[5] biomaterials,[6] biomotors,[7] tissue engineering,[8] and pollution control.[9]

The present chapter covers classification of supramolecular hydrogels, mechanism of formation, and different types of hydrogels for drug delivery and a few key examples of commercial development of pharmaceutical products based on supramolecules.

2. Classification of Supramolecular Hydrogel based on the Formation Mechanism

Supramolecular hydrogels can be classified as either chemical or physical gels.

2.1. *Chemically crosslinked supramolecular hydrogels*

Chemically crosslinked supramolecular hydrogels form three-dimensional interconnecting network structures using covalent bonds between gelator molecules. The chemical bonds are irreversible therefore this type of hydrogels exhibit permanently crosslinked microstructures and do not possess thermo-responsive properties.

2.2. *Physically crosslinked supramolecular hydrogels*

On the other hand, physical crosslinked supramolecular hydrogels form as a result of non-covalent interactions between gelator molecules. Non-polymeric physically crosslinked supramolecular hydrogels have gained interest recently.[2–9]

The aggregation of gelator molecules into micronano fibrous interconnecting networks is driven by multiple weak interactions such as π–π stacking, H-bonding, van der Waals interaction, hydrophobic interactions, salt bridges, and transition metal coordination etc.

The low molecular mass gelator molecules form a continuous three dimensional interconnecting network structure in the solvent, and prevent the liquid from flowing via surface tension.[10]

3. Molecular Structures of Gels

Like protein structure, a gel structure can be classified into a primary, secondary, and tertiary structures which can explain the mechanism of gel formation.[11]

The primary structure (angstrom to nanometer scale) is determined by the molecular level recognition. As a result, gelator molecules are aggregated in one dimension. There must be a balance between the tendency of the molecules to dissolve or to aggregate for the formation of gel. Hydrophobic interaction becomes most important in the aqueous environments of hydrogels. The secondary structure (nano- to micrometer scale) is defined as the morphology of the aggregates like micelles, vesicles, fibers, ribbons or sheets. This secondary structure is directly influenced by the molecular structure of gelators.

Finally, the tertiary structure of a gel (micro- to millimeter scale) involves the interaction of individual aggregates which ultimately determines whether a gel is formed or instead, fibers (or other aggregates) precipitated in the solution without trapping the solution.[12] The aggregation of gelator molecules occurred through the both branched fibers and entangled fibers. It is also found out that long, thin and flexible fibers are better able than shorter fibers to trap the solvent molecules, leading to gelation.[13] Modeling also showed various modes of gelator aggregations in various solvents and hence the gel formation.[14–17]

Most of the gels are formed by techniques similar to crystallization or sol-gel procedures such as heating a solution of gelator molecules to dissolve it (sol) and, then allowing the solution to cool to form a gel.

4. Types of Hydrogelators

Small organic hydrogelators have been categorized by the type and placement of the polar groups, as discussed below. Examples provided below help us to understand the effect of chemical functional groups of the gelator on the gel properties and mechanisms of the formation of the gel. Using this information we can develop novel supramolecules for the drug delivery purpose.

4.1. *Cholesterol-based hydrogelators*

Novel tripodal cholic acid derivatives (Fig. 1) form gels in an aqueous acid medium at very low concentration.[18] The gels derived from 1 in acetic acid/water medium are

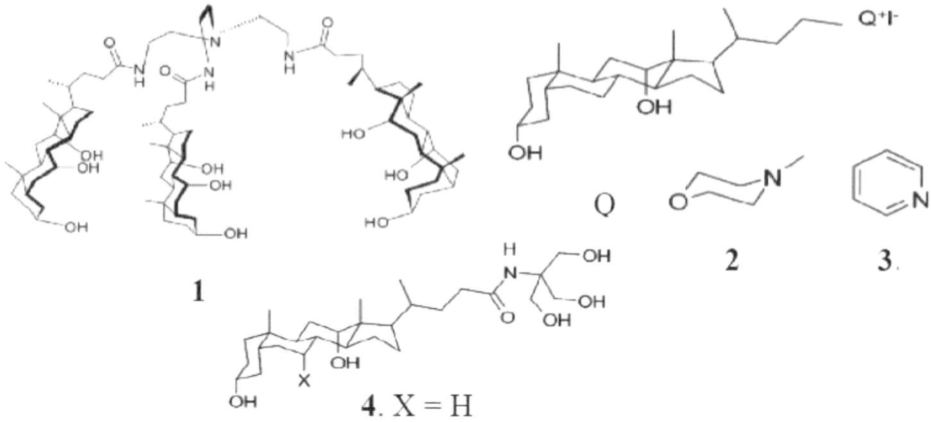

Fig. 1. Cholesterol derivative based supramolecular gelators.
Copyright permission from supramolecular hydrogels, *J. Phys. Sci.*, **12**, USM Press.

very strong and transparent whereas the gel derived from hydrochloride salt of 1 in water medium is weak and turbid. But the addition of organic cosolvent to the hydrochloride salt of 1 in water led to formation of strong and transparent gel.

The gel-to-sol transition temperature (T_{gel}) of the gelator was measure at various concentrations of the gelator. Additionally, it was also found that T_{gel} decreased at higher acetic acid concentrations.

The microstructure analysis using the transmission electron microscopy (TEM) has showed that the network structure was composed of submicrometer fibers.

Novel cationic and neutral analogues of bile acids 2 and 3 formed thermos-reversible gels in water in presence of sodium chloride (NaCl), whereas neutral 4 formed gels in water in the presence of organic solvents such as ethanol, methanol, DMSO, and DMF. Additionally, the bile acids 2 and 3 exhibited different aggregation behaviors which can be attributed to the difference in their chemical structures. Compound 2 (1% w/v) formed stable thermo-reversible gels in 0–4 M aqueous NaCl solution and also formed gel in 1M aqueous solution of NaBr, Na_2SO_4, Na_2CO_3, NaN_3, $NaNO_3$, KCl, LiCl, and $BaCl_2$. The thermal stability of the gel derived from 2 (1% w/v of NaCl) was increased by both the increase in the concentration of 2 and NaCl. Additionally, the scanning electron microscopy (SEM) of xerogel derived from compounds 3 and 4 showed the interconnecting submicron size fibrous network structures (that can hold aqueous solutions).

4.2. *Sugar based hydrogelators*

Compounds 5–8 (Fig. 2) are the efficient hydrogels at very low concentrations.[19–23] In these gelators, the sugar moiety helps to improve solubility in aqueous medium

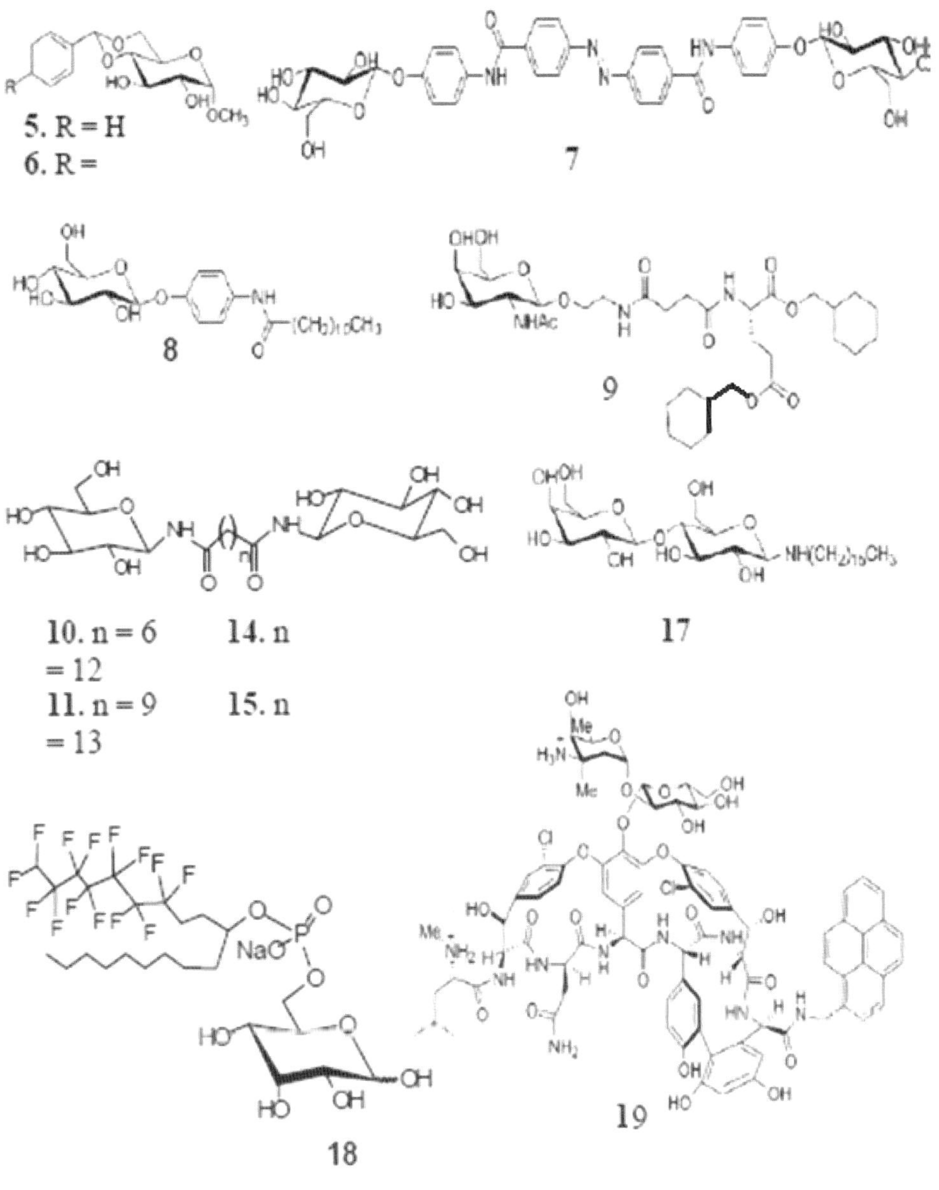

Fig. 2. Sugar derivative based supramolecular gelators.

and the aromatic group helps to promote π–π aggregation in water. Particularly, the compound 5 reveals highly flexible gelation properties.

It is capable to form gels of a broad variety of organic solvents and water at the same time. Glycosylated amino acid derivative 9 is a potential gelator for aqueous solvent and has exhibited the thermo-responsive phase transition.[24,25] This gelator could be useful for the encapsulation and release of drugs.[24] The glycosylated gelator have found to stabilize myoglobin in its active state for long times without significant leakage into buffer.[25]

The bola-amphiphilic glycolipid[26] and bola-amphiphilic sugar derivatives (10–16) found to forming efficient gels in an aqueous medium.[27,28] The aggregation properties of sugar derivatives (10–16) have been attributed to the presence of different linking chain between the two sugar units. A disaccharide amphiphiles (17) containing one alkyl chain is also found to be an efficient hydrogelator in the presence of alcohol as a cosolvent.[29]

Sugar containing fluorocarbons- (18) based hydrogel have been prepared[30,31] where the fluorocarbon unit has been used as strong hydrophobic force for the aggregation. Xing *et al.*[32] have reported that the vancomycin-pyrene derivative containing sugar moiety (19) is an efficient gelator for aqueous solvent. This is the first claimed example of an antibiotic hydrogel.

4.3. *Amino acid based hydrogelators*

Amino acid surfactants with alanine, aspartic acid and glutamic acid as head groups (20–23) (Fig. 3) have found to form hydrogels at very low gelator concentrations.[33,34]

The formation of gel is pH dependent due to the presence of carboxylic acid in the gelator. The pKa values of the acid carboxylic acid can be moderated by varying the length of the alkyl chain. An increase in the chain length leads to the increase in the pKa values by 2 pH units. The cryo-TEM image of the gelator exhibited the rod like morphology.

Additionally, lysine derivatives (24 and 25) have also formed hydrogels at the low gelator concentration. Again, the hydrophobic effects is the major driving force for the formation of hydrogel.[35–37]

Franceschi *et al.* have been reported the gel-forming amino acid bolaamphiphiles (26–28).[38] They showed that depending on the linker length, different microstructures such as micelles, vesicles, or fibers can be formed.

Light harvesting hydrogels are formed by two-headed amphiphiles (29–31)[39] where the charge–charge interactions and hydrophobic effects are key driving forces to form the gel with water. The compound (32) with bulky (phenyl or isopropyl) side chain is also found to be effective gelators for both water and organic solvents.[40,41] In water, the hydrophobic interactions of the side chains promote linear aggregation, and then hydrogen bonding of the adjacent carboxylic acid further strengthens the assembly. Similarly, Serine-based amphiphiles (33) also form hydrogel based on above-said mechanisms.[42]

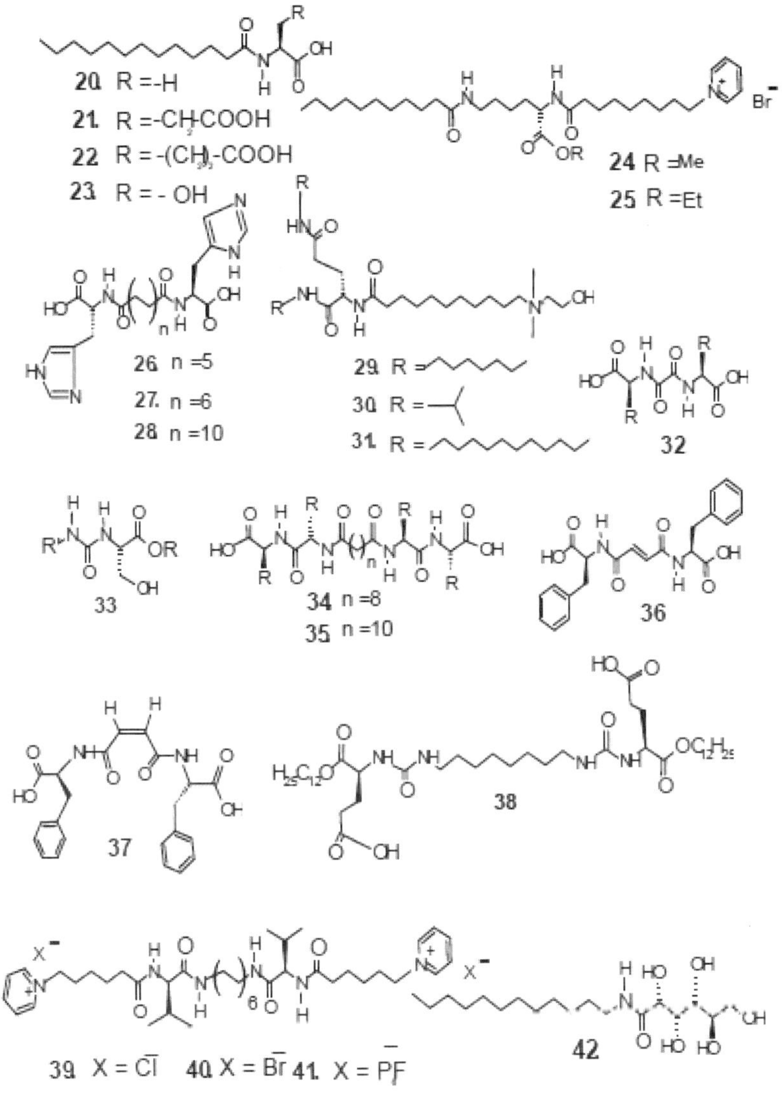

Fig. 3. Amino acid derivative based supramolecular gelators.
Copyright permission from supramolecular hydrogels, *J. Phys. Sci.*, **12**, USM Press.

The amino acid derivatives (34, 35) containing terminal carboxylic acid are aggregated in aqueous solution. The aggregation properties are found to be pH-dependent.[43,44]

In an interesting example, the photoisomerization of the double bond from 37 to 36 has found led the gel formation.[45] Only trans double bond in compound 36, projects the hydrophobic side chains in an orientation similar to the oxalyl group, helps promoting stacking which led to the gelator fiber formation, and subsequent hydrogelation.

The compound (38) is found be self-assembled during the formation of hydrogel at very low concentration.[46] The gelation is found to dependent on the pH and ionic strength.[46] It is also found that this molecule forms giant vesicles at low pH.

New L-valine derivatives having positively charged group formed transparent hydrogels of pure water as well as aqueous solutions of inorganic acids and salts at 0.2 wt% of the gelator.[47] In pure water, the hydrogelation abilities significantly depended on the concentrations. Among L-valine derivatives, the compound 39 has no hydrogelation ability at 70 g/L, whereas compounds, 40 and 41, have formed hydrogels of pure water. Particularly, compound 41 is able to gel pure water at 2 g/L. Additionally, the hydrogelation abilities of compounds 39 and 40 have found to be increased in the presence of inorganic acids and salts. Furthermore, the hydrogelation ability of compound 40 in inorganic salts solution is 12 times higher than that in pure water. Compared with 40, the thermal stability of 41 is greater. The hydrogels based on 41 and 40, which contain an inorganic acid and salts have the similar T_{gel} values, and their $MgCl_2$ and $CaCl_2$ hydrogels have good thermal stabilities. The T_{gel} of hydrogels based on 41 increases in the presence on inorganic salts. Enantiomerically pure, N-(l-alkyl)-aldonamides (42) is aggregated in water solution during the formation of hydrogels.[48] As aldonamides have demonstrated antileishmanial activity,[49] its hydrogels can be used for the treatment of *Leishmania amazonensis.*

The recemic mixture did not form gel in water. Different diastereomers forms different morphologies, indicating that the geometry of the head group is important for the packing.

4.4. *Nucleotides based hydrogelators*

Shimizu and co-workers have synthesized various hydrogelators based on nucleotides.[50–52] The molecular organization of the nucleotide is controlled by hydrogen bonding of the end groups (43) (Fig. 4).

It is found that solubilizing deoxyribose and phosphodiester groups have formed hydrogels. The terminal sugar part helps the molecules to aggregate during hydrogelation.[50,51]

4.5. *Gemini surfactants based hydrogelators*

The surfactant molecule (44) (Fig. 5) forms hydrogels at very low concentration. It is found that reducing the length of one alkyl chain from 18 methylenes to 10 methylenes resulted in the formation of a coaservate (a spongy, water-insoluble mass)

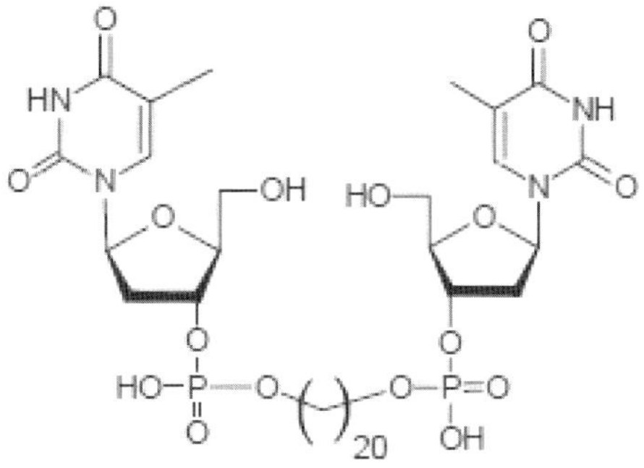

Fig. 4. Nucleotide based supramolecular gelators.
Copyright permission from supramolecular hydrogels, *J. Physl. Sci.*, **12**, USM Press.

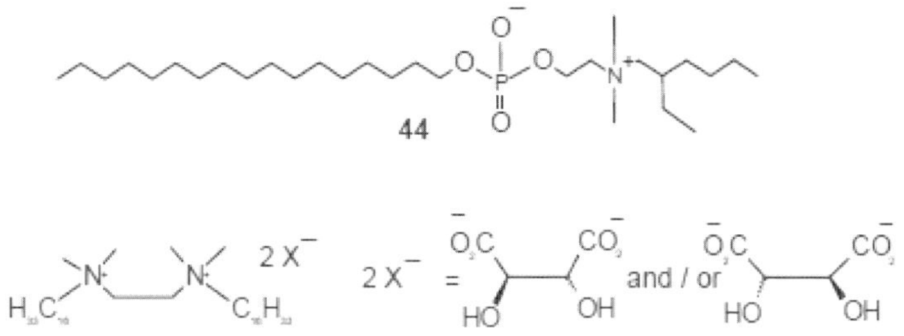

44

45

Fig. 5. Nucleotide-based supramolecular gelators.
Copyright permission from supramolecular hydrogels, *J. Phys. Sci.*, **12**, USM Press.

rather than a gel.[53,54] Hydrophobicity plays a determinant role during the formation of hydrogels. Oda and Hue have reported a family of Gemini surfactants with chiral counter ions (45) to form hydrogels.[55–57] The TEM images indicated that the gels are formed from twisted ribbon. Direct interactions of the dications and chiral tartrate anions are found using the 1H NMR spectroscopy. The chiral structure of the aggregate may be induced by the chiral anion.

4.6. *Cyclodextrin-based hydrogelators*

The supramolecular structures of cyclodextrins (CDs) and polymers have inspired interesting developments of novel supramolecular biomaterials.[58] The development of the supramolecular biomaterials through inclusion complexation has opened up a new approach for designing novel drug and gene delivery systems, which may have many advantages over the systems based on the conventional polymeric materials. Li and Loh[58] have reviewed such structures and their applications for the drug delivery.

Supramolecular hydrogels based on the self-assembly of the inclusion complexes between CDs with biodegradable block copolymers can be used as injectable drug delivery systems for the sustained release of macromolecular drugs (58). Biodegradable polyrotaxanes with drug-conjugated CDs threaded on a polymer chain with degradable end-caps form supramolecular prodrugs for controlled and targeting delivery of drugs.[58] These CD-containing cationic polymers have been used as gene carriers. These polymers are less cytotoxicity than non-CD-containing polymer counterparts. Additionally, the polyplexes of CD-containing cationic polymers with DNA can be pegylated through a supramolecular process using inclusion complexation between the CD moieties and a modified PEO.

4.7. *Miscellaneous hydrogelators*

Haines and Harrison have developed and demonstrated that resorcinarene is an efficient pH–reversible gelator for water.[59] The gelator aggregation and gelation occur at the solution pH below 2.5 which is corresponding to the pKa value for the protonation of one of the carboxylates and the molecule is neutral. The gel can be dissolved by raising pH and reformed again by lowering the pH. However, gelation is inhibited in the presence of divalent cations, in particular Cu^{2+}, due to strong interactions of the iminodiacetate ligand and cation. The SEM images revealed that the gel is composed of fibers that form random interconnecting networks.

Hanabusa *et al.*[60] have demonstrated the use of sodium salts of terpyridine-containing carboxylic acid (47–50) to form hydrogels above critical micelle concentration (CMC). The hydrogel is most likely composed of rod-like micelles.

Menger and co-workers[61] have synthesized various derivatives of aroyl L–cystines to study the gelation capability and found that these gelators formed thermally stable gels.

Also a large bolaamphiphiles-based molecules forms gels in aqueous medium.[62] In these bolaamphiphile-based gelators, a large number of hydroxyl head groups provides solubility and the hydrophobic core promotes the aggregation of the gelator.

5. Supramolecular Hydrogels for Drug Delivery

This section describes a diverse variety of supramolecular hydrogels used for different types of drug delivery to treat different diseases.

5.1. *CD-based supramolecular drug delivery systems*

CD and its derivatives exhibit excellent biocompatibility and unique inclusion capability, and powerful functionalization capacity. Therefore, CD-based supramolecular systems have been increasing used to achieve drug and gene delivery. In this regard, a recent review by Zhang and Ma[63] is useful to understand state of the art and recent advances in the construction of CD-based assemblies and their applications for controlled drug delivery.

A few interesting examples based on chemistry of CD and drug delivery applications are discussed below.

Long *et al.*[64] have fabricated a supramolecular hydrogel with the complexation of α-CDs and monoend-functionalized low molecular weight methoxy poly (ethylene glycol) (mPEG, Mn = 2000) micelles. Hydrophobic cinnamic acid was immobilized on methoxy poly (ethylene glycol) via L-lysine as linker to prepare amphiphilic mPEG. The monoend-functionalized mPEG self-assembled micelles in aqueous solution with the CMC of 42.5 mg/L. These monodisperse micelles exhibited core-shell structure and the mean diameter of around 40 nanometers. α-CDs were then added in the suspension of above-said micelles to form supramolecular hydrogel with the polypseudorotaxanes complexation. These supramolecular systems have been used for the delivery of a hydrophilic drug doxorubicin hydrochloride.

Marchesan *et al.*[65] have reported the self-assembly of a sparingly soluble antibiotic (ciprofloxacin) and a hydrophobic tripeptide (Leu–Phe–Phe) into supramolecular nanostructures that yield a macroscopic hydrogel at physiological pH. Drug incorporation has changed the morphology and rheological properties of the self-assembled hydrogel. These changes were correlated with intermolecular interactions between the drug and the peptide, using spectroscopic analysis (fluorescence, circular dichroism and IR). The drug was bound within the hydrogel by non-covalent interactions, and retained its activity over a prolonged release timescale. Antimicrobial activity of the ciprofloxacin–peptide self-assembled hydrogel was evaluated against *Staphylococcus aureus*, *Escherichia coli*, and a clinical strain of *Klebsiella pneumoniae*. The peptide hydrogel alone exhibited a mild anti-bacterial activity against Gram-negative bacteria and no major cytotoxicity was seen in haemolysis assays of human red blood cells or in mouse fibroblast cell cultures. This new approach of drug incorporation into the nanostructure of a simple tripeptide hydrogel by self-assembly may find potential applications in cost-effective wound dressings and novel antimicrobial formulations.

Yu *et al.*[66] found that the CD/polymer inclusion system was sensitive to physiological reduction condition caused by glutathione tripepetide (GSH). These novel reduction-sensitive supramolecular hydrogels were, for the first time, fabricated by the inclusion of [poly (ethylene glycol) monomethyl ether]-graft-[disulfide-linked poly (amido amine)] (mPEG-g-SS-PAA) with α-CD in an aqueous solution. The reduction-sensitivity was described to the disulfide linker in the SS-PAA main chain while various physical conjugations contributed to a reversible gel–sol transition under shearing as a key of injectable function. This supramolecular system exhibited a prominent sustained release profile, and the release rate can be regulated by controlling the reduction condition. Additionally, incorporating a low loading-level of reducing agent did not inhibit the formation of hydrogel. Therefore, it is possible to use the reduction-sensitivity to regulate the drug release profile in extracellular milieus and normal tissue. Combined with acceptable cytotoxicity, this kind of reduction-sensitive supramolecular hydrogel based on CD/polymer inclusion showed a great potential as an injectable smart biomaterial for the application of drug controlled-release.

Yhaya *et al.*[67] have synthesized a block copolymer based on poly(*N*–isopropyl acrylamide) (PNIPAAm) and a block with a statistical distribution of poly(2–hydroxyethyl acrylate) (PHEA) and repeating unit with carrying β-CD via reversible addition–fragmentation chain transfer (RAFT) polymerization and click reaction. Addition of poly (2–hydroxyethyl acrylate–*s*–adamantylmethyl acrylate) $P(HEA_{17}$–*s*–$AdMA_7)$ above the lower critical solution temperature (LCST) of the block copolymer led to capture of the micelle structure of 36 nm against disassembly. The drug (albendazole) loaded supramolecular assembly, which was fixed via host–guest complexation between β-CD and adamantane, was evaluated as a drug carrier. Cell viability studies performed using human ovarian carcinoma cell line (OVCAR-3) cell lines showed a higher toxicity of the shell crosslinked micelle compared with the free block copolymer.

Intracellular pH-sensitive supramolecular block amphiphiles based on the host–guest interaction between benzimidazole (BM) modified poly(ε-caprolactone) (BM–PCL) and β-CD terminated dextran (Dex–β–CD) were designed by Zhang *et al.*[68] These supramolecular block amphiphiles can self-assemble into supramolecular micelles and exhibit pH-sensitive behavior in acidic aqueous solution for the pH values below 6. Doxorubicin (DOX), a model anticancer drug, was effectively loaded into the supramolecular micelles *via* hydrophobic interactions. The DOX release from micelles was found to be accelerated in acidic conditions which mimic the endosomal/lysosomal compartments. The enhanced intracellular DOX release in HepG2 cells was observed. Additionally, DOX-loaded intracellular pH-sensitive supramolecular micelles showed higher cellular proliferation inhibition towards HepG2 cells than pH-insensitive micelles. These results suggested that the supramolecular micelles can efficiently load and deliver DOX into tumor cells and enhance the inhibition of cellular proliferation *in vitro* thus providing a powerful device for delivering and releasing drugs at the tumor sites.

The blood–brain barrier (BBB) hinders the accumulation of active compounds in the central nervous system, resulting in decreasing therapeutic effectiveness. To overcome this obstacle, interesting supramolecular nanodevices were developed by Paolino *et al.*[69] These systems have many advantages over the conventional formulations, such as having structures made up of biocompatible and biodegradable materials, the possibility of bypassing the BBB in a non-invasive manner (without structural modifications) and modulate the biopharmaceutical properties of the encapsulated compounds by structural modifications. Polymolecular (liposomes, niosomes, nanogels) and oligomolecular (CD) devices have potential clinical applications in brain drug delivery, being capable of active targeting that can concentrate bioactives in the brain.

Wu *et al.*[70] have developed supramolecular hydrogels of α-CD and methoxypolyethylene glycol–poly (caprolactone)–(dodecanedioic acid)–poly (caprolactone)–methoxypolyethylene glycol (MPEG–PCL–MPEG) triblock polymers. These polymeric hydrogels have exhibited the sustained release of dextran-fluorescein isothiocyanate (FITC) that lasted for more than one month. Additionally, the *in vivo* viability studies and the *in vivo* histological studies have demonstrated that these hydrogels were non-cytotoxic and biocompatible.

A CD-based supramolecular hydrogel system with supramolecularly anchored active cationic copolymer/plasmid DNA (*p*DNA) polyplexes (Fig. 6) used as a sustained gene delivery carrier.[71]

Authors have developed a few biodegradable triblock copolymers of methoxypoly (ethylene glycol)–*b*–poly (ε-caprolactone)–*b*–poly[2–(dimethylamino)ethyl methacrylate] (MPEG–PCL–PDMAEMA) with well-defined cationic block lengths to condense *p*DNA. The MPEG–PCL–PDMAEMA copolymers have exhibited good ability to condense *p*DNA into 275–405 nm polyplexes with hydrophilic MPEG in the outer corona. The MPEG corona imparted greater stability to the *p*DNA polyplexes and served as an anchoring segment when the *p*DNA polyplexes were encapsulated in α-CD-based supramolecular polypseudorotaxane hydrogels. The resultant hydrogels were able to sustain release of *p*DNA for up to 6 days. The *p*DNA was bound electrostatically to the cationic segment of the MPEG–PCL–PDMAEMA copolymers and released as polyplex nanoparticles. Authors found that the protein expression level of *p*DNA polyplexes released over the durations was comparable to that of freshly prepared PEI polyplexes.These hydrogels were thixotropic and easily prepared without using organic solvent therefore, these gels can be used as *in situ* gelling injectable carrier for sustained gene delivery.

A novel injectable supramolecular hydrogel of polymer–CDcomplex–nanoparticles based on polymer–CD complex[72] was developed for the stepwise release of cisplatin to treat cancer.Poly (ethylene glycol)–*b*–poly (acrylic acid) (PEG–*b*–PAA) block copolymer nanoparticles containing cisplatin were released

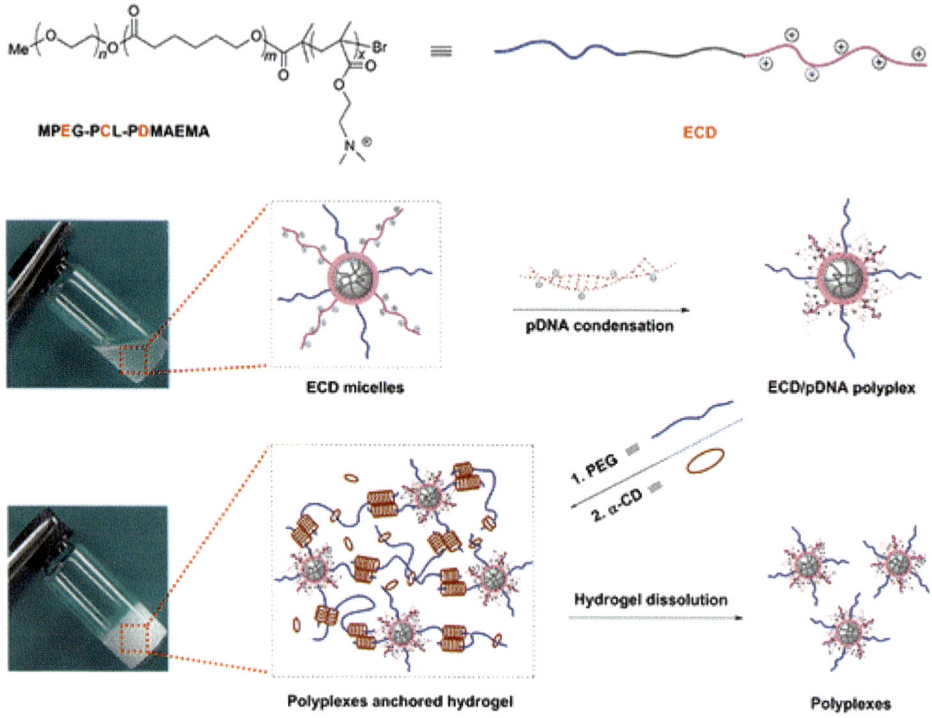

Fig. 6. CD-based supramolecular gelators.

Reprinted (adapted) with permission from (Biocompatible Hydrogels by Oxime Click Chemistry Gregory N. Grover, Jonathan Lam, Thi H. Nguyen, Tatiana Segura, and Heather D. Maynard *Biomacromolecules*, **13**(10), 3013–3017 (2012).

by erosion of the hydrogels and then the cisplatin was released from the nanoparticles by exchanging with chloride ions.

By mixing α-CDs and the PEG-*b*-PAA micelles with their PAA cores loaded with the cisplatin in water, the novel supramolecular hydrogels were generated. α-CDs were threaded onto the PEG segments and formed physical crosslinks of molecular necklaces in these hydrogels. The gelation properties were controlled by controlling concentrations of polymers and cisplatin, and by adding PEG homopolymers or Pluronic copolymers as additives. These hydrogels exhibited the reversible sol-gel transition and thixotropic property. The hydrogel erosion liberated discrete nanoparticles from which cisplatin were released completely in saline. However, it was found that when the hydrogels were eroded into nanoparticles in pure water, cisplatin did not release. Additionally, *in vitro* cytotoxicity studies showed that the cisplatin-loaded hydrogels inhibited the growth of human bladder carcinoma EJ cells with a similar potency as that of the free cisplatin, whereas the hydrogels without cisplatin showed no cytotoxicity. Thus, the cisplatin-coordinated

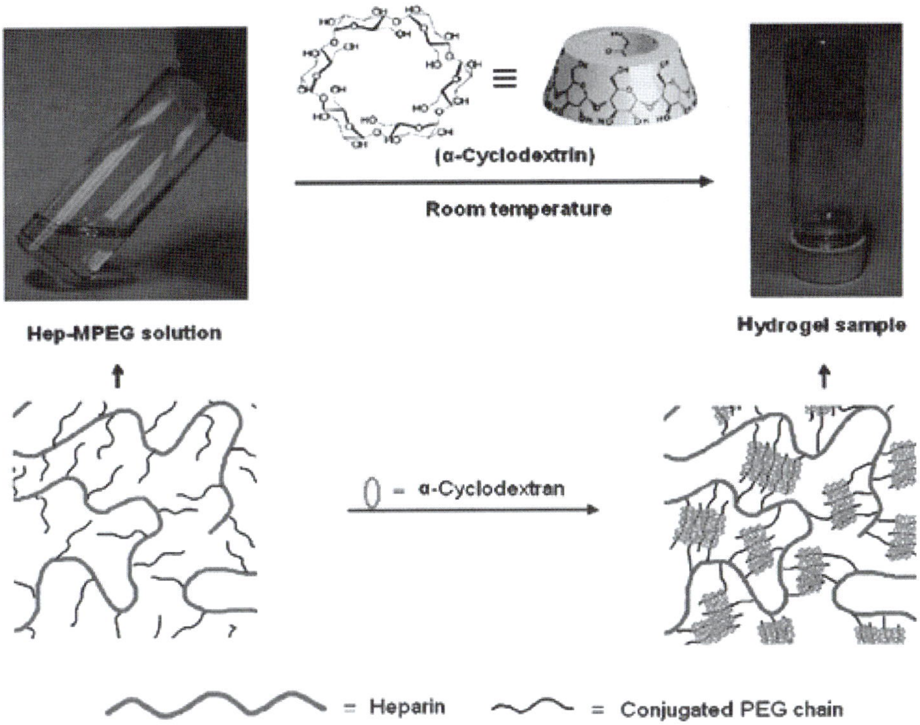

Fig. 7. CD-based supramolecular gelators.

Reprinted (adapted) with permission from Bioactive Supramolecular Hydrogel with Controlled Dual Drug Release Characteristics. Dong Ma, Kai Tu, and Li-Ming Zhang, *Biomacromolecules*, **11**(9), 2204–2212 (2010).

PEG–*b*–PAA/α-CD supramolecular hydrogels were demonstrated as an injectable system for the sustained delivery of cisplatin in the cancer therapy.

Heparin has been used as a powerful anticoagulant and also used for the prophylaxis of both surgical and medical thrombosis. Ma *et al.*[73] have conjugated heparin with amino-terminated poly (ethylene glycol) methyl ether by carbodiimide chemistry and mixed with α-CD in aqueous solution (Fig. 7). The gelation kinetics and hydrogel properties can be modulated by changing the amount of conjugated heparin or α-CD.

The circular dichroism analyses and *in vitro* release studies showed that these supramolecular hydrogels can also be used as an injectable matrix for the encapsulation and sustained release of model protein drug, bovine serum albumin. Additionally, the *in vitro* release, blood clotting, and hemolysis studies of these supramolecular hydrogel have showed a controlled release profile for conjugated heparin and also exhibited good anticoagulant and blood-compatible properties.

5.2. L-phenylalanine derivatives based supramolecular drug delivery systems

Cao *et al.*[74] have prepared supramolecular hydrogels of L-phenylalanine derivative (TC(18)PheBu) as novel hydrogelators. Fu *et al.*[75] have studied the gelation mechanism using SEM, FTIR and fluorescence microscopy and found that intermolecular hydrogen-bonding and hydrophobic interactions are the driving forces for the self-assembly of TC(18)PheBu during the gelation process.

Salicylic acid (SA) was entrapped in the supramolecular hydrogels[74] (Fig. 8). Authors found that the release rate of SA in the supramolecular hydrogels was slightly retarded with an increase of the hydrogelator concentration. Also, the release rate of SA was increased with an increase of temperature and the SA concentration (Fig. 9). Additionally, the release behavior of SA was found to be different at various pH values in buffers as accepting media. The study of the release kinetics indicated that the release behavior of SA was in accord with the Higuchi equation[76] and exhibited the diffusion-controlled mechanism as per the Fickian model.[77]

5.3. Capsosomes based drug delivery systems

Polymer multilayer capsules fabricated using the layer-by-layer (LbL) technique is called "Capsosomes". Capsosomes are biodegradable polymer capsules comprising of thousands of loaded liposomes as subcompartments (78). Capsosomes can be consist of intact, unsaturated, zwitterionic 50 nm liposomes incorporated in a nonbiodegradable polyelectrolyte film assembled from poly (styrene sulfonate) (PSS) and poly(allylamine hydrochloride) (PAH). Fragile small molecule drugs can be trapped in liposomes and liposome are sub-compartments of engineered capsosomes

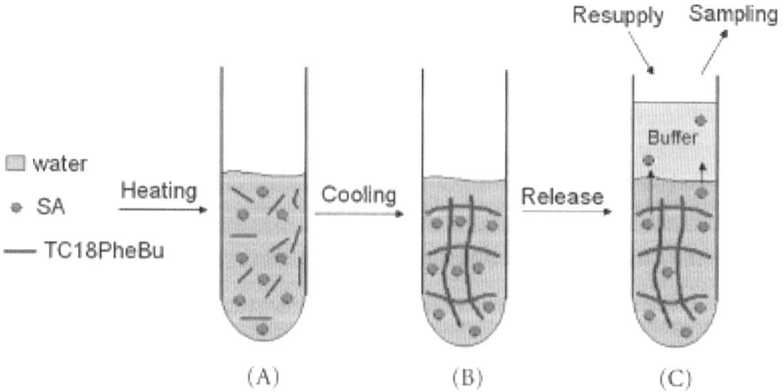

Fig. 8. Scheme of the formation of supramolecular hydrogel and SA released from the hydrogel. (A) Mixture of water, SA and gelator TC_{18}PheBu. (B) Formation of supramolecular hydrogel. (C) SA released from supramolecular hydrogel.

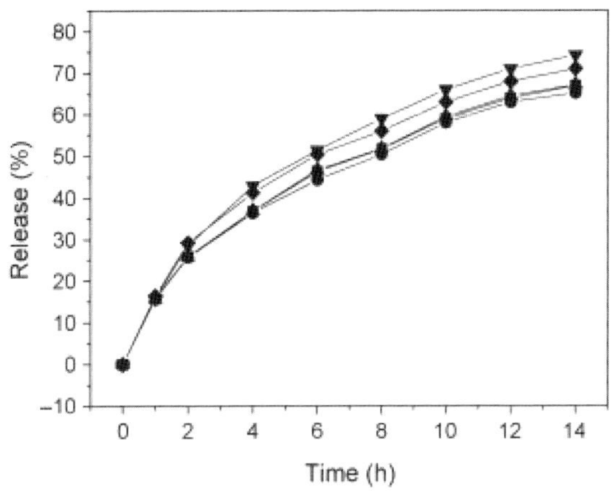

Fig. 9. Release profiles of SA from the hydrogels formed by 4 wt% of TC_{18}PheBu at 25°C. The hydrogels contained 100 (•), 200 (▴), 300 (♦) and 400 (▾) mg/L of SA.

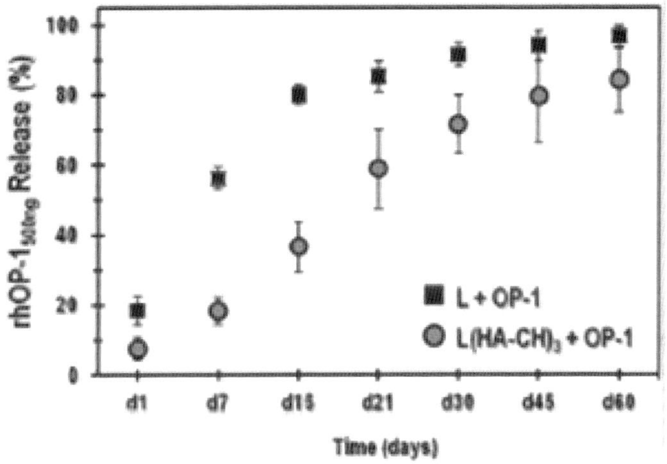

Fig. 10. Cumulative rhOP-1 release profile over 60 days.[81]

therefore capsosomes can be promising systems to perform enzymatic cascade reactions within confined environments. This system opens up a new avenue for targeting, solubilization and controlled delivery of drugs.[79]

Liposomes can incorporate the drug molecules either in core or in bilayer structure via electrostatic/ionic and van der Waal interactions, respectively. This has been used for the delivery of drugs such as amphotericin B and doxorubisine.[80] These liposomes were coated with alternating layers of hyaluronic acid and chitosan by Zoo and Ramasamy[81] to realize a more viscous, sticky, stable, biodegradable and reproducible hydrogel for the sustain drug delivery of rhOp-1 (Fig. 10), an osteoinductive biomaterial that can be used to quickly fill and heal large segmental bone defects. Longer release of rhOp-1 may help improve mechanical strength of the bone.

5.4. *Enzymatically triggered supramolecular drug delivery hydrogels*

Saez *et al.*[82] have demonstrated a novel combination of self-immolative spacers and supramolecular hydrogels for the enzyme triggered drug release. Low-molecular weight drug–gelator conjugates were prepared in which a gel forming lysine moiety linked to model drugs (benzylamine and phenethylamine) through a self-immolating spacer (*p*-aminobenzyloxycarbonyl). Trypsin hydrolyzed the amide linkage between the gelator moiety and the spacer and triggered the release of the model drug (Fig. 11).

The hydrogel microstructure depicted in Fig. 12 shows the interconnecting three dimensional network structures of micro-nano fibers.

The sustain release of phenethylamine and benzylamine (Fig. 13) after trypsan hydrolysis of amide linkage between the gelator moiety and the spacer.

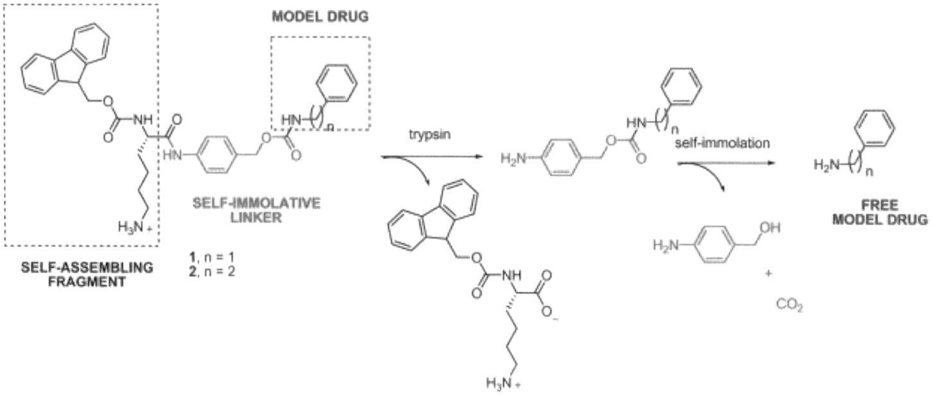

Fig. 11. Enzymatic cleavage and self-immolative degradation mechanism of gelators 1 and 2.

Reprinted from Tetrahedron 66, J. A. Sáez, B. Escuder, and J. F. Miravet, Supramolecular hydrogels for enzymatically triggered self-immolative drug delivery, 2614. Copyright (2010), with permission from Elsevier.

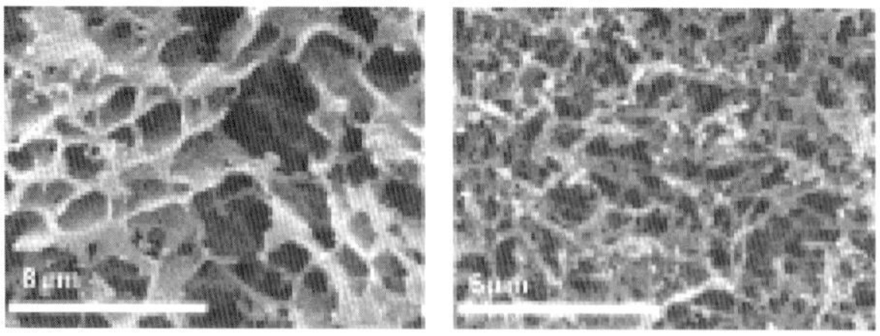

Fig. 12. Cryo-SEM images of hydrogels 1 (left) and 2 (right).

Reprinted from Tetrahedron 66, J. A. Sáez, B. Escuder, and J. F. Miravet, Supramolecular hydrogels for enzymatically triggered self-immolative drug delivery, 2614. Copyright (2010), with permission from Elsevier.

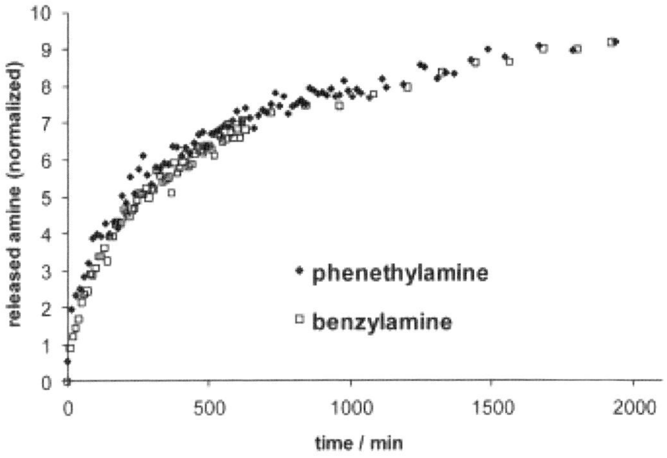

Fig. 13. Kinetic profile of amine release

Reprinted from Tetrahedron 66, J. A. Sáez, B. Escuder and J. F. Miravet, Supramolecular hydrogels for enzymatically triggered self-immolative drug delivery, 2614. Copyright (2010), with permission from Elsevier.

Boekhoven *et al.*[83] have suggested an alternative way where the self-assembly of three-molecular components results in a novel delivery platform where the drug release rate can be tuned after its production. These authors have demonstrated by using a model system where a fluorophore-conjugated gelator is hydrolyzed by an enzyme to release the fluorescent small molecule. To make the release to be tunable, the enzyme is entrapped in liposomes and the system is heated for a short period to release the enzyme which in turn releases the fluorescent small molecule. The heating time controls the amount of enzyme liberated and hence the drug release rate.

5.5. *Self-assembled phage–liposome complex as novel nanocarrier for targeted drug delivery*

Bacteriophage has been exploited as nanocarriers to deliver drugs[84,85] because of it is the cell-specific property. Mao *et al.*[84,85] have used this property and form self-assembled phage–liposome complex of cationic ZnPc-entrapped liposomes and anionic genetically engineered M13 phage and used as a novel nanocarrier for targeted drug delivery and can be used as a potential treatment for cancer.

Figure 14 depicts a schematic representation of the self-assembled complex: (A) cationic ZnPc-entrapped liposome, (B) anionic genetically engineered phage, and (C) rod-like phage–liposome complex. Authors have explained that "M13 phage is a rod-like virus that specifically infects bacteria but is nontoxic to human beings. The side wall of phage is made of ~3000 copies of a genetically modifiable

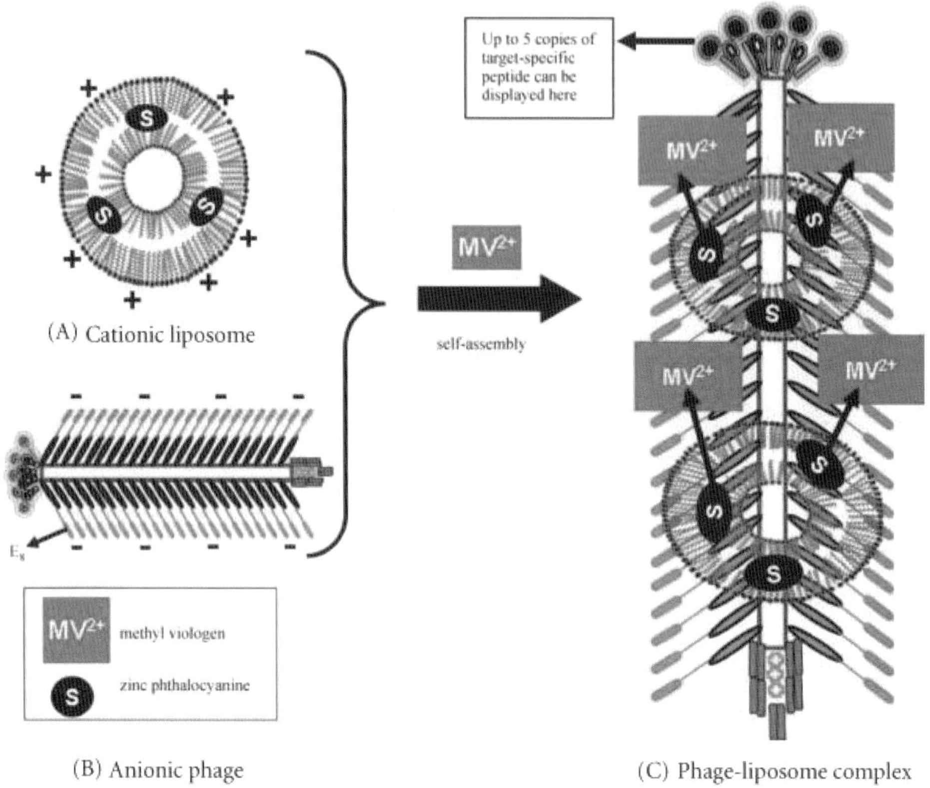

Fig. 14. Schematic representation of the self-assembled complex: (A) cationic ZnPc-entrapped liposome; (B) anionic genetically engineered phage; and (C) rod-like phage–liposome complex.

major coat protein called '*pVIII*'. The two ends of the rod-like virus can be engineered to identify and display a peptide that can recognize a specific target (e.g., cells, tissues, or organs) through phage display." On the other hand, lipid liposomes are known to entrap drugs either within the inner aqueous compartment or within the hydrophobic core of the lipid bilayer. The self-assembled phage–liposome complex formed driven by electrostatic interactions can be used as a target-recognized cargo for a photosensitizer, zinc phthalocyanine (ZnPc), for photodynamic therapy.

6. Commercial Product Development Inspired by Nanostructured Supramolecular Systems

In the pharmaceutical market, only 1 out of 10,000 drug molecules result in a marketable product and only 3 out of 10 drugs generate revenue that meet or exceeds the average R&D cost. The cost of pharmaceutical R&D is exceeding $800 million due to larger and complex clinical studies, and expensive new enabling technologies.[86] Additionally, the introduction of prescription products in markets needs to pass clinical trials and food and drug regulations of different countries. Furthermore, pharmaceutical scientists need to overcome many technological hurdles such as drug solubilization, shelf-life stability of drugs from environment (e.g., sunlight, temperature), targeted delivery of drugs to specific sites within complex human body, and controlling the dosage of drugs to avoid systemic toxicity and increase bioavailability of drugs. Therefore, pharmaceutical industries are careful while investing resources to develop novel drug delivery systems. Supramolecules and its assemblies have offered exciting opportunities for the pharmaceutical industries in this regards.

A few of the following supramolecules based technologies have recently showed a significant progress in the pharmaceutical industry.

6.1. *Captisol technology*

According to Captisol website[87] "Captisol was invented in 1990 by scientists at the University of Kansas Higuchi Biosciences Center for use in drug development and formulation. Six FDA-approved, Captisol-enabled medications are marketed by: Pfizer, Bristol-Myers Squibb and Baxter International. Captisol also has License and Supply Agreements in place with a number of pharmaceutical companies worldwide with Captisol-enabled product candidates. Routes of administration investigated include parenteral, oral, ophthalmic, nasal, topical, oral, and inhalation." A number of articles have been published by the Ligand pharmaceuticals Inc.'s scientist to demonstrate the Captisol for the drug delivery.[88–91] Captisol can be dissolved in aqueous formulations and entrap drugs, and packaged as hydrogels as final products.

6.2. *Dendrimer technology*

Starpharma[92] has developed VivaGel, a veginal microbicide to prevent the transmission of sexually transmitted infections including HIV and genital herpes and bacterial vaginosis.

Dendrimers can protect drugs from degradation and clearance thus an improved dosing regimen can be pursued, for example, replacing a daily infusion with a once weekly injection. Dendrimers have been used to actively or passively target a drug to the selected site in the body. Conversely dendrimers can also be used to restrict the circulation of the drug in the body to avoid side effects, or to allow a higher dose of the drug to be administered than the regular acceptable level to improve efficacy. Drug solubility can be achieved by complexing the drug to a dendrimer construct. An increase in solubility can enhance the "bioavailability" of the drug. According to Starpharma website[92] B recently completed animal studies demonstrated that the Starpharma dendrimer–docetaxel formulation was significantly more efficacious than docetaxel in a breast cancer model. The improved efficacy is understood to be due to a longer circulating half-life, extended release of the docetaxel and targeting of the dendrimer-docetaxel construct to the tumor tissue." Dendrimer-drug constructs are also packaged as hydrogels to realize final products.

6.3. *Liposome technology*

Liposomes, spherical lipid vesicles with a bilayered membrane structure composed of natural or synthetic amphiphilic lipid molecules,[93,94] have been used for drug delivery.[80]

It encapsulates hydrophobic molecules into its hydrophobic bilayers and hydrophilic drugs in aqueous inner core thus liposomes can carry both hydrophobic and hydrophilic drugs. While delivering drugs to sites of action, the lipid bilayers fuse with other bilayers such as the cell membrane and transfer the drugs.

Eleven drugs with liposomal delivery systems have been approved by United States food and drug administration (US-FDA) and six additional liposomal drugs were in clinical trials by 2008.[95]

There has been a progressive increase in the number of commercially available nanoparticle-based therapeutic products in past few decades. A global survey conducted by the European Science and Technology Observatory in 2006 has showed that more than 150 companies are developing nanoscale therapeutics.[96] So far, 24 nanotechnology-based therapeutic products have been approved for clinical use, with total sales exceeding $5.4 billion. Among these products, liposomal drugs and polymer–drug conjugates are two dominant classes, accounting for more than 80% of the total amount (Table 1).

Some of the liposomal products approved by US-FDA in the past 15 years are provided in Table 1.

Doxil was the first liposomal drug formulation approved by US-FDA for the treatment of AIDS associated with Kaposi's sarcoma in 1995. Doxorubicin (an anticancer chemotherapeutic drug) was encapsulated into stealth liposome carriers comprised of hydrogenated soy phosphatidylcholine, cholesterol, and PEGylated phosphoethanolamine to prolong doxorubicin circulation half-life and enhanced drug deposition in the tumor tissue. Liposome based products can also be packaged as hydrogels.

Table 1. List of clinically approved liposomal drugs.

Name	Trade name	Company	Disease
Liposomal amphotericin B	Abelcet	Enzon	Fungal infections
Liposomal amphotericin B	Ambisome	Gilead Sciences	Fungal and protozoal infections
Liposomal cytarabine	Depocyt	Pacira (formerly SkyePharma)	Malignant lymphomatous meningitis
Liposomal daunorubicin	DaunoXome	Gilead Sciences	HIV-related Kaposi's sarcoma
Liposomal doxorubicin	Myocet	Zeneus	Combination therapy with cyclophosphamide in metastatic breast cancer
Liposomal IRIV vaccine	Epaxal	Berna Biotech	Hepatitis A
Liposomal IRIV vaccine	Inflexal V	Berna Biotech	Influenza
Liposomal morphine	DepoDur	Skye Pharma, Endo	Post-surgical analgesia
Liposomal verteporfin	Visudyne	QLT, Novartis	Age-related macular degeneration, pathologic myopia, ocular histoplasmosis
Liposome–PEG doxorubicin	Doxil/Caelyx	Ortho Biotech, Schering-Plough	HIV-related Kaposi's sarcoma, metastatic breast cancer, metastatic ovarian cancer
Micellular estradiol	Estrasorb	Novavax	Menopausal therapy

7. Conclusions and Outlook

The present chapter summarized recent development in supramolecular hydrogels and their applications in the drug delivery therapeutics. Some of the supramolecular gels exhibited physically crosslinked three-dimensional inter-connecting networks of micronano fibers. The microstructure dictates the physico-chemical and drug delivery properties of supramolecular gels. Different classes of hydrogelators and their structures were discussed in the present chapter. In general, number of literature focusing on the synthesis of supramolecules and their drug delivery applications is growing every year. This demonstrates the importance and promise of the use of supramolecules for the drug delivery purposes.

A few pharmaceutical products inspired by supramolecules to treat variety of diseases were also discussed in the present chapter.

So far, more than 150 companies worldwide are engaged in developing nanoscale therapeutics based on supramolecules. 24 nanotechnology-based therapeutic products have been approved for the clinical use by the US-FDA.

Liposomal drugs and polymer–drug conjugates are the two dominant classes that accounting for more than 80% of the total supramolecular drug delivery systems. A few liposomal products were approved by US-FDA in the past 15 years to treat diseases like influenza, fungal infections, cancer and AIDS.

However, the availability of a few products based on supramolecular hydrogels at the market place shows that this field is still emerging. There is a lot of scope for scientists from industry and academia to collaborate and develop novel supramolecular hydrogels for therapeutic treatments.

References

1. D. P. Patel and B. G. Chaudhari, *J. Cur. Pharm. Res.*, **9**, 01 (2012).
2. G. C. Maity, *J. Phys. Sci.*, **12**, 173 (2008).
3. J. Li, *NPG Asia Materials*, **2**, 112 (2010).
4. P. Pakhomov, S. Khizhnyak, M. Ovchinnikov, and P. Komarov, *P. Macromol. Symp.*, **316**, 97 (2012).
5. J. S. Shen, Y. L. Chen, J. L. Huang, J. D. Chen, C. Zhao, Y. Q. Zheng, T. Yu, Y. Yang, and H. W. Zhang, *Soft Matter*, **9**, 2017 (2013).
6. T. Rossow, S. Bayer, R. Albrecht, C. C., and S. Seiffert, *Macromol. Rapid Commun.*, (2013).
7. S. Yamaguchi, S. Matsumoto, K. Ishizuka, Y. Iko, K. V. Tabata, H. F. Arata, H. Fujita, H. Noji, and I. Hamachi, *Chemistry*, **14**, 1891 (2008).
8. Y. Hu, H. Wang, J. Wang, S. Wang, W. Liao, Y. Yang, Y. Zhang, D. Kong, and Z. Yang, *Org Biomol Chem.*, **21**, 3267 (2010).
9. S. Kiyonaka, K. Sugiyasu, S. Shinkai, and I. Hamachi, *J. Am. Chem. Soc.*, **124**, 10954 (2002).
10. T. Tanaka, *Am. Sci.*, **244**, 110 (1981).
11. B. A. Simmons, C. E. Taylor, F. A. Landis, V. T. John, G. L. McPherson, D. K. Schwartz, and R. Moore, *J. Am. Soc.*, **123**, 2414 (2001).
12. A. D. Hamilton and L. A. Estroff, *Chem. Rev.*, **104**, 1201 (2004).
13. P. J. Flory, *Faraday Discuss*, **57**, 7 (1974).
14. S. Bhattacharya and S. N. G. Acharya, *Chem. Mater.*, **11**, 3504 (1999).
15. J. Makarevic, M. Jokic, B. Peric, V. Tomisic, B. Kojic-Prodic, and M. Zinic, *Chem. Eur. J.*, **7**, 3328 (2001).
16. F. S. Schoonbeek, J. H. Van Esch, R. Hulst, R. M. Kellogg, and B. L. Feringa, *Chem. Eur. J.*, **6**, 2633 (2000).
17. K. D. M. Harris, M. Tremayne, and B. M. Kariuki, *Angew. Chem., Int. Ed. Engl.*, **40**, 1626 (2001).

18. U. Maitra, S. Mukhopadhyay, A. Sarkar, P. Rao, and S. S. Indi, *Angew. Chem.*, **113**, 2341 (2001).
19. J. H. Jung, G. John, M. Masuda, K. Yoshida, S. Shinkai, and T. Shimizu, *Langmuir*, **17**, 7229 (2001).
20. J. H. Jung, S. Shinkai, and T. Shimizu, *Chem. Eur. J.*, **8**, 2684 (2002).
21. O. Gronwald and S. Shinkai, *J. Chem. Soc., Perkin Trans.*, **2**, 1933 (2001).
22. K. Yoza, N. Amanokura, Y. Ono, T. Akao, H. Shinmori, M. Takeuchi, S. Shinkai, and D. Reinhoudt, *Chem. Eur. J.*, **5**, 2722 (1999).
23. K. Yoza, Y. Ono, K. Yoshihara, T. Akao, H. Shinmori, M. Takeuchi, S. Shinkai, and D. N. Reinhoudt, *Chem. Commun.*, **8**, 907 (1998).
24. S. Kiyonaka, K. Sugiyasu, S. Shinkai, and I. Hamachi, *J. Am. Chem. Soc.*, **124**, 10954 (2002).
25. S. Kiyonaka, S. Shinkai, and I. Hamachi, *Chem. Eur. J.*, **9**, 976 (2003).
26. R. Ochi, K. Kurotani, M. Ikeda, S. Kiyonaka, and I. Hamachi, *Chem. Commun.*, **49**, 2115 (2013).
27. M. Masuda, Y. Okada, T. Hanada, K. Yase, M. Asai, and T. Shimizu, *Langmuir*, **15**, 4757 (1999).
28. T. Shimizu and M. Masuda, *J. Am. Chem. Soc.*, **119**, 2812 (1997).
29. S. Bhattacharya and S. N. G. Acharya, *Chem. Mater.*, **11**, 3504 (1999).
30. T. Imae, K. Funayama, M. P. Krafft, F. Giulieri, T. Tada, and T. Matsumoto, *J. Colloid Interface Sci.*, **212**, 330 (1999).
31. S. F. Pang and D. B. Zhu, *Chem. Phys. Let.*, **358**, 479 (2002).
32. B. Xing, C. W. Yu, K. H Chow, P. L. Ho, D. Fu, and B. Xu, *J. Am. Chem. Soc.*, **124**, 14846 (2002).
33. T. Imae, Y. Takahashi, and H. Muramatsu, H. *J. Am. Chem. Soc.*, **114**, 3414 (1992).
34. T. Imae, N. Hayashi, T. Matsumoto, T. Tada, and M. Furusaka, *J. Colloid Interface Sci.*, **225**, 285 (2002).
35. M. Suzuki, M. Yumoto, M. Kimura, H. Shirai, and K. Hanabusa, *Chem. Commun.*, **8**, 884 (2002).
36. M. Suzuki, M. Yumoto, M. Kimura, H. Shirai, and K. Hanabusa, *Chem. Eur. J.*, **9**, 348 (2003).
37. M. Suzuki, M. Yumoto, M. Kimura, H. Shirai, and K. Hanabusa, *New J. Chem.*, **26**, 817 (2002).
38. S. Franceschi, N. de Viguerie, M. Riviere, and A. Lattes, *New J. Chem.*, **23**, 447 (1999).
39. T. Nakashima and N. Kimizuka, *Adv. Mater.*, **14**, 1113 (2002).
40. J. Makaveric, M. Jokic, B. Peric, V. Tomisic, B. Kojic-Prodic, and M. Zinic, *M. Chem. Eur. J.*, **7**, 3328 (2001).
41. M. Jokic, J. Makarevic, and M. Zinic, *Chem. Commun.*, **17**, 1723 (1995).
42. G. Wang and A. D. Hamilton, *Chem. Commun.*, **3**, 310 (2003).
43. M. Kogiso, T. Hanada, K. Yase, and T. Shimizu, *Chem. Commun.*, **17**, 1791 (1998).
44. M. Kogiso, S. Ohnishi, K. Yase, M. Masuda, and T. Shimizu, *Langmuir*, **14**, 4978 (1998).
45. L. Frkanec, M. Jokic, J. Makarevic, K. Wolsperger, and M. Zinic, *J. Am. Chem. Soc.*, **124**, 9716 (2002).

46. A. Estroff and A. D. Hamilton, *Angew. Chem., Int. Ed. Engl.*, **39**, 3447 (2000).
47. M. Suzuki, S. Owa, M. Yumoto, M. Kimura, H. Shirai, and K. Hanabusa, *Tetradedron Let.*, 45 (2004).
48. J. H. Fuhrhop, S. Svenson, C. Boettcher, E. Rossler, and H. M. Vieth, *J. Am. Chem. Soc.*, **112**, 4307 (1990).
49. E. S. Coimbra, C. G. Almeida, W. V. Júnior, R. C. N. Dos Reis, A. C. F. De Almeida, L. S. M. Forezi, M. V. De Almeida, and M. Le Hyaric, *The Scientific World J.*, **8**, 752 (2008).
50. R. Iwaura, K. Yoshida, M. Masuda, M. Ohnishi-Kameyama, M. Yoshida, and T. Shimizu, *Angew. Chem., Int.*, **42**, 1009 (2003).
51. R. Iwaura, K. Yoshida, M. Masuda, K. Yase, and T. Shimizu, *Chem. Mater.*, **14**, 3047 (2002).
52. T. Shimizu, R. Iwaura, M. Masuda, T. Hanada, and K. Yase, *J. Am. Chem. Soc.*, **123**, 5947 (2001).
53. F. M. Menger, H. Zhang, K. L. Caran, V. A. Seredyuk, and R. P. Apkarian, *J. Am. Chem. Soc.*, **124**, 1140 (2002).
54. F. M. Menger, V. A. Seredyuk, R. P. Apkarian, and E. R. Wright, *J. Am. Chem. Soc.*, **124**, 12408 (2002).
55. R. Oda, I. Huc, J. C. Homo, B. Heinrich, M. Schmutz, and S. Candau, *Langmuir*, **15**, 2384 (1999).
56. D. Berthier, T. Buffeteau, T. J.-M. Leger, R. Oda, and I. Huc, *J. Am. Chem. Soc.*, **124**, 13486 (2002).
57. R. Oda, I. Huc, M. Schmutz, S. J. Candau, and F. C. MacKintosh, *Nature*, **399**, 566 (1999).
58. J. Li and X. J. Loh, *Adv. Drug Deliv. Rev.*, **60**, 1000 (2008).
59. S. R. Haines and R. G. Harrison, *Chem. Commun.*, **23**, 2846 (2002).
60. K. Hanabusa, T. Hirata, D. Inoue, I. Kimura, and H. Shirai, *Colloid Surf.*, A, **169**, 307 (2000).
61. F. M. Menger and K. L. Caran, *J. Am. Chem. Soc.*, **122**, 11679 (2000).
62. G. R. Newkome, G. R. Baker, S. Arai, M. J. Saunders, P. S. Russo, K. J. Theriot, C. N. Moorefield, L. E. Rogers, J. E. Miller, T. R. Lieux, M. E. Murray, B. Phillips, and L. Pascal, *J. Am. Chem. Soc.*, **112**, 8458 (1990).
63. J. Zhang, and P. X. Ma, *Adv. Drug Del. Rev., 11 May (2013) (Online).*
64. Y. Long, H. Song, B. He, Y. Lai, R. Liu, C. Long, and Z. Gu, *J. Biomater. Appl.*, **27**, 333 (2012).
65. S. Marchesan, Y. Qu, L. J. Waddington, C. D. Easton, V. Glattauer, T. J. Lithgow, K. M. McLean, J. S. Forsythe, and P. G. Hartley, *Biomat.*, **34**, 3678 (2013).
66. J. Yu, H. Fan, J. Huang and J. Chen, *Soft Matter*, **7**, 7386 (2011).
67. F. Yhaya, S. Binauld, Y. Kim, and M. H. Stenzel, *Macromol. Rapid Commun.*, **33**, 1868 (2012).
68. Z. Zhang, J. Ding, X. Chen, C. Xiao, C. He, X. Zhuang, L. Chen, and X. Chen, *Polym. Chem.*, **4**, 3265 (2013).

69. D. Paolino, D. Cosco, R. Molinaro, C. Celia, and M. Fresta, *Drug Discov. Today*, **16**, 311 (2011).
70. D. Q. Wu, T. Wang, B. Lu, X. D. Xu, S. X. Cheng, X. J. Jiang, X. Z. Zhang, and R. X. Zhuo, *Langmuir*, **24**,10306 (2008).
71. Z. Li, H. Yin, Z. Zhang, K. L. Liu, and J. Li, *Biomacromol.*, **13**, 3162 (2012).
72. W. Zhu, Y. Li, L. Liu, Y. Chen, C. Wang, and F. Xi, *Biomacromol.*, **11**, 3086 (2010).
73. D. Ma, K. Tu, and L. M. Zhang *Biomacromol.*, **11**, 2204 (2010).
74. S. Cao, X. Fu, N. Wang, H. Wang, and Y. Yang, *Internat. J. Pharma.*, **357**, 95 (2008).
75. X. Fu, N. Wang, S. Zhang, H. Wang, and Y. Yang, *J. Col. Interf. Sci.*, **315**, 376 (2007).
76. J. Siepmann and N. A. Peppas, *Int. J. Pharm.*, **418**, 6 (2011).
77. A. Peppas, *Pharm. Acta Helv.*, **60**, 110 (1985).
78. B. Stadler, R. Chandrawati, K. Goldie, and F. Caruso, *Langmuir*, **25**, 6725 (2009).
79. R. Chandrawati, B. Stadler, A. Postma, L. A. Connal, S. F. Chong, and A. N. Zelikin, *J. Biomat.*, **30**, 5988 (2009).
80. A. Chonn and P. R. Cullis, *Curr. Op. Biotech.*, **6**, 698 (1995).
81. V. Joo, T. Ramasamy, and Z. S. Haider, *Polymers*, **3**, 967 (2011).
82. J. A. Sáez, B. Escuder, and J. F. Miravet, *Tetrahedron*, **66**, 2614 (2010).
83. J. Boekhoven, M. Koot, T. A. Wezendonk, R. Eelkema, and J. H. Van Esch, *J. Am. Chem. Soc.*, **134**, 12908 (2012).
84. P. Ngweniform, G. Abbineni, B. R. Cao, and C. Mao, *Small*, **5**, 1963 (2009).
85. P. Ngweniform, D. Li, and C. Mao, *Soft Matter.*, **5**, 954 (2009).
86. O. Gassmann, G. Reepmeyer, and M. Von Zedtwitz, 'Innovation: Key to success in the pharmaceutical industry,' *Leading Pharmaceutical Innovation: Trends and Drivers for Growth in the Pharmaceutical Industry*, Springer-Verlag, Berlin, pp. 1 (2008).
87. Available at http://www.captisol.com/captisol-technology/.
88. O. Zimmerer and J. D. Pipkin, *Resp. Drug Deliv.*, **9**, 501 (2008).
89. O. Zimmerer, J. D. Pipkin, S. Somaraju, J. D. Suman, G. Su, and R. N. Dalby, *Resp. Drug Del.*, **9**, 461 (2004).
90. C. Fulk, S. Ghate, G. Mosher, and D. Thompson, *Proceed. Int'l Symp Control. Rel. Bioact. Mater.*, (27), Controlled Release Society; Mira Publishing, St. Louis, MO; 654 (2000).
91. T. Järvinen, K. Järvinen, A. Urtti, D. Thompson, and V. J. Stella, *J. Ocular Pharm. Ther.*, **11**, 95 (1995).
92. Available at http://www.starpharma.com/.
93. L. Zhang and S. Granick, *Nano Lett.*, **6**, 694 (2006).
94. V. P. Torchilin, *Nat. Rev. Drug Discov.*, **4**, 145 (2005).
95. L. Zhang, F. X. Gu, J. M. Chan, A. Z. Wang, R. S. Langer, and O. C. Farokhzad, *Clin Pharmacol Ther.*, **83**, 761 (2008).
96. V. Wagner, A. Dullaart, A. K. Bock, and A. Zweck, *Nat. Biotechnol.*, **24**, 1211 (2006).

Chapter 5

BioHybrid Hydrogels as Environment-Sensitive Materials for Systematic Delivery of Therapeutics

Sara Faraji Dana and Jing Pan
Department of Pharmacy,
National University of Singapore, 18 Science Drive 4, Singapore 117543, Singapore

Maziar S. Ardejani
Department of Chemistry,
King's College London, London, SE1 1DB, UK
maziar.ardejani@kcl.ac.uk

In this chapter we discuss the concepts of environment-sensitive hydrogels and physical phenomena governing their properties. The focus is on how hybridization of hydrogels with biomolecules can improve the performance of this class of soft materials for systematic delivery of therapeutic agents.

1. Introduction

Through millions of years of evolution, organisms have developed delivery mechanisms through which biologically important chemicals are efficiently transported to specific cellular destinations. A common feature of all these mechanisms is the utilization of a multitude of functionalized materials that are tailored to perform a specific delivery task. A goal of pharmaceutical sciences is to engineer delivery systems that can emulate the specificity and efficiency of such biological materials.

Most biological materials are complex in that they are composed of many components held together by relatively weak non-covalent interactions. Such complexity makes the biological materials adaptive in a way that they are able to modify their dynamic properties according to the physiological stimuli. This has inspired the creation of polymeric materials that are complex enough to give emergence to noticeable degrees of adaptivity. Prominent examples of such materials are environment-sensitive biomaterials whose properties change in response to mechanical, physical or chemical variations of their environment.

Elevated temperatures, altered pH, mechanical tensions, chemical potential gradients are all environmental cues that are often caused by disease and can be used to activate the release of the drug from an environment-sensitive carrier. Therefore, application of environment-sensitive materials in drug delivery systems potentially allows for the temporal and spatial modulation of active concentration of the drug in the targeted part of a diseased organ. In other words, ideally such delivery systems will release the drug at the appropriate time and location and otherwise will stop the release. In a sense these smart delivery systems will aid the homeostatic toolkit of the body to maintain and regulate the internal conditions of cells, tissues, and organs.

To act as a homeostatic agent, the smart material must possess at least three functional modules (Fig. 1(A)): a sensor to detect the relevant changes in the environment, an effector to regulate the detected change and a feedback mechanism that connects the last two parts. An ideal feedback loop interprets the input signal using a comparator to adjust the quantity of the signal sent to the effector. In a way, a comparator, as a part of feedback module, couples the physiological requirement with the amount of drug released. The key challenge in the development of environment-sensitive drug delivery systems is to engineer molecular units that can act as these basic components and the calibration of their functional parameters.

A more simplistic model for a smart drug delivery system is an on–off gating system, where an environment-sensitive part of the drug carrier controls the release of a pharmacon (Fig. 1(B)). In the "off" state, the gatekeeper moiety is in its extended configuration and blocks the exit. Binding of the reporter molecule or reception of a physicochemical signal at the vicinity of the gatekeeper moiety causes it to collapse into a more compact configuration hence opening the exit and releasing the pharmacon.

Materials capable of encompassing all the components required for a smart drug delivery system — sensor, processor (feedback module), and effector modules — must fulfill a number of various functional requirements that makes them inevitably complex. In fact such smart materials have to be a hybrid of many components collectively assembled through non-covalent interactions in order to exhibit a reversible responsiveness to the environmental cues. Depending upon the design of such hybrid materials, a vast variety of biomolecules may potentially be embedded in them to play the roles of sensors or effectors. Therefore the delivery

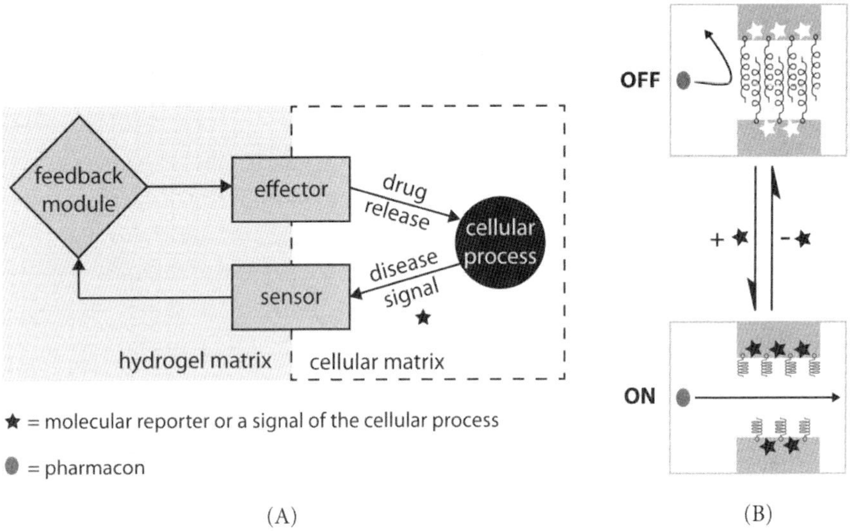

Fig. 1. (A) A proposed model for a smart drug delivery system composed of three major components: a sensor, a feedback loop, and an effector. The feedback loop includes a processing unit that receives, modulates and sends the chemical signal to the effector unit. A classical controlled-release system is distinguished from an environment-sensitive one by the feedback loop. A drug delivery system is environment-sensitive if the effector module is fed by the chemical information received from a sensor. (B) A schematic illustration of an on–off gating system whose permeability is controlled by the local concentration of molecular reporters of a typical cellular process.

matrix that encases these biomolecules must provide a mild ambience to maintain the functional integrity of these molecules. The delivery matrix must also be able to withstand the biochemical and mechanical stress during the delivery process. Moreover, these materials directly interact with biological environment hence must be biocompatible. The complexity and biocompatibility required for the environment-sensitive drug delivery systems directs us toward polymeric materials with high water content.

Hydrogels, due to their high water absorption capacity and structural integrity, are one of the suitable scaffolds for such hybrid biomaterials. Many hydrogel polymers have been developed whose degradation products do not show any detectable detrimental reactions in body and hence are considered to be biocompatible. The tissue-like water content of hydrogels improves their biocompatibility. One significant advantage of hydrogels is their permeability for small molecules such as metabolites, nutrients, and drugs. The highly hydrated matrix of hydrogels can stably host biomolecules without affecting their native structure and activity. Such water-filled three-dimensional (3D) polymeric networks protect the embedded molecules from the hostile factors like degrading enzymes and low-pH of the gastrointestinal tract.

Hybridization of water-absorbing polymer networks with various sensor and effector modules of biological or chemical origin yields "smart" hydrogels. Biohybrid hydrogels have the chance to inherit the useful features of both synthetic hydrogels (solubility, processability, and elasticity) and biological molecules (hierarchical structure, chirality, functionality, and biocompatibility). Fast and cooperative but reversible phase transition in response to changes in the environment is what makes biohybrid hydrogels deserve the title "smart". By this salient feature (sensitivity to physiological cues/signals), smart hydrogels prove useful in micromanagement of varying metabolic and physiological requirements. Moreover, the kinetics of pharmacon release may be modulated through modification of the shape, size, and composition of the hydrogels during their assembly process.

Smart hydrogels are fabricated in different forms based on their target delivery. The major forms include particles and films. Particulate hydrogels are suitable for applications where a flow is available to transport the particle to the target. The surface of particulate hydrogel may display targeting tags to improve the site-specific delivery. Film-like hydrogel formulations suit the topical delivery. In both of these dosage forms, the outermost layer of the hydrogel construct, which is in contact with biological matrix, can be made of environment-sensitive hydrogel to control the drug diffusion (Fig. 2). An important difference of these two types of dosage is the directionality of drug release. The drug release from a particulate formulation lacks directionality, whereas, it can be made directional in film dosage form by having an impermeable backing layer. Using a directional film dosage form, the adsorption of pharmacon can be confined to the contact area and hence the effect of active ingredients can be restricted to the very site of application.

The film forms of environment sensitive hydrogels have been used for many drug delivery applications. Patches for transdermal and transmucosal delivery are examples of film dosage forms that are for the most part based on hydrogels.

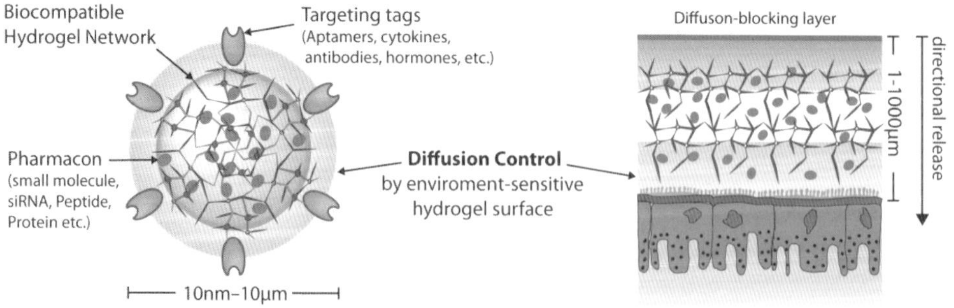

Fig. 2. Hydrogel-based hybrid biomaterials depending on their application can be assembled into two different forms — particle or film. For example, pressure-sensitive hydrogels are usually fabricated as films and have larger dimensions suitable for the topical application.

Delivering pharmacons through skin and mucous have many advantages, among which are relatively large surface area (e.g., the skin area is about 1.5–2.0 square metres for an average adult), accessibility, ease of use, low metabolic activity, low variability, robustness, prolonged retention, and distinct possibility of achieving a zero-order controlled-release. The application of stimuli-responsive hydrogels in the delivery systems of this route is ever expanding. For example, the pressure-sensitive adhesive (PSA) films that are commonly used for transdermal delivery are, by and large, based on pressure-sensitive hydrogels.

2. Environment-Sensitive Hydrogels

Hydrogels are insoluble 3D crosslinked hydrophilic polymeric networks, which can retain a large amount of water. Presence of hydrophilic functional groups in the polymeric chain is responsible for hydrogel's ability to absorb water, while hydrogel resistance to dissolution arises from crosslinks between network chains.

The resemblance of hydrogels to the natural living tissues more than any other synthetic biomaterials and their biocompatibility due to their high water content, has made them an excellent candidate for pharmaceutical (drug delivery) applications. Hydrogels with ionizable groups or unfoldable moieties are defined as environment-sensitive because their structure undergoes a sharp transition induced by changes in the surrounding conditions, such as pH changes, temperature fluctuations, etc.

Hydrogels can play the roles of sensors and effectors in a delivery system (see Fig. 1) if they can respectively respond to environmental changes and regulate the drug release. This control is achieved through a change in the structure of the gel that may enhance or decrease the release rate. Some specially functionalized hydrogels are able to couple this structural change with the environmental stimuli. To have such sensor properties, the chemical structure of the gel must be sensitive to the changes in its surrounding. Variations in environmental condition cause the hydrogel to undergo a reversible volume or sol–gel phase transition as a result of which the release rate of the embedded pharmacon is altered.

A wide variety of hydrogel systems have been developed to be responsive to different physicochemical and mechanical stimuli. These phase-change hydrogels so called "intelligent" or "smart" hydrogels, which release drug in the response to the environmental stimuli, are of the most interest in the field since they may command the site of delivery and also the timing and an interval of drug release. These hydrogels with "sensor" properties can endure either sol–gel phase transition or reversible volume phase transition in the response to the environmental changes. Four fundamental forces that control the behavior of most of environment-sensitive hydrogels are ionic interactions, hydrophobic interactions, hydrogen bonding, and

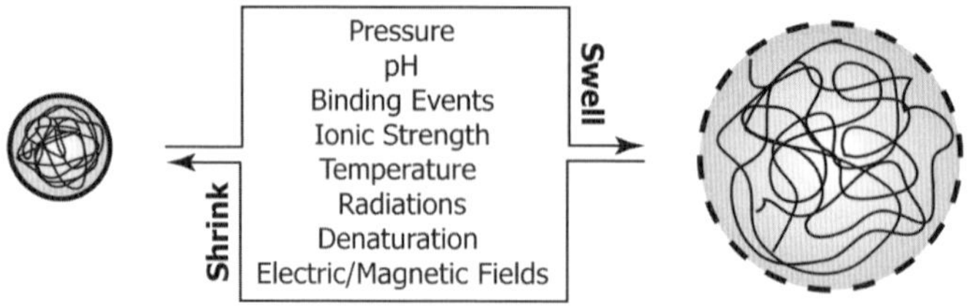

Fig. 3. Various physicochemical stimuli can affect shrinking and swelling behavior of environment-sensitive hydrogels. Swollen hydrogels can release the entrapped drug.

van der Waals interactions. Variation of one or all of these forces enables hydrogels to inter-convert between two phases. These forces can vary due to physical stimuli (pressure, temperature, light, electric fields, ultrasound, magnetic fields), or chemical, and biochemical stimuli (chemical concentrations such pH and ionic strength and specific molecular recognition events of molecules such as glucose or antigen), see Fig. 3. The hydrogels capable of exhibiting structural responses to these stimuli have proven useful in engineering smart delivery systems.[1–3] Some examples of such hydrogel systems those of biohybrid origin are reviewed in the following sections of this chapter.

2.1. *PSA hydrogels*

PSA hydrogels are a special class of tacky viscoelastic polymers that adhere to substrates of various chemical nature under application of slight external pressures over a short period of time (1–2 s).[4,5] To be qualified as a PSA, the polymer needs a balance of elasticity and viscosity.[6] It should possess both relative viscous flow under applied bonding pressure to form a proper adhesive contact, and elastic cohesive strength, which are necessary for resistance to debonding stresses.[7]

Generally, for the tight interaction of adhesive with the surface of a substrate, it should be able to viscously flow into the surface cavities of the substrate.[8] When the adhesive makes a close contact with the surface of substrate because of its viscoelastic properties then it will be able to make a greater amount of molecular interactions such as van der Waals with the substrate e.g., skin. Following the initial adhesion, the adhesive–substrate bonds can be additionally enhanced by spatially tighter molecular interactions (i.e., hydrogen bonding, hydrophobic interactions etc.). [8,9]

Most of the biomedical substrates are comprised of complex arrays of biomolecules with co-localized display of various interaction chemistries. Therefore, development of polymeric systems capable of simultaneously forming multiple types of interactions with substrates will extend the current scope of pharmaceutical applications of PSAs.

PSAs have found ever-expanding potential in drug delivery applications during the recent years. They have been proposed to be utilized in transdermal[10] and transmucosal[11] therapeutic systems for programmed drug delivery,[12] transdermal patches[10,13–16] and moisture-insensitive orthodontic primers.[17] Tailoring PSAs for various pharmaceutical applications however, requires in depth understanding of physiology, chemistry, and physics of the substrates as well as precise engineering of the PSA film.

There are three main factors that determine the characteristics of a substrate for adhesion. The first parameter is the chemical composition and structure of the substrate surface that contributes to the thermodynamics of the adhesion. Second, mechanical properties of substrate's contact volume govern the dynamics of adhesion process while as a third factor, surface morphology of the substrate controls the effective contact area. Although all these three factors contribute to the work of adhesion, however, when comparing PSAs for wet and dry surfaces, it is the chemical composition of the substrate that plays the main role.[18]

Biological surfaces greatly vary in their hydration levels. The main difference between skin and mucous membranes is that the latter is non-keratinized and is highly moist because continuously produces mucus to prevent itself from becoming dry. This makes the mucous membrane to behave as a rather hydrophilic substrate for adhesives. Whereas, stratum corneum, the outer layer of the skin, is hydrophobic in nature to effectively act as a barrier to transepidermal water loss. Although there are considerable levels of morphological and mechanical differences between the skin and mucous membranes, the main factor to be considered for the development of specific adhesives for each type of these substrates is their surface chemical composition i.e., water content.[19]

As for skin applications, the performance pre-requisites of PSAs for drug delivery applications are challenging because they must be able to exhibit appropriate gel strength and sufficient adhesiveness against varying skin types[20] and at the same time they should be easily removable from the skin surface without causing excessive irritation and leaving no residues behind. Since the PSA hydrogels for drug delivery purposes are usually in direct contact with skin, thus the biocompatibility and non-toxicity are two main factors to consider.[21,22] Various functional hydrogels for use in transdermal drug delivery systems (TDDS) have been prepared with biocompatible, non-toxic monomers/macromers, such as poly(ethylene glycol) diacrylate (PEGDA)[21,23,24], polyvinylpyrrolidone (PVP),[24–27] polyethylene glycol (PEG),[21,24,26] etc.

2.1.1. *Pressure-sensitive bioadhesives*

Most of the conventionally developed PSAs intended for adhesion to the skin are basically hydrophobic in nature. These PSAs are based on natural and/or synthetic hydrophobic monomers like polyisobutylenes, silicones and natural or synthetic rubbers. The main disadvantage of PSAs solely made of hydrophobic components is that they lose their tackiness upon presence of moisture on the substrate. This is a major problem for the application of these PSAs on skin because the moisture accumulated due to sweating or other dermal secretions in the adhesive–skin interface will cause loss of adhesion.[28]

To approach this problem, researchers have developed a class of adhesive polymers named bioadhesives that are defined as adhesives capable of adhering to highly hydrated biological surfaces such as mucosal tissues. To be considered bioadhesive, a PSA must plasticize in the presence of water and remain adhered to the hydrated surface. This requires the bioadhesive film to be made of hydrophilic elastomers.[29] As a more specific class of bioadhesives, those materials of this type that are designed to directly interact with mucosal surface are referred as mucoadhesives. Since mucous membranes cover a significant portion of the body's available surface, mucosal path is a major direction for the development of novel drug delivery systems.[30]

Similar to adhesion, an initial step in the process of bioadhesion is formation of a series of interactions between surfaced molecular moieties of bioadhesive and the biological substrate. Subsequently, polymeric chains of the bioadhesive interpenetrate into the bio-substrate. It has been shown that by incorporation of specific ligands into the bioadhesives, they can be guided to directly bind to the receptors on the cell surface rather than mucous gel membrane.[31] This enables targeted delivery of active pharmaceutical agents into the cells since binding to cell surface receptors often results in endocytosis and internalization.

From another point of view, bio/mucoadhesion can be considered as a pressure-sensitive character of adhesives toward hydrated biological substrates, which provides several advantages in using them in drug delivery systems. These advantages include but are not limited to the followings:

(1) Longer residence time of the formulation at the delivery site due to close contact and adhesion. This will result in higher bioavailability at lower concentrations of drug.
(2) Possibility of targeted drug delivery to particular tissues or parts in the body by incorporation of target-specific ligands in the bioadhesive.
(3) Controlled-release of the active pharmaceutical agent which in combination with extended residence time may result in lower administration frequency.
(4) Possibility of avoiding the first-pass metabolism.
(5) Reduction in cost and dose-related side effects due to efficiency and localization of the drug delivery.[32,33]

A bioadhesive PSA must be able to absorb a considerable amount of moisture to avoid adhesion loss due to the accumulation of interfacial water. Being highly hydratable is the characteristic property of hydrogels. Therefore, hydrogels are the candidates for the synthesis of bioadhesive PSAs if they can be modified to show an appropriate degree of viscoelasticity. Hydrogel polymers have been used to produce PSAs for drug delivery applications.[7] The major chemical systems used for these PSAs are acrylate-based hydrogels, due to their suitable adhesive properties and low levels of skin irritation. Other polymer types, used as PSAs, include silicone-based adhesives, polyvinyl ether-based adhesives and PVP-based adhesives.[7,34]

2.2. *Hydrogels sensitive to changes in chemical concentration*

2.2.1. *pH-sensitive hydrogels*

pH-sensitive hydrogels respond to changes in pH of environment. Possessing pendant ionizable acidic (such as carboxylic or sulfonic acids) or basic (such as ammonium salts) functional groups, these gels responds to environmental pH changes by either releasing or accepting protons. Based on their nature, they can be classified as ionic (i.e., anionic or cationic hydrogels) or neutral hydrogels. Anionic hydrogels have negatively charged moieties in their structure, cationic hydrogel have positively charged moieties and neutral hydrogel have both positively and negatively charged moieties.

The pH-sensitive hydrogels consisted of crosslinked polyelectrolytes (i.e., polymer with a large number of ionizable groups) show significant changes in the swelling behavior once their pendant acidic or basic groups undergo ionization as a function of the environmental pH. For instance, crosslinked anionic (e.g., poly(acrylic acid) (PAA) and cationic polyelectrolytes (e.g., poly(N,N′-diethylaminoethyl methacrylate) (PDEAEM)) swell due to ionization at high and low pH respectively. The ionization of polyelectrolytes however, is rather difficult compared to the ionization of individual constituent monoacids or monobases because of the stabilizing electrostatic interactions between the adjacent ionizable groups.[2,35,36]

The hydration of these ionized groups causes the swelling of the polyelectrolyte hydrogels to be more extensive than that of the non-electrolyte polymer hydrogels. The swelling behavior of polyelectrolyte hydrogels correlates with the amount of electrostatic repulsion between the same-charge functional groups of the polymeric chain. Therefore, any condition that decreases the electrostatic repulsion will negatively affect the extent of swelling. These conditions include but are not limited to changes in counterion type, ionic strength, and pH.[35,37]

The swelling and pH-responsiveness of a hybrid hydrogel depends on the amount of polyelectrolyte monomers present in its structure. For example, neutral co-monomers, such as 2-hydroxyethyl methacrylate, methyl methacrylate and maleic anhydride can be added to adjust the swelling and pH-dependent behavior of a hydrogel. A wide range of pH-sensitive behaviors can be achieved by including various co-monomers each of which providing a different level of hydrophobicity/-philicity to the polymeric chain. The pH-sensitive behavior of interpenetrating polymer networks (IPN) systems can be manipulated through a similar principle where, two or more different types of polymer chains interact.

Once all the ionizable functional groups are ionized, the swelling process stops. Additional increase in pH, enhances the ionic strength which leads to the declining of osmotic pressure and shrinkage of the gel.[35,38]

2.2.2. *Glucose-sensitive hydrogels*

Development of auto-regulated insulin delivery systems has been one of the greatest challenges in controlled dug delivery. As a certain amount of insulin has to be delivered at a precise timing, the glucose sensing ability and automatic cut-off mechanism are necessary for the systematic regulation of the insulin delivery.[2]

Hydrogels with a built-in glucose sensor have been fabricated for the modulation of insulin delivery. These glucose-sensitive hydrogels can couple the insulin release with the local glucose concentration. For instance, the embedded glucose oxidase of the cationic pH-sensitive polymers correlates hydrogel's swelling and insulin release with the blood glucose level. Once the glucose level of blood is high, glucose diffuses in the hydrogel and is converted to gluconic acid, which elevates the pH causing the hydrogel to swell. When the hydrogel is swelled, the trapped insulin is discharged with a pulsating trend and thus the glucose level in the blood is controlled.[2,39,40]

Based on a similar approach, glucose-sensitive hydrogels can be fabricated to endure sol–gel phase transformations depending on the environmental glucose concentration. A glucose-dependent crosslinking is required for the reversible sol–gel phase transformations. One of such pharmaceutical hydrogel systems is prepared through conjugation of monomethoxy poly(ethylene glycol) with glucose containing polymer which undergoes reversible sol–gel phase transition in response to the changes of glucose concentration.[39]

Besides the sol–gel approach, glucose-sensitive hydrogels can be developed based on binding of insulin–glucose conjugates to the binding sites of hydrogels carrying Concanavalin A (Con A), which is a glucose-binding protein. In the presence of free glucose, the insulin conjugate is released from the gel due to the competitive binding of glucose to Con A.[41,42]

2.3. *Temperature-sensitive hydrogels*

Temperature-sensitive or thermo sensitive hydrogels are capable of swelling and de-swelling due to changes in the environmental temperature. The temperature-dependent phase behavior of hydrogels is an important property that can be used for the design of temperature-sensitive delivery systems. For example, the partial miscibility of some hydrogels is temperature-dependent. The chemical structure of temperature-sensitive hydrogels is comprised of hydrophobic groups, such as methyl, ethyl, and propyl groups. These hydrogels are classified into negative thermo sensitive, positive thermo sensitive, and thermally reversible hydrogels.

2.3.1. *Negative thermo sensitive hydrogels*

The water-solubility of some thermo sensitive hydrogels decreases at temperatures above lower critical solution temperature (LCST). In other words, such hydrogels start to shrink when the temperature rises beyond the LCST because of the inter-polymer chain association through hydrophobic interactions. Therefore, due to this inverse swelling behavior these types of hydrogels are known as negative thermo sensitive. Negative thermo sensitive hydrogels are made of polymeric chains containing moderately hydrophobic groups or a mixture of hydrophilic and hydrophobic groups. Such hydrogels can be fabricated as copolymers of hydrophobic (e.g., N-isopropylacrylamide (NIPAAm)) and hydrophilic (e.g., acrylic acid) monomers. By adjusting the ratio of hydrophilic and hydrophobic segments of the polymer, LCST can be fine-tuned. One of the most extensively used temperature-sensitive hydrogels is poly(N-isopropylacrylamide) (pNIPAAm). Polymers based on NIPAAm can also be copolymerized with other monomers such as butylmethacrylate (BMA) in order to alter the LCST. Derivatives of poly(N,N-diethylacrylamide) (pDEAAm) are also widely used as temperature-sensitive hydrogels due to their LCST range (25–32°C) which is close to the body temperature.

2.3.2. *Positive thermo sensitive hydrogels*

Positive thermo sensitive hydrogels behave opposite to negative thermo sensitive hydrogels. This means they swell at high temperatures and shrink at low temperatures. Water-solubility of positive thermo sensitive hydrogels increases as the temperature increase. For instance, the swelling of IPNs of PAA and polyacrylamide (pAAm) or p(AAm–co-BMA) are positively correlated to the temperature.

2.3.3. *Thermally reversible hydrogels*

Thermally reversible hydrogels are composed of non-covalently crosslinked polymer chains. These types of temperature-sensitive hydrogels may undergo sol–gel phase transitions instead of swelling–shrinking transitions because of the non-covalent crosslinking. Such hydrogels exhibit an inverse temperature dependent sol–gel behavior in that they become sol as the temperature increases. Block copolymers of poly(ethylene oxide) (PEO) and poly(propylene oxide) (PPO) show this type of behavior. Having LCSTs close to the body temperature, these polymers become increasingly attractive for the development of delivery systems that release the drug based on a sol–gel phase conversion around the body temperature.[1,2]

2.4. *Widely used monomers in environment-sensitive hydrogels*

The number of synthetic monomers that can be used for the fabrication of environment-sensitive hydrogels has expanded to include monomers with tailor-made properties suitable for specialized applications. In that sense, hydrogel monomers are selected based on their physicochemical properties, biocompatibility, ease of delivery, as well as availability and cost. For example, some of the most commonly used monomers intended for delivery of proteins are biodegradable conjugates of lactic and glycolic acid. However, polymers of such monomers are hydrophobic with a tendency to denature the encapsulated protein. They also cause inflammation when degraded. These problems can be tackled by using hydrophilic monomers. Acrylic acid, PEG, and methacrylic acid are among hydrophilic monomers that are useful for therapeutic applications. Monomers such as PNIPAAm, polyvinyl alcohol (PVA) are also being tailored using synthetic techniques to fulfill the requirements of new applications.

Following are monomers used in the synthesis of environment-sensitive hydrogels for pharmaceutical applications: hydroxyethoxyethyl methacrylate (HEEMA), hydroxydiethoxyethyl methacrylate (HDEEMA), methoxyethyl methacrylate (MEMA), methoxyethoxyethyl methacrylate (MEEMA), methoxydiethoxyethyl methacrylate (MDEEMA), Ethylene glycol dimethacrylate (EGDMA), N-vinyl-2-pyrrolidone (NVP), N-isopropyl AAm (NIPAAm), vinyl acetate (VAc), acrylic acid (AA), methacrylic acid (MAA), N-(2-hydroxypropyl methacrylamide) (HPMA), ethylene glycol (EG), PEG acrylate (PEGA), PEG methacrylate (PEGMA), PEGDA, PEG dimethacrylate (PEGDMA).[43]

3. Encapsulating Hybrid Biomaterials for Therapeutic Delivery

Encapsulation is often achieved using vesicles and vesicle-like assemblies. Vesicles are semispherical assembly of molecules that like a membrane enclose a hollow volume.

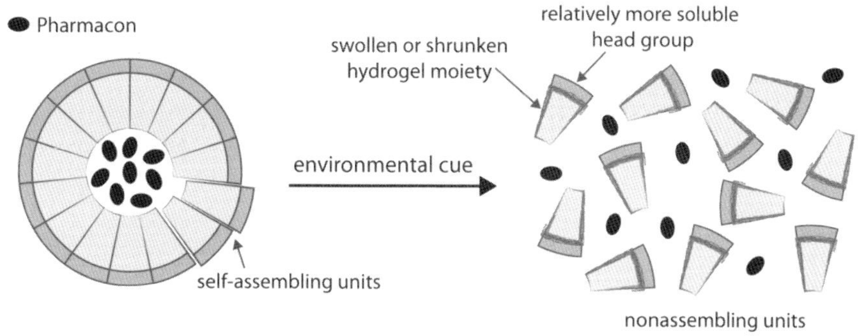

Fig. 4. A general schematic showing how swelling or shrinking of an environment-sensitive hydrogel can control the release of pharmacons from encapsulating assembly.

Biological and synthetic molecules and their conjugates may be used to form wedge-shaped units that are capable of self-assembling into vesicles. For a structural unit to be self-assembling, its corona or wider part of the wedge has to be more hydrophilic than its vertex sides. Hydrogels whose hydrophobicity can change in response to environmental cues can act as a switch to convert a self-assembling unit into non-assembling unit and hence trigger the disassembly of the vesicle (Fig. 4) and release the cargo.

Encapsulation of therapeutic agents in a membranous biomaterial offers few advantages for their effective delivery. A pharmacon protected by a biomaterial layer is more stable and less prone to degradation. Encapsulation into vesicle-like assemblies also provides the opportunity of gaining control over the formulation's particle size — a distribution range that can vary from nano- to micro-scale. Moreover, the ability to diversify the chemical composition of the enclosing material will add to the functional modularity of the final encapsulated assembly.

3.1. *Peptide/protein-hydrogel hybrid vesicles*

Proteins and their shorter forms, peptides, can be engineered to self-assemble into nanoscale vesicle-like particles. Mimicry of naturally polymerizing proteins is one of the approaches for the development of such self-assembling systems with potential application as environment-sensitive drug delivery vesicles. In an early example of such systems, Lee *et al.* developed an amphiphiles block copolymer based on the structure of elastin. They showed that this elastin-mimicking polypeptide forms micelle-like nanoscale aggregates.[14] Triggered disassembly of such vesicles after they have been localized in the site of interest is a challenge and yet to be addressed. In an approach opposite to encapsulation, similar elastin-like polypeptides (ELPs) are able to bear a drug molecule on their hydrophilic part. Thermally-triggered self-assembly of these ELP block copolymers into micelles will cause multivalent localization of the drug molecule displayed on the outer surface of the particle.[45]

Peptosomes are another type of hybrid vesicles assembled from amphiphilic copolymers. At least one of the blocks of the constituent copolymer is a peptide with completely distinct properties than the other blocks that may or may not be synthetic. Peptosomes can be regarded as the proteinaceous counterparts of liposomes. They however, offer prolonged clearance time and useful properties such as rational designability, cell adhesion and recognition, targeted delivery, controllable nanoscale order, and adjustable environment sensitivity.

Early examples of environment-sensitive peptosomes were based on polybutadieneb-poly(L-glutamate) copolymers.[46] These peptosomes exhibited a pH-dependent swelling behavior that varied by ionic strength. Relatively more biocompatible peptosomes have been made using polysacrosine and poly(γ-methyl-L-glutamate) respectively as hydrophilic and hydrophobic blocks.[47] The self-assembly, morphology and size of the peptosomes has been found to be dependent on the hydrophobicity to hydrophilicity ratio of the blocks.

Cage-like structures with hydration levels comparable to those of hydrogels have been also fabricated using *de novo* designed peptides.[48] Fletcher *et al.* have developed peptide components that assemble into enclosed, hollow nanostructures, the morphology of which is defined by specific rules governing the inter-peptide interactions (Fig. 5).[49] An advantage of *de novo* design is that the manipulation of these rules results in the controlled alteration of the morphology

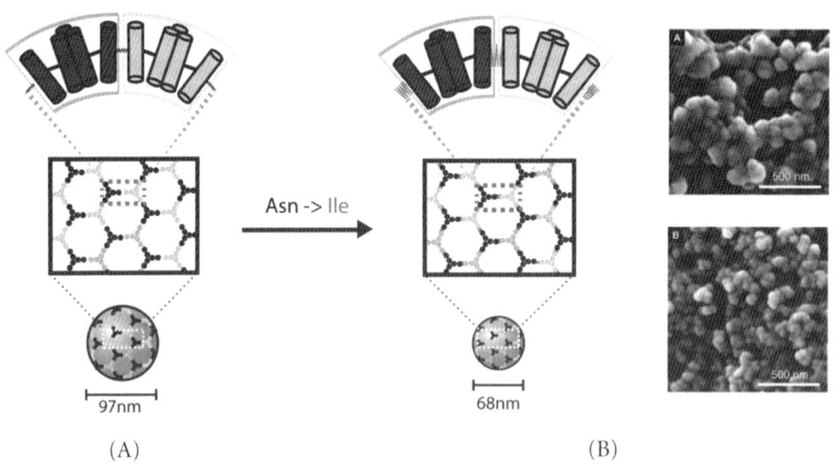

Fig. 5. Highly hydrated capsular morphologies are made from peptide building blocks. Subtle manipulation of the helix–helix binding through a single amino acid change (Asn $\rightarrow$ Ile) causes a significant change in nanostructure size, from around 100 nm in (A) to about 70nm in (B). The scanning electron micrographs are reprinted from Fletcher *et al.*, *Science* 340 (6132), 2013.[49]

(Fig. 5). Because the system is so precisely defined up from the molecular level, it provides easily testable sequence–structure relationships and controlled manipulation of structure to generate new cages that can be used for encapsulation and targeted therapeutic delivery.

Environment-sensitive vesicles can also be assembled using large protein–polymer hybrid amphiphiles. Resembling an almost wedge-shaped geometry, the head group of such hybrid amphiphiles is a protein or enzyme and the tail group is composed of a relatively more hydrophobic but extended polymeric chain. These two groups can be linked either covalently for example through click reactions or non-covalently through methods like reconstitution (Fig. 6(A)). It has been shown such biohybrid block copolymers can self-assemble into vesicles that retain the biological activity of the protein group. For instance, a triblock copolymer prepared by non-covalent linkage of a PEG-based di-block polymer with myoglobin or horseradish peroxidase (Fig. 6(B)) was shown to have the ability to self-assemble into vesicle-like architectures. The di-block copolymer, poly(ethylene oxide)-block-polystyrene–protoporphyrin IX (Fig. 6(C)), was first conjugated with a heme-like group and then reconstituted with apo-protein to from the triblock. The chemical and functional diversity of such biohybrid constructs expands their applicability.[1]

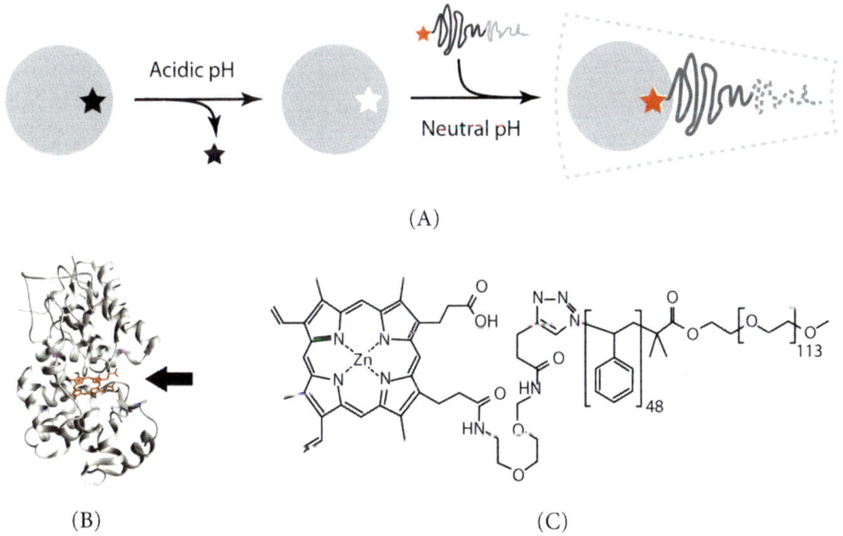

Fig. 6. (A) A schematic illustrating how reconstitution can be used to non-covalently attach a polymeric chain of hydrogel to a protein to form a wedge-shaped self-assembling unit. The cofactor of the protein is removed at low pH and then the polymeric chain conjugated with a cofactor-like molecule is placed back in its binding pocket. (B) The structure of Horseradish peroxidase with heme cofactor in red. The arrow shows the heme positioning. (C) PEO-*block*-polystyrene–protoporphyrin IX (PEO-*b*-PS-PPIXZn).

3.2. *Compacted DNA nanoparticles*

Environment-sensitive nanoparticles of hydrogels loaded with DNA molecules are candidates for efficient and sustained release gene delivery systems mainly because they can rapidly escape from the degradative endo-lysosomal to the cytoplasmic compartment. Much attention has been given to poly(D,L-lactide-co-glycolide), PLGA, and polycationic polymers such as polyethylenimines and chitosan for the development of non-viral gene delivery vectors. Polycationic hydrogels are able to condense large DNA plasmids into small nanoparticles while protecting them against degradative enzymes such as nucleases. Biohybrid hydrogels such as PEG-substituted poly-lysine peptides and cationic bovine serum albumin conjugated with PEG–poly(lactide) nanoparticle has also shown to be promising delivery systems.[50,51] Cationic dipeptides are also able to condense plasmid DNA into spherical nanostructures.[52] It has been shown that such nanoparticles can be loaded into environment-sensitive dosage forms. For instance, temperature sensitive contact lenses loaded with nanoparticles of ophthalmic drugs has shown triggered and extended delivery.[53]

4. Biomolecular Hydrogel Conjugates

In addition to synthetic polymers, natural polymers, such as peptides, DNA and polysaccharides, have also been used to constitute the polymeric networks in hydrogels. Due to their low toxicity, biocompatibility and functionality, these biological polymers, have been used as molecular building blocks of hydrogels in diverse applications of drug delivery and tissue engineering.

4.1. *DNA-hydrogel conjugate networks*

The double helix structure of DNA was discovered by Francis Crick and James D. Watson in 1953.[54] Since then, it is known that base pairs are the building blocks of the DNA double helix. The precisely defined complementary nature of this base-paired structure has expanded the role of DNA from biology to other fields, such as material science, nanotechnology and etc.[55,56] Currently, based on the unique characteristics of nucleic acids, DNA is also used in the formation of smart hydrogels capable of responding to external stimuli. As a hydrophilic, biocompatible and highly programmable natural polymer, DNA can self-assemble into hydrogel networks and also be incorporated in synthetic polymers to form DNA–polymer biohybrids.[57]

4.1.1. *Temperature-sensitive DNA hydrogels*

As mentioned before, the temperature-sensitive hydrogels undergo an often-reversible sol–gel phase transition at a critical solution temperature. It is also well known that the hydrogen bonds between bases on opposite strands of DNA can break due to thermal fluctuations. This denaturation results in untwisting of DNA double helix. In addition, because of the complementary nature of the base-paired structure, the temperature-sensitive DNA hydrogel can be designed based on the hybridization of double strands and thermal dissociation feature. One of the early examples of thermo sensitive DNA hydrogel was reported in 1996. In this study, 5′-terminal-amino-modified DNAs were conjugated to the side chains of succinimide copolymers. The DNA–polymer conjugate was water-soluble. DNA strands complementary to the DNA–polymer were added to the DNA–polymer solution to act as a crosslinker. Upon hybridization of the crosslinker DNA with DNA–polymers, the state of polymers undergoes a transition from solution to hydrogel. With an increase in temperature, the double strands of hybridized DNA become unstable and denature as a result of which the hydrogel transforms from gel into a solution.[58]

Kang and co-workers also reported the fabrication of a nanostructure composed of Au–Ag nanorods, oligonucleotides and polymers (Fig. 7). In this study,

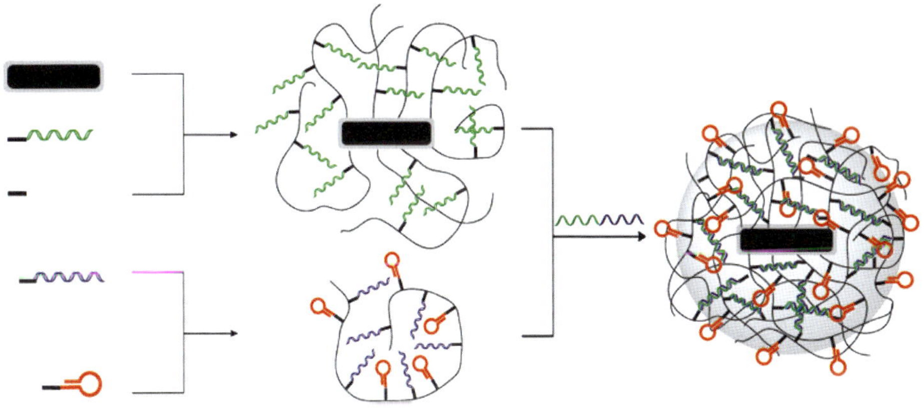

(A)

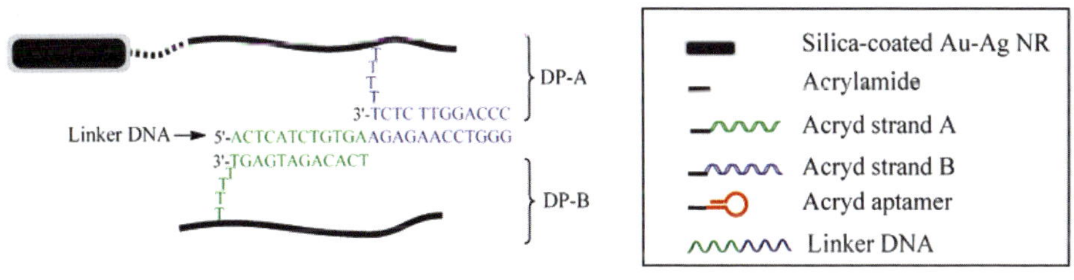

(B)

Fig. 7. (A) Schematic diagram illustrating the formation of an aptamer-functionalized core-shell nanogel. (B) DNA sequences and linkages in the nanogel (not to scale). Reprinted with permission from Kang *et al.*[59]

the methacryl groups on the Au–Ag nanoparticles initiate the growth of polymeric chains through co-polymerization with acrylamide monomer and acrydite-modified oligonucleotides. These DNA grafted polymers are crosslinked by complementary DNA sequences to form a nanogel. Due to the photosensitive nature of Au–Ag nanorods, once irradiated by a near-infrared laser beam, the encapsulated nanorods absorb the photo-energy and convert it to heat with high efficiency. Produced heat elevates the local temperature of the gel, which leads to the dissociation of hybridized DNA strands and finally resulting in a gel-to-sol transition. These nanogels could be developed for targeted cancer therapy by functionalizing nanogels with aptamers, since there are aptamers shown to have excellent binding affinity for different types of tumor cells. Similarly, they may be used for controlled released drug delivery by controlling the heat generated through Au–Ag nanorods.[59]

4.1.2. *Ion-responsive DNA hydrogels*

Metal cations serve as cofactors in most DNA/RNA self-cleavage reactions. This role of metal ions can be harnessed for the development of ion-responsive hydrogels. DNAzymes are DNA molecules capable of catalyzing DNA/RNA self-cleavage reactions.[60] Metal cations are able to specifically recognize secondary structures of the DNAzymes and subsequently cause DNA/RNA to irreversibly cleave at the cleavage site.[61–64] Lin and co-workers reported a Cu^{2+} DNAzyme sequence in which DNA substrate and DNAzyme were incorporated into pAAm, respectively, to form the side chains of linear polymers. When these DNA-grafted polymers were mixed in the buffer solution, they were crosslinked into 3D hydrogels as a result of hybridization between the complimentary parts of DNA substrate and DNA enzyme. The DNAzyme/substrate complex consisted of two different domains, triplex domain and duplex domain. When Cu^{2+} ions were added to the hydrogel, the DNA substrate was cleaved at the cleavage site. This breaks the crosslinks at triplex domain and causes the DNA side chains to dissociate. As a result, the DNAzyme crosslinked hydrogel underwent gel-to-sol transition in response to copper ions.[65]

Similar to metal ions, protons are also able to induce the formation of DNA hydrogels. One structural feature of DNA useful for this purpose is that the fluctuation of environmental pH may cause conformational changes in a cytosine-rich DNA.[66] Cheng and co-workers has reported the creation of a pH-triggered DNA hydrogel based on this effect of cytosine protonation.[67] In their study, a Y-shaped DNA nanostructure was fabricated through co-assembly of three single-stranded DNAs with cytosine-rich tails. At high pH, Y-shaped DNA nanostructures were dissociated as separate random coils due to electrostatic repulsion.

At an acidic pH, cytosines located at the tail of each DNA strand became partially protonated. This protonation causes the formation of hydrogen bonds between protonated and unprotonated cytosines. This leads to the crosslinking of Y-shaped DNA nanostructures into a 3D complex. An increase in environmental pH slightly beyond 7 causes the crosslinks of hydrogel to collapse due to the deprotonation of cytosines. As a consequence of this, the hydrogel was transformed into a liquid solution. Moreover, it is reported that gel transition took place in less than 1 min after increasing pH to 8. Hence, this pH-sensitive DNA hydrogel may have potential application in immediate release dosage forms and drug delivery systems for cancer therapy since tumor sites are usually more acidic than other parts of human body.

4.1.3. *Biochemically-responsive DNA hydrogels*

In addition to temperature- and ion-responsive hydrogels, bio-responsive hydrogels have also attracted much attention because of their advantages in the biomedical application. Biomolecules (antibodies, enzymes, aptamers etc.) have been incorporated into hydrogels to diversify the responsiveness and expand the use of hydrogels.[68,69] Yang and co-workers designed a selective target-responsive hydrogel using a DNA aptamer.[70] The design is based on the competitive binding of targets to the aptamer. There were two strands of acrydite-modified DNA (strand A and strand B), which were separately copolymerized with acrylamide to form the side chains of pAAm. Upon the addition of a DNA linker, two acrydite-modified DNA strands were hybridized with the linker. As a consequent of this, the two polymers were crosslinked into a DNA-based hydrogel. Nevertheless, the DNA linker contained an aptamer sequence for adenosine, which partially hybridized with strand B. Once adenosine was added in the hydrogel, adenosine competitively bound the aptamer, which destabilized the hybridization between strand B and the DNA linker, and in turn led to the dissociation of strand B from the DNA linker. As a result of this, the hydrogel was converted into liquid state.

Another biohybrid hydrogel was developed by Christen and co-workers (Fig. 8). Firstly, they designed two types of linear pAAm of which one was covalently conjugated with acrylating amine-functionalized double-stranded DNA containing the tetO motif and the other one was coupled to a protein, single-chain TetR. Mixing these two types of linear polymers resulted in the formation of a biohybrid hydrogel via TetR/tetO-interaction. In the presence of tetracycline, competitive binding of TetR to tetracycline dissociated the link between TetR–tetO, thereby destabilizing the networks of hydrogel. Consequently, the hydrogel was dissolved and transformed into liquid phase.[71]

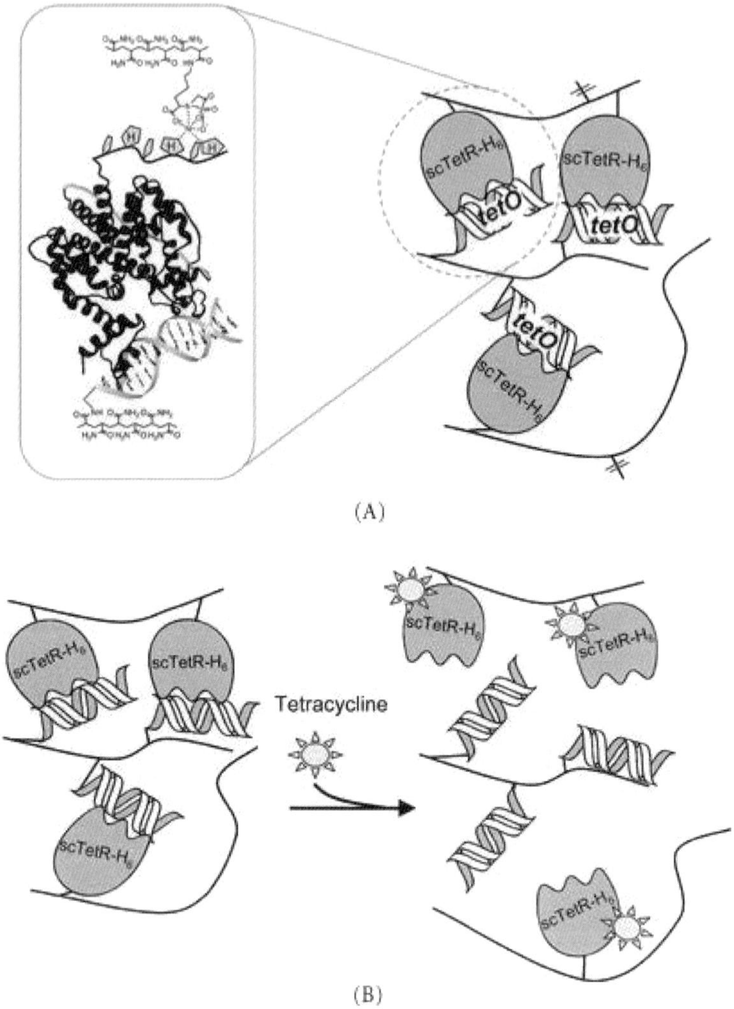

Fig. 8. Design concept for the Tetracycline repressor (TetR)-based tetracycline-responsive hydrogel: (A) Schematic representation of the hydrogel design. The hydrogel is composed of two derivatives of pAAm, one functionalized with the double-stranded *tetO* DNA operator and one coupled to the single-chain TetR, scTetR-H_6, via Ni^{2+}-mediated electrostatic interactions between nitrilotriacetic acid and the hexa-histidine tag (H_6). (B) Binding of scTetR-H_6 to *tetO* crosslinks the polymers, thereby resulting in the formation of a hydrogel, whereas the addition of tetracycline triggers the dissociation of scTetR-H_6–*tetO* to loosen the network structure. Reprinted from Christen *et al.*[71]

4.2. *Peptide-hydrogel conjugate networks*

Peptides can form a hydrogel by self-assembly under certain conditions, and they can also be a part of hybrid molecules that constitute hydrogels. For example, peptides can be conjugated with synthetic polymers to form hydrogels.

4.2.1. *Alpha-helical peptide hydrogels*

The alpha helix (α-helix) is a common secondary structure of proteins and it is a right-handed coiled conformation. Two or more α-helices twine parallel to each other to form coiled–coil structure. The coiled–coil structure has been widely used as a motif in peptide hydrogel design. It is well known that coiled–coil motifs possess a heptad repeat $(abcdefg)_n$ of polar and non-polar amino acid residues (Fig. 9).[72] The positions "a" and "d" usually contain hydrophobic amino acids. They direct the dimer formation due to the hydrophobic interactions. The positions g and e which often contain charged and polar amino acids configure the dimer interface together with the position a and d. Position b, c, and f that are exposed on the surfaces of coiled–coil structures have been less well studied but generally are hydrophilic to enhance the peptide solubility in water.[73]

In several studies, the fabrication of peptide-hydrogels has been developed based on the coiled–coil assembly. For example, Dong and co-workers used 21 amino acid peptides to form self-assembled nanofibers with a diameter of 4 nm.[74] They incorporate positively charged residues in the positions b, c and f and controlled the length of nanofibers via the selection of amino acids with different helix propensities. Hydrogels were engineered with a relatively high peptide concentration at low pH value (~3.5). In addition, Woolfson and co-workers chose alanine and glutamine residues in the positions b, c and f to promote weak hydrophobic interfibril interactions and hydrogen bond formation.[75] As a result, hydrogels consisting of bundles of small and flexible fibers were self-assembled. In this study, they also found that peptide sequence could affect the hydrogel properties. Increasing temperature destabilized the hydrogels with hydrogen-bonded crosslinks, whereas strengthened hydrophobic interactions of fibrils.

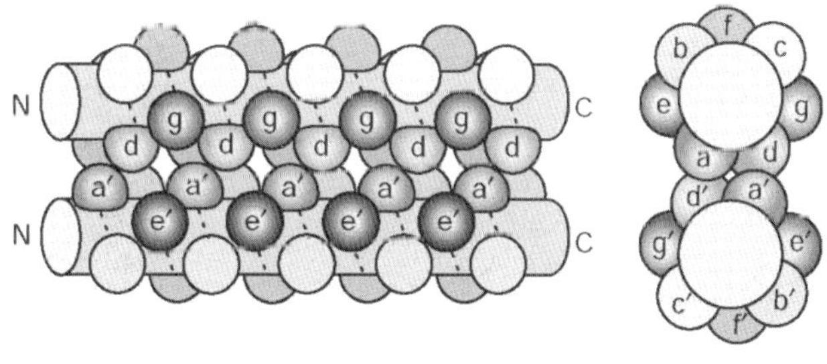

(A) (B)

Fig. 9. Cartoon of a parallel two-stranded coiled–coil (a) side view and (b) top view. The interface between the α-helices in a coiled–coil structure is formed by residues at the core positions **a**, **d**, **e** and **g**. Positions in the two helices are distinguished by the prime notation; for example, **a** and **a'** are analogous positions in the two helices. N, amino terminus; C, carboxy terminus. Reprinted from Fong *et al.*[72]

4.2.2. *Beta-sheet peptide hydrogels*

Similar to α-helix, β-sheet structure is also ubiquitous in natural and disease-related proteins, including silk fibroins and fibrillar amyloids of Alzheimer's disease.[76,77] It is characterized by two parallel or anti-parallel peptide chains connected by hydrogen bonds (Fig. 10).[78] In particular, the repetitive GAGAGS, GAGAY, and GAGAGVGY sequences were found to be responsible for the formation of anti-parallel structures in silk fibers.[79] Besides, there are several factors including polarity of the solvent, which may influence the process of gelation with β-sheet conformation. It has been shown that K24 and K27 domains of IsK protein could be fabricated into fibril-rich hydrogels with a high β-sheet conformation ($\geq$70%) in the presence of moderately polar solvents such as methanol.[80] Gelation of peptides in more polar solvents (e.g., water) can be achieved by incorporation of more hydrophilic residues in the peptides.[81]

Ionic interactions have also been observed to induce the formation of hydrogels based on peptides with β-sheet conformation. Zhang and co-workers designed peptide sequences to form hydrogels with regular β-sheet structures in water.[82] These peptide-based self-assembling structures were stabilized through self-complementary ionic interactions. In this study, positively charged lysines and negatively charged glutamines were complementary ionic pairings in the molecular structure of peptide, which undergo spontaneous association in the process of gelation. If the positively charged residues are displaced with negatively charged residues, like glutamines, hydrogels may not be self-assembled.[83]

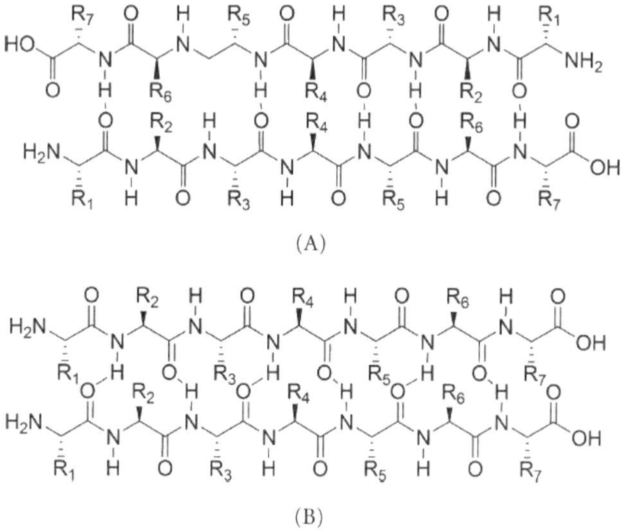

Fig. 10. (A) Antiparallel and (B) parallel β-sheet structures of two peptide chains connected by hydrogen bonds. Reprinted from Altunbas *et al.*[78]

Schneider and Pochan laboratories discovered a class of β-sheet peptides, named as β-hairpins, which can be assembled into hydrogels through intramolecular folding and intermolecular association induced by external stimulus including changes in pH, ion concentration or temperature.[84–86] The folded β-hairpins are amphiphilic with hydrophobic valine and hydrophilic lysine residues on opposite sides. When many hairpins suspend in the water, the valine residues of two β-hairpins collapse through hydrophobic interactions to form a hairpin sandwich which also forms intermolecular interactions through hydrogen bonds of deprotonated lysine residues at the end of two adjacent β-hairpins. These intermolecular interactions result in the self-assembly of fibrils.[87] Peptide hydrogels consist of a series of these crosslinked fibrils.[88]

4.2.3. *Hydrogels based on peptide amphiphiles*

Hydrogels can also be synthesized using hybrid monomers composed of peptides and other organic functional groups. Jun and co-workers assembled nanofibers using amphiphilic peptide molecules consisting of a hydrophobic alkyl tail, a short peptide sequence, a hydrophilic moiety and a short bioactive headgroup.[89] Peptide sequences were employed to form hydrogen bonds with other peptide sequences and in turn, self-assemble into β-sheet or α-helix structures. The hydrophobic alkyl tails were used to form the fiber core based on hydrophobic interaction. The hydrophilic moieties are employed to improve the water solubility of the peptide amphiphiles. Under proper conditions, nanofibers composed of peptide amphiphiles assemble into hydrogel networks. Furthermore, moieties with different biochemical functions can be incorporated to facilitate the assembly of peptide amphiphiles. For examples, incorporation of cysteine residues in the peptide sequences or adding diacetylene groups in the hydrophobic block enables the formation of crosslinking networks by disulfide bonds or UV irradiation.[90,91]

4.2.4. *Hydrogels based on polymer–peptide conjugates*

Peptide sequences can also be incorporated into the synthetic polymer monomers for synthesis of polymer–peptide conjugated hydrogels. Such hydrogels possess the advantages of both polymer and peptide moieties. Peptide sequences impart biological, chemical, thermal responsiveness to the hydrogel systems. The solubility and stability of hydrogel systems are also improved by the addition of some synthetic polymers, like PEG. One widely used peptide sequence in the preparation of cell-compatible hydrogels is arginine–glycine–aspartic acid (RGD). The RGD sequence is commonly known as a cell binding domain of the extracellular matrix (ECM) and is implicated in cellular attachment via integrins.[92] Hern and co-workers incorporated

RGD sequence into non-adhesive hydrogels.[93] The N-terminus of the RGD sequence was functionalized with an acrylate moiety and then, the peptide sequence was mixed with PEG diacrylate to facilitate the formation of hydrogels upon UV polymerization.

As discussed earlier, peptides can also provide thermal responsiveness to the polymer–peptide conjugates. Song and co-workers synthesized a thermo sensitive poly (organophosphazene)–RGD conjugate through a covalent amide linkage between a peptide sequence, GRGDS, and poly (organophosphazene) through carboxylic acid terminal group.[94] This peptide–polymer conjugate exhibited a solution state at room temperature and underwent a solution–gel transition at body temperature.

4.3. *Polysaccharide-based hydrogels*

Polysaccharides have also been widely utilized for hydrogel preparations since they are usually abundant, biocompatible, and non-toxic. Moreover, they show complex physical and chemical properties that may allow multiple functional modifications to meet the requirements of specific applications. As a class of natural macromolecules, polysaccharides are capable of forming hydrogels by themselves under certain conditions. Polysaccharides may be combined with other natural or synthetic polymers in order to crosslink into 3D complex. This section will introduce two polysaccharides that are well-studied for the development of hydrogel dosage forms.

4.3.1. *Alginate-based hydrogels*

Alginate is extracted from marine brown algae or produced by bacteria. It is consisted of a backbone of 1, 4 linked β-D mannuronic acid (M) and α-L-guluronic acid (G) residues of different sequences (Fig. 11(A)). These homopolymeric M and G blocks are covalently linked with alternating structure. The physicochemical properties of alginate are affected by the M/G ratio and the structure of alternating zones as well. The formation of alginate-based hydrogel is mainly based on the interactions between the carboxylic acid moieties and bivalent counterions (i.e., Cu^{2+}, Ca^{2+}, Pb^{2+}). However, these physically associated hydrogels are unstable due to the inhomogeneity of networks and their low mechanical strength, which restrict the use of alginate-based hydrogels in the drug delivery applications. To improve the mechanical strength and to increase the diversity of hydrogels for responding to different environmental stimuli, "smart" alginate-based hydrogels have been developed through chemical crosslinking of alginate with other natural or synthetic polymers. Here, we will briefly introduce several "smart" alginate–polymer hybrid hydrogels.

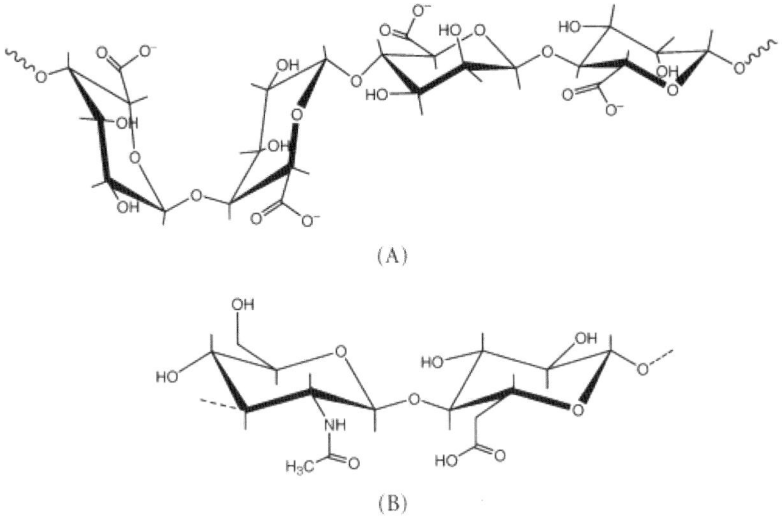

Fig. 11. Structures of (A) alginate and (B) hyaluronic acid. Reprinted from Coviello *et al.*[95]

Muniz and co-workers reported a thermo sensitive alginate–Ca^{2+} hydrogels interpenetrated with pNIPAAm networks.[96,97] As mentioned before, polymers of NIPAAm have been extensively used in the preparation of thermo sensitive hydrogels as its phase-transition happens near body temperature.[98] Muniz and co-workers found that the mechanical strength of hydrogels composed of alginate–Ca^{2+} and pNIPAAm networks was much higher than that of plain alginate–Ca^{2+} and pNIPAAm hydrogels. Besides, the hybrid hydrogels exhibited prominently reduced water uptake and decreased pore size when temperature was increased from 25 to 40°C. It may be due to the collapse of pNIPPAm chains at higher temperature, which leads to the contraction of gel. The permeation of orange II, which served as a model drug was also evaluated in this hydrogel system and a remarkable decrease of orange II diffusion was detected with the temperature increase.

Chen and co-workers blended alginate and N,O-carboxymethyl chitosan (NOCC) together to create a pH sensitive hydrogel in the presence of genipin, working as a crosslinker in the hydrogel system.[99] As a water-soluble chitosan derivative, NOCC possesses carboxymethyl substituents on some of both the amino (N-position) and primary hydroxyl sites (O-position) of the modified chitosan structure. The authors suggested that genipin might form intermolecular crosslinks among NOCC molecules with a heterocyclic structure. Alginate entangles through the NOCC network, leading to the formation of alginate–NOCC interpenetrating polymeric hydrogel (Fig. 12). They also found that the swelling of

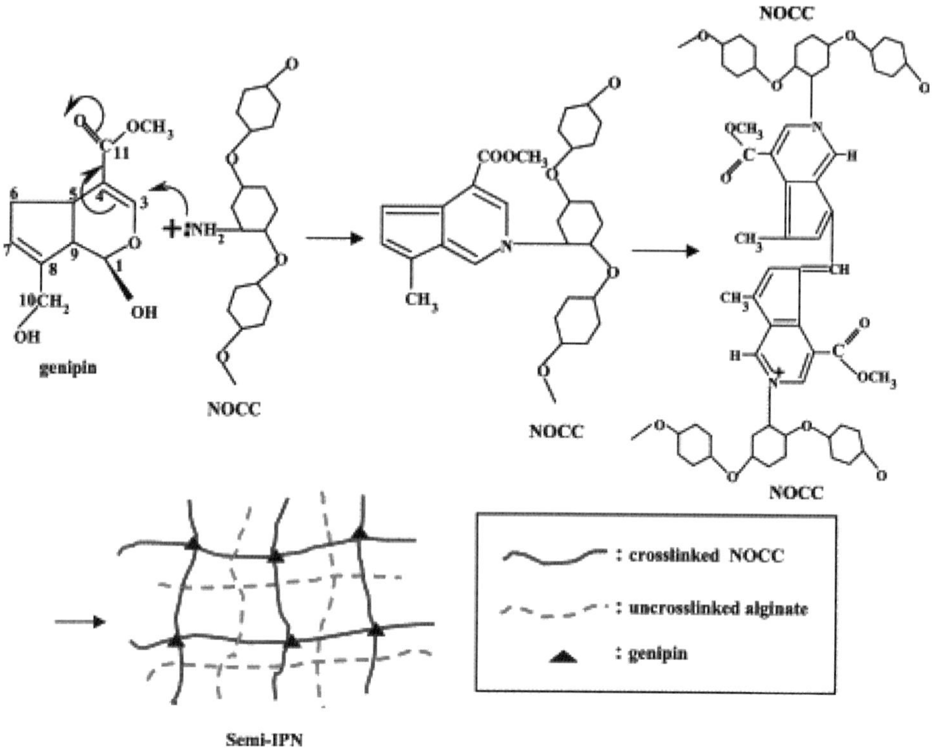

Fig. 12. Presumable reaction mechanism of genipin with NOCC and schematic illustration of the semi-IPN formation of NOCC and alginate crosslinked with genipin. Reprinted from Chen *et al.*[99]

the hydrogel was related to the environmental pH. In an acid environment (pH=1.2), the swelling of the hydrogel was limited due to the formation of hydrogen bonds between NOCC and alginate. When the pH was increased to 7.4, the carboxylic acid groups became ionized, thereby resulting in significant swelling of the hydrogel due to the electrostatic repulsion between the ionized acid groups. Because of the pH sensitivity, the authors suggested that the NOCC/alginate hydrogel could serve as a carrier for site-specific drug delivery in the intestine.

Photopolymerization is also commonly utilized to synthesize chemically associated networks for the biomedical applications, since it is a generally fast process and several biocompatible photoinitiators are available for *in vivo* use.[100–102] Pescosolido and co-workers mixed alginate solution and dextran methacrylate derivatives (DexMA) together, followed by addition of $CaCl_2$ solution to allow the formation of calcium–alginate networks.[103] After UV curing, the fully interpenetrating polymeric networks were developed through the radical polymerization of DexMA in the presence of the initiator Irgacure 2959. It was reported that the mechanical property of the hydrogel could be tuned by adjusting the molecular weight (MW) and derivatization degree of dextran chains. When DexMA with higher MW and more methacrylic groups was added in the polymer solution, the synthesized hydrogel became stiffer and the release of Horseradish peroxidase, the model drug, turned slower. On the contrary, the release of Horseradish peroxidase was faster from the loose networks composed of calcium–alginate and a DexMA with a lower MW and derivatization degree.

4.3.2. *HA-based hydrogels*

Hyaluronic acid (HA) is an unbranched, non-sulphated glycosaminoglycan composed of β-1,4-linked disaccharide units of β-1,3-linked glucuronic acid and N-acetyl-D-glucosamine (Fig. 11(B)).[95] HA is one of the major components of the ECM and it is also present in human connective tissues. Due to its biocompatibility and biodegradability, HA is often considered as a suitable material for preparation of hydrogels for the drug delivery applications.

As discussed above, pNIPAAm is a temperature responsive polymer that is used in the preparation of alginate-based hydrogels. Likewise, the usage of pNIPAAm can be extended to the formation of thermo sensitive HA-based hydrogels. Santos and co-workers combined HA with crosslinked pNIPAAm to create a pH and temperature responsive hydrogel system.[104] They discovered that at pH 7.4, the hydrogels exhibited an improved de-swelling behavior at 37°C and an enhancement in water-uptake and swelling kinetics at 25°C by the addition of HA to plain pNIPAAm networks. Moreover, since the pKa of HA is about 2.9, HA was in the anionic form at pH 7.4 which increased the hydrophilicity of polymeric networks and in turn, speeded up the water-uptake and swelling process. On the other hand, at pH 2.1, HA is in its neutral form and had no additional contribution to the swelling of hydrogel.

Due to the poly-anionic nature of HA, it also can be introduced to other synthetic polymers to produce polyelectrolytes. La Gatta and co-workers reported an interpenetrating polymeric network in which HA was combined with poly(2-hydroxyethylmethacrylate-co-2-methacryloxyethyltrimethylammonium) (poly(HEMA-co-METAC)).[105] Positively charged poly(HEMA-co-METAC) could be partially neutralized by anionic HA in physiological solution and the degree of neutralization may result in the changes of swelling and mechanical properties of hydrogel. In their study, it was discovered that the water-uptake of this polymeric network decreased with increasing ratio of HA in the matrix. This HA/poly(HEMA-co-METAC) matrix proved biocompatible and appropriate for cell adhesion and proliferation.

HA also can be chemically modified with small molecular moieties and the HA derivatives can form hydrogels by self-polymerization in the presence of crosslinking reagent. Messager and co-workers prepared HA-based nanogels using water-in-oil emulsions as nanoreactors.[106] HA was initially reacted with methacrylic anhydride to obtain methacrylated HA, which is photocrosslinkable. Then, the solution of methacrylated HA, containing Irgacure 2959 as a photoinitiator, underwent UV polymerization to yield nanogels. The swelling behavior of these nanogels was controlled by the conditions of photopolymerization, the extent of HA methacrylation and the environmental conditions. The authors also suggested that these nanogels could be used as polymeric carriers of insulin for the pH-dependent delivery.

5. Applications of Directed Therapeutic Delivery Using Environment-Sensitive Hydrogels

5.1. *Encapsulation of living cells with environment-sensitive hydrogels*

Bio-orthogonal polymerization can be harnessed to encapsulate the microscale tissues and cells through layered hydrogel assembly on the surface of the living cells. This technique has been used to encapsulate pancreatic islets. Although the viability of the cells partially affected however, it has been shown the viability of the encapsulated cells can be improved by changing the cationic strength of the coating polymer.[107]

Encapsulation within hydrogel shells provides the possibility to stabilize, culture, and deliver living cells. These applications require gaining spatial and temporal control over the release rate of encapsulated cells. Due to the size of the cells, the only mechanism through which the encapsulated cell can be released is through the degradation of hydrogel sheath. Some chemical bonds of interpenetrated polymer network of environment-sensitive hydrogels can be broken to dissolute the gel and release the content. The cleavage of the polymer network can be promoted by different mechanisms such as photolysis, pH-dependent hydrolysis and enzymatic degradation. For instance, cells encased within microgels made of dendritic polyglycerol (dPG) and linear PEG can be released from the hydrogel matrix by varying the pH without any negative effect of their viability.[108]

6. Conclusions and Outlook

In the view of above review, it can be concluded that a variety of biomolecule-based hydrogels have been developed including DNA-, peptide-, and polysaccharide-based hydrogels. Although the formation rationales of these hydrogels are diverse based on their different natural features, they possess a great potential to be candidate materials for drug delivery applications.

Although the concept of environment-sensitive hydrogels has been subject of vast investigation, the practical applications require significant improvement of the hydrogel properties. The most significant weakness of all these stimulisensitive hydrogels is that their response time is too long. Thus, fast-acting hydrogels are necessary and the easiest way of achieving this goal is to make thinner and smaller hydrogels. This usually makes the hydrogel system too fragile and they do not have the mechanical strength necessary for many applications. Another approach for shortening the response time is to employ relatively more cooperative biomolecular events as sensor and effector processes in the hydrogel. Furthermore, cooperative coupling of biological activity of biomolecules with the hydrogel phase-transition behavior enables the construction of specific stimuli-responsive soft materials capable of sensing a variety of biomarkers.[109]

References

1. K. S. Soppimath, T. M. Aminabhavi, A. M. Dave, S. G. Kumbar, and W. E. Rudzinski, *Drug Dev. Ind. Pharm.*, **28**(8), 957–974 (2002).
2. Y. Qiu and K. Park, *Adv. Drug Del. Rev.*, **64**, 49–60 (2012).
3. E. M. Ahmed, *J. Adv. Res.*, **6**(2), 105–121 (2015).
4. J. T. Padding, L. V. Mohite, D. Auhl, W. J. Briels, and C. Bailly, *Soft Matter*, **7**(10), 5036–5046 (2011).
5. M. M. Feldstein, V. G. Kulichikhin, S. V. Kotomin, T. A. Borodulina, M. B. Novikov, A. Roos, and C. Creton, *J. Appl. Polym. Sci.*, **100**(1), 522–537 (2006).
6. T. Wang, P. J. Colver, S. A. F. Bon, and J. L. Keddie, *Soft Matter*, **5**(20), 3842–3849 (2009).
7. Z. Czech and A. Kowalczyk, *Wide Spectra of Quality Control*, Ed. I. Akyar, Rijeka, Croatia; In Tech, pp. 309–332 (2011).
8. I. Benedek, *Pressure-Sensitive Adhesives and Applications*, 2nd edn., Marcel Dekker New York, NY, U.S.A., pp. 1–5 Inc. (2004).
9. A. M. Wokovich, S. Prodduturi, W. H. Doub, A. S. Hussain, and L. F. Buhse, *Eur. J. Pharm. Biopharm.*, **64**(1), 1–8 (2006).
10. S. Venkatraman and R. Gale, *Biomaterials*, **19**(13), 1119–1136 (1998).
11. N. A. Peppas and J. J. Sahlin, *Biomaterials*, **17**(16), 1553–1561 (1996).
12. C. L. Silva, J. C. Pereira, A. Ramalho, A. A. C. C. Pais, and J. J. S. Sousa, *J. Membrane Sci.*, **320**(1–2), 268–279 (2008).
13. K. L. Ulman and L. Chi-Long, *J. Control. Release*, **10**(3), 273–281 (1989).
14. N. Baït, B. Grassl, C. Derail, and A. Benaboura, *Soft Matter*, **7**(5), 2025–2032 (2011).
15. X. Chen, W. Liu, Y. Zhao, L. Jiang, H. Xu, and X. Yang, *Drug Deliv. Ind. Pharm.*, **35**(6), 704–711 (2009).
16. L. Li, L. Fang, X. Xu, Y. Liu, Y. Sun, and Z. He, *Biopharm. Drug Dispos.*, **31**(2–3), 138–149 (2010).

17. M. Mizanur Rahman and H. D. Kim, *J. Appl. Polym. Sci.*, **104**(6), 3663–3669 (2007).
18. G. Cleary, M. M. Feldstein, and P. Singh, *Technology of Pressure-Sensitive Adhesives and Products*, CRC Press: Boca Raton, FL, U.S.A., Chapter 7, pp. 7-1–7-80 (2008).
19. S. Leung Sau-Hung and R. Robinson Joseph, *Polyelectrolyte Gels*, American Chemical Society: U.S.A., Vol. 480, pp. 269–284 (1992).
20. Y. Onuki, M. Hoshi, H. Okabe, M. Fujikawa, M. Morishita, and K. Takayama, *J. Control. Release*, **108**(2–3), 331–340 (2005).
21. M. Bae, R. Divan, K. J. Suthar, D. C. Mancini, and R. A. Gemeinhart, *J. Vac. Sci. Technol. B*, **28**(6), 24–29 (2010).
22. J. S. Boateng, K. H. Matthews, H. N. E. Stevens, and G. M. Eccleston, *J. Pharm. Sci.*, **97**(8), 2892–2923 (2008).
23. A. S. Sawhney, C. P. Pathak, and J. A. Hubbell, *Macromolecules*, **26**(4), 581–587 (1993).
24. S. F. Dana, D.-V. Nguyen, J. S. Kochhar, X.-Y. Liu, and L. Kang, *Soft Matter*, **9**(27), 6270–6281 (2013).
25. S. D. Barhate, *J. Pharm. Res.*, **2**(4), 663–665 (2009).
26. A. A. Chalykh, A. E. Chalykh, M. B. Novikov, and M. M. Feldstein, *J. Adhesion*, **78**(8), 667–694 (2002).
27. O. Z. Higa, S. O. Rogero, L. D. B. Machado, M. B. Mathor, and A. B. Lugão, *Radiat. Phys. Chem.*, **55**(5–6), 705–707 (1999).
28. K. Mahrag Tur and H.-S. Ch'ng, *Int. J. Pharm.*, **160**(1), 61–74 (1998).
29. J. Woodley, *Clin. Pharmacokinet.* **40**(2), 77–84 (2001).
30. J. M. Gu, J. R. Robinson, and S. H. S. Leung, *Crit. Rev. Ther. Drug.*, **5**(1), 21–67 (1988).
31. C. M. Lehr and J. Haas, *Expert Opin. Biol. Th.*, **2**(3), 287–298 (2002).
32. G. P. Andrews, T. P. Laverty, and D. S. Jones, *Eur. J. Pharm. Biopharm.*, **71**(3), 505–518 (2009).
33. H. Li, Y. Yu, S. Faraji Dana, B. Li, C.-Y. Lee, and L. Kang, *J. Drug Target.*, **21**(7), 611–629 (2013).
34. I. Webster, *Int. J. Adhes. Adhes.*, **17**(1), 69–73 (1997).
35. A. Richter, G. Paschew, S. Klatt, J. Lienig, K.-F. Arndt, and H.-J. Adler, *Sensors*, **8**(1), 561–581 (2008).
36. M. Falamarzian and J. Varshosaz, *Drug Dev. Ind. Pharm.* **24**(7), 667–669 (1998).
37. B. A. Firestone and R. A. Siegel, *J. Appl. Polym. Sci.*, **43**(5), 901–914 (1991).
38. N. A. Peppas and J. Klier, *J. Control. Release*, **16**(1–2), 203–214 (1991).
39. S. Amin, S. Rajabnezhad, and K. Kohli, *Sci. Res. Essay*, **3**(11), 1175–1183 (2009).
40. T. Traitel, Y. Cohen, and J. Kost, *Biomaterials*, **21**(16), 1679–1687 (2000).
41. J. J. Kim and K. Park, *J. Control. Release*, **77**(1–2), 39–47 (2001).
42. A. A. Obaidat and K. Park, *Biomaterials*, **18**(11), 801–806 (1997).
43. A. K. Gupta and A. W. Siddiqui, *J. Drug Deliv. Th.*, **2**(1), 81–88 (2012).
44. T. A. T. Lee, A. Cooper, R. P. Apkarian, and V. P. Conticello, *Adv. Mater.*, **12**(15), 1105–1110 (2000).
45. M. R. Dreher, A. J. Simnick, K. Fischer, R. J. Smith, A. Patel, M. Schmidt, and A. Chilkoti, *J. Am. Chem. Soc.*, **130**(2), 687–694 (2007).

46. F. Chécot, S. Lecommandoux, Y. Gnanou, and H.-A. Klok, *Angew. Chem. Int. Edit.*, **41**(8), 1339–1343 (2002).
47. H. Tanisaka, S. Kizaka-Kondoh, A. Makino, S. Tanaka, M. Hiraoka, and S. Kimura, *Bioconjugate Chem.*, **19**(1), 109–117 (2007).
48. M. S. Ardejani and B. P. Orner, *Science*, **340**(6132), 561–562 (2013).
49. J. M. Fletcher, R. L. Harniman, F. R. H. Barnes, A. L. Boyle, A. Collins, J. Mantell, T. H. Sharp, M. Antognozzi, P. J. Booth, N. Linden, M. J. Miles, R. B. Sessions, P. Verkade, and D. N. Woolfson, *Science*, **340**(6132), 595–599 (2013).
50. X. Chen and P. B. Davis, *Mol Ther*, **13**(S1), S152–S152 (2006).
51. J. Han, Q. Wang, Z. Zhang, T. Gong, and X. Sun, Small, **10**(3), 524–535 (2014).
52. J. Panda, A. Varshney, and V. Chauhan, *J. Nanobiotechnol.*, **11**(1), 18 (2013).
53. H. J. Jung and A. Chauhan, *Biomaterials*, **33**(7), 2289–2300 (2012).
54. J. D. Watson and F. H. Crick, *Nature*, **171**(4356), 737–738 (1953).
55. J. J. Storhoff and C. A. Mirkin, *Chem Rev.*, **99**(7), 1849–1862 (1999).
56. N. C. Seeman, *J. Theor. Biol.*, **99**(2), 237–247 (1982).
57. X. Xiong, C. Wu, C. Zhou, G. Zhu, Z. Chen, and W. Tan, *Macromol. Rapid Commun.*, **34**(16), 1271–1283 (2013).
58. S. Nagahara and T. Matsuda, *Polym. Gels Netw.* **4**(2), 111–127 (1996).
59. H. Kang, A. C. Trondoli, G. Zhu, Y. Chen, Y. -J. Chang, H. Liu, Y. -F. Huang, X. Zhang, and W. Tan, *ACS Nano.*, **5**(6), 5094–5099 (2011).
60. G. M. Emilsson and R. R. Breaker, *Cell. Mol. Life. Sci.*, **59**(4), 596–607 (2002).
61. J. Elbaz, B. Shlyahovsky, and I. Willner, *Chem. Commun. (Camb).*, **13**, 1569–1571 (2008).
62. J. Liu and Y. Lu, *Angew. Chem. Int. Ed. Engl.*, **46**(40), 7587–7590 (2007).
63. J. Liu and Y. Lu, *J. Am. Chem. Soc.*, **129**(32), 9838–9839 (2007).
64. J. Liu, A. K. Brown, X. Meng, D. M. Cropek, J. D. Istok, D. B. Watson, and Y. Lu, *Proc. Natl. Acad. Sci. U S A*, **104**(7), 2056–2061 (2007).
65. H. Lin, Y. Zou, Y. Huang, J. Chen, W. Y. Zhang, Z. Zhuang, G. Jenkins, and C. J. Yang, *Chem. Commun.*, **47**(33), 9312–9314 (2011).
66. K. Gehring, J. L. Leroy, and M. Gueron, *Nature*, **363**(6429), 561–565 (1993).
67. E. Cheng, Y. Xing, P. Chen, Y. Yang, Y. Sun, D. Zhou, L. Xu, Q. Fan, and D. Liu, *Angew. Chem. Int. Ed. Engl.*, **48**(41), 7660–7663 (2009).
68. T. Miyata, N. Asami, and T. Uragami, *Nature*, **399**(6738), 766–769 (1999).
69. M. Kurisawa, Y. Matsuo, and N. Yui, *Macromol. Chem. Phys.*, **199**(4), 705–709 (1998).
70. H. Yang, H. Liu, H. Kang, and W. Tan, *J. Am. Chem. Soc.*, **130**(20), 6320–6321 (2008).
71. E. H. Christen, M. Karlsson, M. M. Kämpf, R. Schoenmakers, R. J. Gübeli, H. M. Wischhusen, C. Friedrich, M. Fussenegger, and W. Weber, *Adv. Funct. Mater.*, **21**(15), 2861–2867 (2011).
72. J. Fong, A. Keating, and M. Singh, *Genome Biol.*, **5**(2), R11 (2004).
73. H. Dong and J. D. Hartgerink, *Biomacromolecules*, **8**(2), 617–623 (2007).
74. H. Dong, S. E. Paramonov, and J. D. Hartgerink, *J. Am. Chem. Soc.*, **130**(41), 13691–13695 (2008).

75. E. F. Banwell, E. S. Abelardo, D. J. Adams, M. A. Birchall, A. Corrigan, A. M. Donald, M. Kirkland, L. C. Serpell, M. F. Butler, and D. N. Woolfson, *Nat. Mater.*, **8**(7), 596–600 (2009).
76. O. S. Makin, E. Atkins, P. Sikorski, J. Johansson, and L. C. Serpell, *Proc. Natl. Acad. Sci. USA*, **102**(2), 315–320 (2005).
77. U. Slotta, S. Hess, K. Spiess, T. Stromer, L. Serpell, and T. Scheibel, *Macromol. Biosci.*, **7**(2), 183–188 (2007).
78. A. Altunbas and D. Pochan, *Peptide-Based Materials*, Ed. T. Deming, Berlin, Heidelberg: Springer, Vol. 310, pp. 135–167 (2012).
79. A. Matsumoto, J. Chen, A. L. Collette, U. J. Kim, G. H. Altman, P. Cebe, and D. L. Kaplan, *J. Phys. Chem. B*, **110**(43), 21630–21638 (2006).
80. A. Aggeli, M. Bell, N. Boden, J. N. Keen, T. C. B. McLeish, I. Nyrkova, S. E. Radford, and A. Semenov, *J. Mater. Chem.*, **7**(7), 1135–1145 (1997).
81. J. J. Yang, M. Pikeathly, and S. E. Radford, *Biochemistry*, **33**(23), 7345–7353 (1994).
82. S. Zhang, T. C. Holmes, C. M. DiPersio, R. O. Hynes, X. Su, and A. Rich, *Biomaterials*, **16**(18), 1385–1393 (1995).
83. S. Zhang, *Biotechnol. Adv.* **20**(5–6), 321–339 (2002).
84. B. Ozbas, J. Kretsinger, K. Rajagopal, J. P. Schneider, and D. J. Pochan, *Macromolecules*, **37**(19), 7331–7337 (2004).
85. J. P. Schneider, D. J. Pochan, B. Ozbas, K. Rajagopal, L. Pakstis, and J. Kretsinger, *J. Am. Chem. Soc.*, **124**(50), 15030–15037 (2002).
86. D. J. Pochan, J. P. Schneider, J. Kretsinger, B. Ozbas, K. Rajagopal, and L. Haines, *J. Am. Chem. Soc.*, **125**(39), 11802–11803 (2003).
87. K. Rajagopal, B. Ozbas, D. J. Pochan, and J. P. Schneider, *Eur. Biophys. J.*, **35**(2), 162–169 (2006).
88. T. Yucel, C. M. Micklitsch, J. P. Schneider, and D. J. Pochan, *Macromolecules*, **41**(15), 5763–5772 (2008).
89. J. M. Anderson, A. Andukuri, D. J. Lim, and H. W. Jun, *ACS Nano.*, **3**(11), 3447–3454 (2009).
90. J. D. Hartgerink, E. Beniash, and S. I. Stupp, *Science*, **294**(5547), 1684–1688 (2001).
91. J. C. Stendahl, M. S. Rao, M. O. Guler, and S. I. Stupp, *Adv. Funct. Mater.*, **16**(4), 499–508 (2006).
92. E. Ruoslahti and M. D. Pierschbacher, *Science*, **238**(4826), 491–497 (1987).
93. D. L. Hern and J. A. Hubbell, *J. Biomed. Mater. Res.*, **39**(2), 266–276 (1998).
94. C. Chun, H. J. Lim, K. Y. Hong, K. H. Park, and S. C. Song, *Biomaterials*, **30**(31), 6295–6308 (2009).
95. T. Coviello, P. Matricardi, C. Marianecci, and F. Alhaique, *J. Control. Release*, **119**(1), 5–24 (2007).
96. M. R. Guilherme, M. R. de Moura, E. Radovanovic, G. Geuskens, A. F. Rubira, and E. C. Muniz, *Polymer*, **46**(8), 2668–2674 (2005).
97. M. R. de Moura, M. R. Guilherme, G. M. Campese, E. Radovanovic, A. F. Rubira, and E. C. Muniz, *Eur. Polym. J.*, **41**(12), 2845–2852 (2005).
98. D. Schmaljohann, *Adv. Drug Deliv. Rev.*, **58**(15), 1655–1670 (2006).

99. S.-C. Chen, Y.-C. Wu, F.-L. Mi, Y.-H. Lin, L.-C. Yu, and H.-W. Sung, *J. Control Release*, **96**(2), 285–300 (2004).
100. N. E. Fedorovich, M. H. Oudshoorn, D. van Geemen, W. E. Hennink, J. Alblas, and W. J. Dhert, *Biomaterials*, **30**(3), 344–353 (2009).
101. K. T. Nguyen and J. L. West, *Biomaterials*, **23**(22), 4307–4314 (2002).
102. S. R. Van Tomme, G. Storm, and W. E. Hennink, *Int. J. Pharm.*, **355**(1–2), 1–18 (2008).
103. L. Pescosolido, S. Miatto, C. Di Meo, C. Cencetti, T. Coviello, F. Alhaique, and P. Matricardi, *Eur. Biophys. J.*, **39**(6), 903–909 (2010).
104. J. R. Santos, N. M. Alves and J. F. Mano, *J. Bioact. Compat. Pol.*, **25**(2), 169–184 (2010).
105. A. L. Gatta, C. Schiraldi, A. D'Agostino, A. Papa, and M. D. Rosa, *J. Funct. Biomater.*, **3**(2), 225–238 (2012).
106. L. Messager, N. Portecop, E. Hachet, V. Lapeyre, I. Pignot-Paintrand, B. Catargi, R. Auzely-Velty, and V. Ravaine, *J. Mater. Chem. B*, **1**(27), 3369–3379 (2013).
107. K. M. Gattás-Asfura and C. L. Stabler, *ACS Appl. Mater. Interfaces*, **5**(20), 9964–9974 (2013).
108. D. Steinhilber, T. Rossow, S. Wedepohl, F. Paulus, S. Seiffert, and R. Haag, *Angew. Chem. Int. Ed.*, **52**(51), 13538–13543 (2013).
109. M. Ikeda, T. Tanida, T. Yoshii, K. Kurotani, S. Onogi, K. Urayama, and I. Hamachi, *Nat. Chem.* **6**, 511–518 (2014) **advance online publication** (2014).

Chapter 6

Application of PEG in Drug Delivery System

Bramasta Nugraha
Department of Biosystem Science and Engineering,
Swiss Federal Institute of Technology (ETH Zürich), Switzerland
Pharmaceutical Research and Early Development,
Department Discovery Technology, Roche Innovation Center Basel,
F. Hoffmann-La Roche Ltd, Switzerland
bramasta.nugraha@bsse.ethz.ch & bramasta.nugraha@roche.com

This chapter describes the application of polyethylene glycol (PEG) in drug delivery system. It begins with an overview of the fundamental and benefits of shielding drugs with hydrophilic polymer like PEG, through PEGylation, in altering the uptake of a drug. Characteristics of PEG to be used in pharmaceutical application will be elaborated and examples of drug-modified PEG conjugates will also be discussed. Lastly, possible limitations of PEG that could hamper its use in pharmaceutical applications will be described.

1. Applications of PEG Hydrogels in Drug Delivery System: An Overview

In disease treatment, preferably a minimum dose of pharmaceutical agents is needed to impose a maximum effect. However during drug administration, delivered drug will circulate in the body through blood before reaching the specific target site thus the drug concentration theoretically needs to be in excess to compensate this long circulation and yet reach the specific site at the required therapeutic concentration, even though this process could lead to non-specific tissue exposure to this drug, which may cause side effects.[1] This phenomenon clearly enlightens the needs to design a drug carrier that is capable to deliver drug into specific site without imposing too much exposure on non-specific site.[1–3]

Shielding a drug with polymer conjugate is required to avoid fast recognition of the delivered drug in circulating blood by the body's immune system followed by rapid clearance from the body.[4] Shielding a drug from its free form prevents opsonin, a blood protein that causes immune response, from recognizing this drug thus limiting the phagocytosis by the reticulo endothelial system (RES) cells and increasing the duration of systemic drug circulation from hours to even days.[5] Among materials that have been used to modify pharmaceutical carriers and drugs, polyethylene glycol (PEG) is a known material in use due to its excellent hydrophilicity and extended blood circulating property.[6–9] The concept of the drug shielding is achieved through PEGylation as the modifier, a term implies covalent and possibly non-covalent conjugation of PEG onto an object, as illustrated in Fig. 1. The drug that has been modified with PEG is denoted as prodrug. Figure 1 depicts a drug that has been covalently PEGylated through a linker that acts as cell targeting ligand.

Drug delivery vector can be also coated with PEG to enhance its water solubility. PEG also reduces the tendency of particles to aggregate by providing steric stabilization thus producing more stable particles during storage and application.[4] PEG offers the advantage from its non-toxicity, non-immunogenicity and FDA approval for use in humans.[11]

The conjugated PEG functions to alter the mechanism of how a drug is delivered into the target site. Solubility barrier i.e., endothelial cell layer will normally block a delivered drug if it is in its free form due to hydrophobicity of most of the drugs, while a drug that is delivered in the form of prodrug will have an altered delivery mechanism due to an enhanced hydrophilicity (illustrated in Fig. 2). PEGylated drug could have a better solubility to pass through this barrier. PEG that shields the drug will also suppress the interaction with the body inducing activation of the complement system causing a reduced blood clearance of the prodrug and drug carriers.[4] The ability of PEG to influence the pharmacokinetics behaviors of certain drugs is now widely utilized in pharmaceutical industry.

One important influence of drug shielding with PEG is a prolonged blood circulation time which then increases the probability of a drug to reach its site of

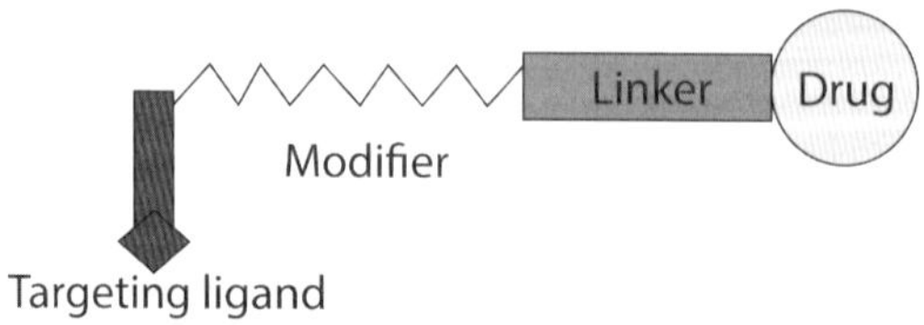

Fig. 1. Prodrug sketch: PEG as modifier in drug PEGylation.[10]

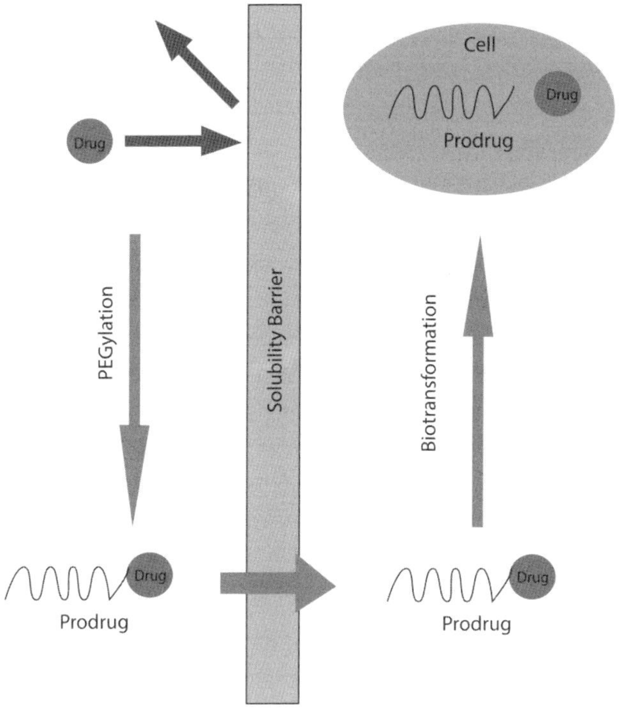

Fig. 2. Comparison of the uptake of free form drug and prodrug.[12]

action before being recognized as a foreign object and cleared from the body. When the presence of drug in blood circulation is prolonged, the possibility events in which the drug is successfully entrapped will be also higher. No other synthetic polymer could reach comparable benefit yet. Various kinds of drugs that have used PEG for shielding are listed in Table 1.

However, not all kinds of PEG can provide optimum drug shielding function, they have to be carefully selected based on several criteria as follows:

(a) Molecular weight

Molecular weight of PEG used in various pharmaceutical and medical applications varies from 400 Da to ~50,000 Da. It has been shown that PEG with molecular weight up to 20,000 Da is excreted through renal system, while larger molecular weight PEG will be excreted through fecal excretion.[13] PEG with molecular weight larger than 20,000 Da is known as poly (ethylene oxide) (PEO).[14] PEG with molecular weight in the range 20,000 Da–50,000 Da are used mostly in conjugation of small molecular weight drugs such as oligonucleotides and small interfering ribonucleic acid (siRNA). This conjugation helps in avoiding quick renal clearance by increasing the size of conjugates

Table 1. List of commercialized prodrugs.[4]

PEG drug description	Company	Indications
Adagen (mPEG adenosine deaminase)	Enzon Inc. (Europe & USA)	Severe combined immunodeficiency
Oncospar (mPEG L-asparaginase)	Enzon Inc. (USA) & Rhône-Poulenc Rorer (Europe)	Acute lymphoblastic leukimia
Doxyl/Caelyx (stabilized liposome of doxorubicin)	Alza Corp. (USA) & Schering Plough Corp. (Europe)	Karposi's sarcoma, ovarian cancer, breast cancer, multiple myeloma
PEG-Intron (mPEG interferon- α-2a)	Schering Plough Corp. (USA & Europe)	Chronic hepatitis C
Pegasys (mPEG interferon- α-2b)	Hoffmann-La Roche (USA & Europe)	Chronic hepatitis C
Neulasta (mPEG granulocyte colony stimulating factor)	Amgen Inc. (USA & Europe)	Febrile neutropenia
Somavert (mPEG human growth receptor antagonist)	Pfizer (USA & Europe)	Acromegaly
Macugen (mPEG anti Vascular Endothelial Growth Factor (VEGF))	Pfizer (USA & Europe) & OSI Pharm Inc. (USA)	Age-related macular degeneration
Cimzia (mPEG anti Tumor Necrosis Factor α (TNFα)	UCB S.A. (USA & Europe)	Crohn's disease, rheumatoid arthritis

above renal clearance threshold. Smaller molecular weight PEGs in the range of 1,000 Da–5,000 Da are used in conjugation of larger molecular weight drugs i.e., antibodies. In this case, opsonization, a recognition process of antigen by phagocytes and subsequent elimination by RES is avoided.[4]

(b) Biodegradability
Biodegradable polymer is preferred in medical application because of the avoidance in achieving complete excretion and/or secondary removal, despite some possible issues in degradation byproducts toxicity and polymer shelf life.

(c) Low polydispersity index
Pharmaceutical applications requires low polydispersity index (PDI), a ratio of weight average molecular weight (MW) to number average molecular weight (Mn) of a polymer. PDI lower than 1.1 displays homogeneity, tolerated body immunogenic reaction and good body residence time. PEG with smaller molecular weight fulfill this requisite with having PDI ~1.01.[15]

(d) High solubility in organic solvents
Although PEG has good water solubility thus providing low intrinsic toxicity and is ideal for biological applications, it is also important to be well soluble in organic solvents in facilitating the ease of PEG end-group chemical modification.[15] Hydrophobic compounds attached into PEG received benefits in having enhanced solubility in aqueous media.

2. Drug Delivery Applications: PEG Prodrug Conjugate in Drug Delivery System

(a) PEG–protein conjugates

The benefits of therapeutic proteins such as antibodies, cytokines, growth factors and enzymes have been recognized for years, however the potential use of these proteins as drug is hampered by the rapid proteolytic degradation property in the blood, poor stability, immunogenic reaction and recognition by the cells in RES. Modifying these proteins with hydrophilic polymer could impose steric hindrance that limits the interaction with proteolytic degradation enzyme and receptor of the RES cells.[8] The hydrophilic part of the polymer prevents the absorption of blood proteins including opsonin onto the surface of the therapeutic proteins. PEG as hydrophilic polymer has been used to modify protein as drug. Modification of the proteins surface by PEGylation enhanced the stability of these proteins towards antigens processing cells and proteolytic enzymes.[16] The coupling of protein to PEG was first described in 1977 by Abuchowski *et al.*, where he covalently conjugated bovine serum albumin (BSA) with PEG to alter its immunological properties.[17] The increase in the molecular weight of proteins after PEGylation also results in longer protein half-life during blood circulation thus reduces the needs to repeatedly administer these proteins.

(b) PEG–drug conjugates

Small molecular weight drug could diffuse into tissue through endothelial cell layers in blood capillaries.[12,18] PEG conjugation into these molecules eventually yields larger molecular weight molecules, named prodrugs. This significantly changes the mechanism of how these prodrugs being internalized by the cells, from diffusion into slower endocytosis process. Shielded drugs in these prodrugs also display lower drug activity compared to the free form drugs.[12] This has been shown by reduced cytotoxic effect of the anticancer prodrugs compared to its free form. To counterbalance this drug activity reduction, one has to target the prodrugs into the specific organs with specific vector and with an optimized concentration. Targeting these prodrugs into tissue can be achieved by two methods: passive and active targeting (depicted in Fig. 3).

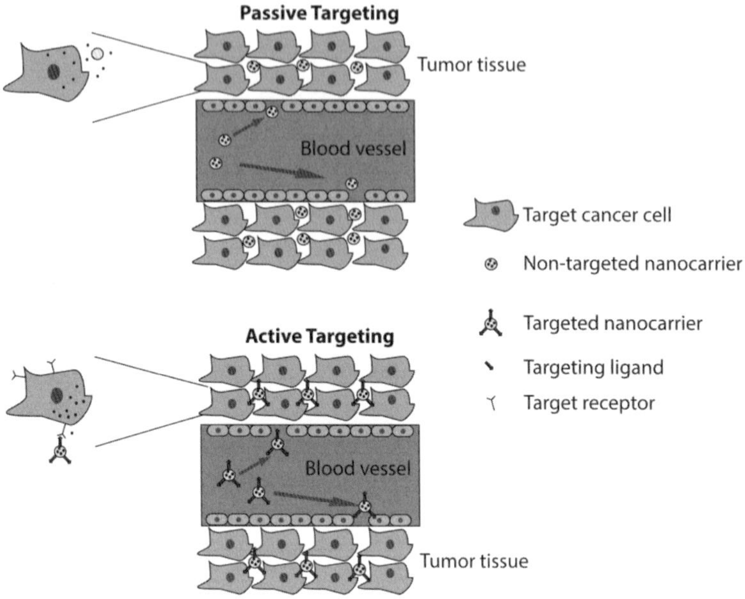

Fig. 3. Comparison between passive and active drug targeting.[19]

2.1. *Passive targeting*

The leaky vasculature and poor lymphatic drainage of the tumor make the endothelial layer of the tumorous tissue easily entraps the substances flowing into. Matsumura and Maeda firstly observed, in 1986, this phenomenon named enhanced permeability and retention (EPR) effect.[20] The presence of large gaps between neighboring endothelial cells in tumor vasculature allows for passive targeting into the tumor site. The size of the pore in most peripheral human tumor range from 200–600 nm, this creates a sieve control for the drugs that could access these gaps to be up to 40,000 Da MW and liposome in the range of 50–400 nm in diameter.[21] Defective lymphatic drainage also helps in retaining these drugs longer than in normal healthy tissue. Although EPR effect offers an advantage in drug targeting, however, drug delivery vectors need to be present in the blood circulation long enough before reaching its target. Opsonin, a plasma protein, could still bind these circulating vectors and remove them from circulation quickly through the RES.

2.2. *Active targeting*

Active targeting involves further chemical conjugation of prodrugs with targeting ligand that binds to specific cell receptor thus helps in selection of the main target cell type within tumor and cell internalization into greater extend.[14] This concept comes from specific biological pairs e.g., ligand receptor, antigen–antibody, or enzyme substrate.[22]

Mostly used targeting ligands are peptides such as Arg–Gly–Asp (RGD),[23,24] small molecule ligands such as folate,[25] proteins such as transferrin,[26] antibodies and or their fragments,[27,28] which are expressed in the targeted cells. Regardless which ligands are used, the principle is the same, to improve tumor cell recognition, tumor uptake, and reduce recognition in non-specific site. Chemical conjugation of PEG products with targeting ligand is performed through PEG functionalization with reactive carboxylic acid, amine or sulfhydryl groups, creating amide bond or disulfide bridge formation.

Folate is an example of a targeting ligand used vastly in cancer therapy, because of its recognition to vitamin folic acid, which is commonly over-expressed in various tumors.[29] Tumor targeting with transferrin-modified PEG nanoparticles has also been demonstrated in transferrin-receptor positive tumor.[30] Small peptide RGD is used to exclusively bind to particular integrins, which are highly over-expressed on the surface of endothelial cells in tumor vasculature.[24] In addition, aptamers, DNA or RNA-based oligonucleotide strands have the ability to recognize antigens with high affinity and specificity through their tertiary folding pattern.[31]

Prodrugs will eventually result in uptake of the drugs in two main approaches: (a) internalization of the whole prodrugs or (b) internalization of only the drug part in prodrug into the targeted cells by endocytosis and phagocytosis.[32] For the (a) mechanism, the process is illustrated in Fig. 3(B). The interaction of the targeted prodrug with relevant cellular receptor triggers receptor-mediated endocytosis by formation of an endocytic vesicle and endosome-membrane limited transport vesicle.[16] During this process, the drug activity is preserved due to drug degradation prevention from cellular detoxification enzymes by the membrane-coated endosome. The process follows with fusion of endosomes with lysosomes forming secondary lysosomes. The drug could be released from prodrug by lysosomal enzymes and exit lysosome by diffusion. In (b) the prodrug is cleaved externally of the targeted cells possibly due to the acidic microenvironment in tumor. Drug activity may be compromised in this process.

These observations showed that active targeting of the prodrugs to specific tumor/cellular compartment has been more precise with cell-specific recognition ligand. PEG shielding in prodrugs in addition to active targeting gives advantages in tumor accumulation through passive targeting with versatile incorporation of specific targeting agent, thereby improving the anti-cancer drug therapy efficacy.

3. Drug Delivery Applications: Non-Covalent Conjugation of PEG

Drug modification by PEG can also be achieved by the non-covalent conjugation of PEG chain with the drug or drug vector. This can be performed for example with hydrophobic–hydrophobic interaction.[33]

A study demonstrated the ease and effective of non-covalent conjugation of PEG, through blending, in developing new material is shown by Legashvili *et al.*[33] In this study non-covalent conjugation of PEG is effective in encapsulating drug through multiple interaction of methylene groups of PEG and co-poly(ester amide (co-PEA) to obtain water-stable nano adducts. The presence of both hydrophobic (co-PEA) and hydrophilic PEG offers unique application similar to those found in macromolecules or liposomes containing hydrophobic and hydrophilic sites.

Non-covalently bound PEG is also currently explored in Borchard group in the University of Geneva, in which hydrophobic interaction of PEG is utilized. Hydrophobic head of PEG chain interacts with hydrophobic parts of the proteins. In partially unfolded protein, the exposed hydrophobic parts of the PEG and proteins will lead to aggregation.[34] Furthermore, upon attachment of the hydrophobic part of PEG, PEG tail will shield the partially folded protein from each other thus interrupting the aggregation process. Examples of the studied PEG in this groups are dansyl-PEG and Trp-PEG.[35]

Eventually, these examples of non-covalent conjugation of PEG demonstrated the advantages of non-covalent conjugation over covalent conjugation of PEG in terms of shorter time in experimentation, less expenses in conjugation without involving organic synthesis, less effort involved and possibly better preserving of proteins functionality upon conjugation that is sensitive to organic solvents.

4. Limitations of PEG Hydrogel in Drug Delivery System

Application of hydrophilic polymer PEG in drug delivery system is not without limitations. There are different features that could possibly limit the applications of PEG thus need to be taken care of.

4.1. *Administration induced immunological response*

PEG property in preventing opsonization, a binding between antigen and its complement molecule, and could lead to specific and non-specific immune recognitions in the body.[4] PEG adverse reaction in the body occurs through complement activation (C activation) in which further leads to hypersensitive reaction.[36,37] Complement activation system is a part of the immune system, a biochemical cascade that is started by C3 hydrolysis (a blood protein). This hydrolysis further triggers biochemical cascade that results in generation of different fragments that label the identified foreign body. Hydroxyl groups of PEG that are surface-exposed provide molecular sites where C can covalently bind to the surface and initiate complement activation.[37]

However, as of now, the majority of PEGylated products are methoxy-PEG (mPEG)-based that could minimize this effect.[4]

4.2. *Non-biodegradability*

Non-degradable property of PEG also imposes a drawback in PEG usage. When the molecular weight of PEG lies below 400 Da, it is found to be toxic in humans due to sequential oxidation result into diacid and hydroxyl acid metabolites by alcohol and aldehyde dehydrogenase.[38,39] On the other hand, PEG molecular weight should not be too high thus exceeding the renal clearance threshold. PEG with molecular weight 20,000 Da will be secreted easily through urine, while larger molecular weight in the range of 40,000–60,000 Da will undergo liver accumulation.[40,41] Multiarm or branched PEG with smaller total molecular weight could be a strategy in designing easily excreted PEG or PEGylated systems. However, deep investigations are still needed to observe the effects that follows, once PEG is disintegrated from the prodrug upon drug release.[42] Long-term study on how PEG affects the site where it accumulates is needed.

4.3. *Side product toxicity*

PEG synthesis process produces a side product known as 1,4-dioxane, a cyclic dimer form of ethylene oxide. International Agency for Research on Cancer (IARC) classified dioxane as a group 2b member, a compound that is possibly carcinogenic in humans. European Pharmacopoeia (Ph. Eur.) has limited the content of dioxane up to 10 ppm for pharmaceutical applications. Other side products from PEG synthesis are ethylene oxide and formaldehyde. These compounds were classified by IARC in group 1 (human carcinogens). As a consequence, Ph. Eur. has regulated the maximum amounts of ethylene oxide and formaldehyde to be up to 1 and 30 ppm, respectively, in pharmaceutical products.[4] These compounds clearly show the needs of using pharmaceutical grade PEG for any biomedical applications.

5. Conclusion

PEGylation could enhance the intrinsic property of a drug or protein to be delivered in terms of hydrophilicity, stability, residence time, and resistance to body clearance. This specifically solves the challenge of delivering hydrophobic drugs, as well as, unstable protein. However, careful consideration is needed to select molecular weight of PEG, which can be used to modify the drug, in order to properly address its non-degradability issue.

References

1. V. P. Torchilin, *Eur. J. Pharm. Sci.*, **11**(2), S81–S91 (2000).
2. C. A. Presant, W. Wolf, V. Waluch, C. Wiseman, P. Kennedy, D. Blayney, and R. R. Brechner, *Lancet*, **343**, 1184–1187 (1994).
3. R. K. Jain, *Sci. Am.*, **271**, 58–65 (1994).
4. K. Knop, R. Hoogenboom, D. Fischer, and U. S. Schubert, *Angew. Chem. Int. Ed. Engl.*, **49**, 6288–6308 (2010).
5. D. E. Owens and N.A. Peppas, *Int. J. Pharm.*, **307**, 93–102 (2006).
6. R. Duncan and F. Spreafico, *Clin. Pharmacokinet.*, **27**, 290–306 (1994).
7. M. J. Roberts and J. M. Harris, *J. Pharm. Sci.*, **87**, 1440–1445 (1998).
8. J. M. Harris, N. E. Martin, and M. Modi, *Clin. Pharmacokinet.*, **40**, 539–551 (2001).
9. R. Duncan, *Nat. Rev. Drug. Discov.*, **2**, 347–360 (2003).
10. J. W. Qihang Sun, M. Radosz, and Y. Shen, Polymer-based prodrugs for cancer chemotherapy. In *Functional Polymers for Nanomedicine*, Ed. Y. Shen, Cambridge UK: The Royal Society of Chemistry, pp. 245–260 (2013).
11. http://www.accessdata.fda.gov/scripts/cder/drugsatfda/index.cfm?fuseaction=Search.SearchAction&SearchType=BasicSearch&SearchTerm=polyethylene+glycol.
12. S. S. Banerjee, N. Aher, R. Patil, and J. Khandare, *J. Drug. Deliv.*, **2012**, 103973 (2012).
13. T. Yamaoka, Y. Tabata, and Y. Ikada, *J. Pharm. Sci.*, **83**, 601–606 (1994).
14. L. E. van Vlerken, T. K. Vyas, and M. M. Amiji, *Pharm. Res.*, **24**, 1405–1414 (2007).
15. V. G. K. Betageri, *Polymers* (2011), **3**(4), 1972–2009; doi:10.3390/polym3041972 (This article belongs to the Special Issue Water-Soluble polymers).
16. J. J. Khandare, P. Chandna, Y. Wang, V.P. Pozharov, and T. Minko, *J. Pharmacol. Exp. Ther.*, **317**, 929–937 (2006).
17. A. Abuchowski, T. van Es, N.C. Palczuk, and F. F. Davis, *J. Biol. Chem.*, **252**, 3578–3581 (1977).
18. H. Ringsdorf, *J. Polym. Sci. C*, **51**, 135–153 (1975).
19. Z. Y. Lei, J. Li, and C. Yu, *J. Mater. Chem. B*, **2**, 452–470 (2014).
20. Y. Matsumura and H. Maeda, *Cancer Res.*, **46,** 6387–6392 (1986).
21. F. Yuan, M. Dellian, D. Fukumura, M. Leunig, D. A. Berk, V. P. Torchilin, and R. K. Jain, *Cancer Res.*, **55**, 3752–3756 (1995).
22. T. Minko, S. S. Dharap, R. I. Pakunlu, and Y. Wang, *Curr. Drug. Targets*, **5**, 389–406 (2004).
23. D. Lochmann, E. Jauk, and A. Zimmer, *Eur. J. Pharm. Biopharm.*, **58**, 237–251 (2004).
24. R. M. Schiffelers, A. Ansari, J. Xu, Q. Zhou, Q. Tang, G. Storm, G. Molema, P. Y. Lu, P. V. Scaria, and M. C. Woodle, *Nucleic Acids Res.*, **32**, e149 (2004).
25. A. Gabizon, H. Shmeeda, A. T. Horowitz, and S. Zalipsky, *Adv. Drug. Deliv. Rev.*, **56**, 1177–1192 (2004).
26. A. Widera, F. Norouziyan, and W. C. Shen, *Adv. Drug. Deliv. Rev.*, **55**, 1439–1466 (2003).
27. F. M. Veronese and G. Pasut, *Drug Discov. Today*, **10**, 1451–1458 (2005).

28. M. Morpurgo, F. M. Veronese, D. Kachensky, and J. M. Harris, *Bioconjug. Chem.*, **7**, 363–368 (1996).
29. M. K. Russell-Jones, G. McEwan, J. Rice, and D. Nowotnik, *J. Inorg. Biochem.*, **98**, 1625–1633 (2004).
30. S. H. Pun, F. Tack, N. C. Bellocq, J. Cheng, B. H. Grubbs, G. S. Jensen, M. E. Davis, M. Brewster, M. Janicot, B. Janssens, W. Floren, and A. Bakker, *Cancer Biol. Ther.*, **3**, 641–650 (2004).
31. K. M. Song, S. Lee, and C. Ban, *Sensors (Basel)*, **12**, 612–631 (2012).
32. S. S. Dharap, B. Qiu, G. C. Williams, P. Sinko, S. Stein, and T. Minko, *J. Control. Release.*, **91**, 61–73 (2003).
33. I. Legashvili, N. Nepharidze, R. Katsarava, B. Sannigrahi, and I. M. Khan, *J. Biomater. Sci. Polym. Ed.*, **18**, 673–685 (2007).
34. M. E. Cromwell, E. Hilario, and F. Jacobson, *AAPS J.*, **8**, E572–E579 (2006).
35. C. Mueller, M. A. Capelle, T. Arvinte, E. Seyrek, G. Borchard, *J. Pharm. Sci.*, **100,** 1648–1662 (2011).
36. A. Chanan-Khan, J. Szebeni, S. Savay, L. Liebes, N. M. Rafique, C. R. Alving, and F. M. Muggia, *Ann. Oncol.*, **14**, 1430–1437 (2003).
37. J. Szebeni, *Toxicology*, **216**, 106–121 (2005).
38. D. A. Herold, K. Keil, and D. E. Bruns, *Biochem. Pharmacol.*, **38**, 73–76, (1989).
39. K. D. Hinds, K. M. Campbell, K. M. Holland, D. H. Lewis, C. A. Piche, and P. G. Schmidt, *J. Control Release*, **104**, 447–460 (2005).
40. L. Sprincl, J. Exner, O. Sterba, and J. Kopecek, *J. Biomed. Mater. Res.*, **10**, 953–963 (1976).
41. L. W. Seymour, R. Duncan, J. Strohalm, and J. Kopecek, *J. Biomed. Mater. Res.*, **21**, 1341–1358 (1987).
42. S. M. Moghimi and J. Szebeni, *Prog. Lipid Res.*, **42**, 463–478 (2003).

Chapter 7

Hydrogels as Actuators for Biological Applications

Hongrui Jiang* and Difeng Zhu
Department of Electrical Engineering, University of Wisconsin Madison
1415 Engineering Drive, Madison, WI 53706, USA
**hongrui@engr.wisc.edu*

In this chapter, we provide a summary on hydrogel based actuators for biological applications. The behaviors and characteristics of various types of hydrogels are discussed. Current work regarding hydrogel actuators employed in microfluidics, bionics, medical systems, biohybrid systems, etc. are reviewed.

1. Introduction

Stimuli-responsive hydrogels, with the capability of significant volumetric change in response to changes in certain external stimuli, such as temperature, pH, photons, and antigens, have drawn considerable attention in both academia and industry. Since 1950 when this "muscle-like functional" polymer was discovered,[21] hydrogel has been serving as functional materials in numerous applications in the areas of sensors and actuators.

Nowadays, along with the rapid development in many areas including drug delivery, microfludics, biomedical systems, and bionics, the potential of hydrogels has been further explored. Thanks to its prominent functionality and compatibility with biological systems and micro-fabrication technology, hydrogels have played more and more important roles in biology related fields.[25]

In this chapter, we will discuss on concepts relevant to hydrogels and their behavior as actuators, and focus on their applications such as in microfludics, bionics, and advanced biomedical systems.

2. Structure and Behavior

A hydrogel is a three-dimensional (3D) network of polymer chains containing water. Possessing mechanical properties similar to natural rubbers, it has high deformability and nearly complete recoverability.[52]

Generally, a hydrogel employed in an actuator design consists of molecular chains, crosslinkers and the solvent: water.[42] The long molecular chains, which are composed of small molecular units, monomers and co-monomers, are bonded together by crosslinks, forming the polymer matrix. The amount of water in the matrix could weigh over 90% of the hydrogel and determines the whole volume, thus actuation.[42] Flory and Rehner[15] investigated on the mechanism of gel network composition and provided the Flory–Rehner equation to interpret the hydrogel behavior, as shown below:

$$-\left[\ln(1-v_2)+v_2+\chi_1 v_2^2\right] = \frac{v_1}{vM_C}\left(1-\frac{2M_C}{M}\right)\left(v_2^{1/3}-\frac{v_2}{2}\right),$$

where v_2 is the volume fraction of polymer in the swollen mass, $\overline{v}$ is the specific volume of the polymer, v_1 is the molar volume of the solvent, n is the number of network chain segments bounded on both ends via crosslinking, χ_1 is the Flory solvent–polymer interaction term, M is the primary molecular mass, M_c is the average molecular mass between crosslinks or the network parameter.[3]

A typical hydrogel has two phases in water solution: a separated phase and a mixed phase, as shown in Fig. 1.[32] The separated phase is dominated by polymer–polymer interactions, at which the gel aggregates and reaches its maximal value of hydrophobicity. Thus, the gel shrinks. The other phase, the mixed phase, is characterized by solvent–polymer interactions, which is the best mixing of polymer and water. Therefore, within the mixed phase the hydrogel gains its maximum of hydrophilicity and swells. The phase transition can be stimulated by various environmental conditions, including solvent composition, pH, temperature, light and electric field.[32]

Photo-initiator is always included in the synthesis of hydrogels, since it could trigger the hydrogel curing process under ultraviolet (UV) light exposure. In this way, the hydrogel actuator could be photopatterned, which enormously widens its applications.[5]

Typically, hydrogel could be classified into temperature responsive hydrogels and pH responsive ones, owing to different tuning mechanisms and characteristics.[42] Other types of hydrogels have also been reported in the past, such as antigen-sensitive hydrogels[30] and glucose-sensitive hydrogels.[7]

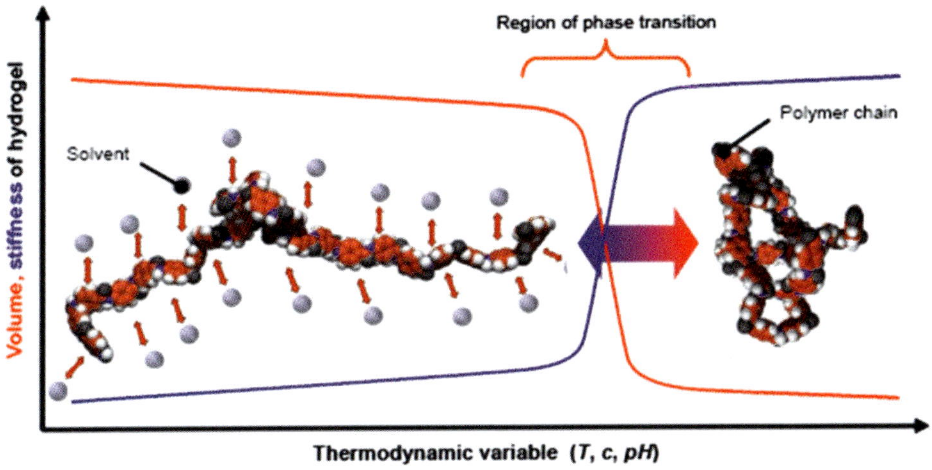

Fig. 1. Phase transition behavior of stimuli-reponsive hydrogels. It undergoes reversible volumetric transitions in response to changes in external environment.[32]

2.1. *Temperature-responsive hydrogel*

Temperature- or thermo responsive hydrogel relies on the N-alkyl acrylamides monomers, whose side-chain could form hydrogen bonds with water. The strength of hydrogen bonds possesses a negative relationship with temperature. When the temperature increases above the lower critical solution temperature (LCST) of the hydrogel, the polymer chains begin to contract and finally become compact. When the temperature decreases and is below the LCST again, the polymer chains repel each other and the hydrogel expands to its original swollen state. Poly-N-isopropylacrylamide (PNIPAAm, or sometimes PNIPAm), a typical thermo responsive hydrogel, consists of N-alkyl acrylamides monomers. The synthesis and mechanism of PNIPAm is illustrated in Fig. 2.[18] The original LCST of PNIPAAm hydrogel is 32°C.[35] An ionizable monomer, 3-(methacryloylamino) propyl trimethylammonium chloride (MAPTAC), can be added into the hydrogel pre-polymer solution to shift the LCST in a certain range.

2.2. *pH-responsive hydrogel*

Hydrogels that respond to pH (or extended to ion concentration) are composed of monomers with weak acidic or weak basic side groups.[42] These side groups are ionizable and their interaction with water is also a function of the pH. Schematic diagram of a pH-responsive hydrogel network interacting with the external environment is illustrated in Fig. 3.[46] Typical weak acidic monomer acrylic acid (AAc or AA) and basic monomer dimethylamino ethylmethacrylate (DMAEMA) are shown in Fig. 4.[42]

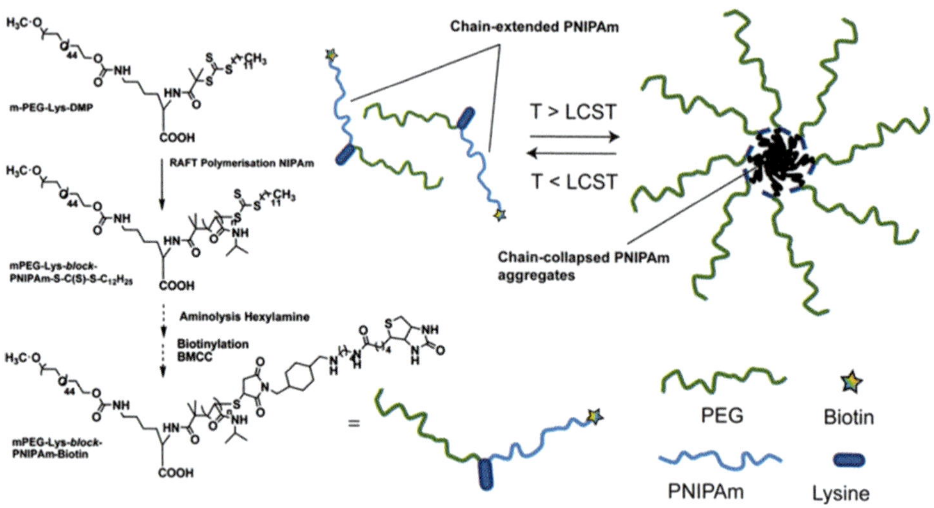

Fig. 2. Synthesis of PNIPAm in block copolymers (left) and schematic representation of the association of PNIPAm in copolymers in response to temperature (right).[18]

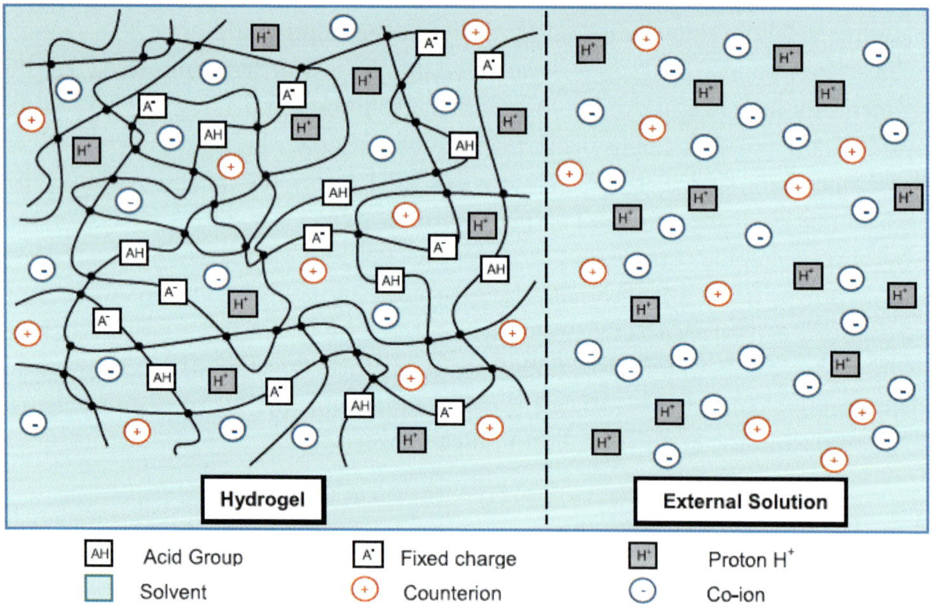

Fig. 3. pH-responsive hydrogel is a polymer network that imbibes a solution. The polymer chains are covalently crosslinked and bear acidic groups AH. Some of the acidic groups dissociate into fixed charges attached to the network and mobile hydrogen ions in the solvent.[46]

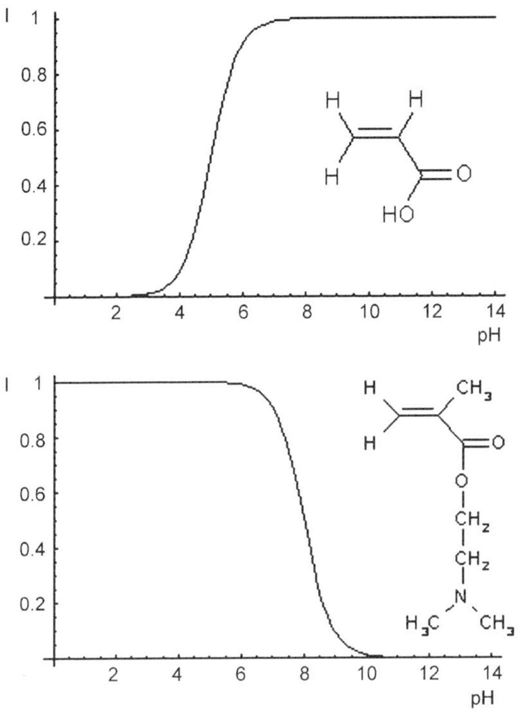

Fig. 4. Ionization behavior of acrylic acid and dimethyl aminomethacrylate. The degree of ionization (I) is plotted Versus the pH value. AA is ionized at high pH and DMAEMA at low pH.[42]

The swelling and shrinking of a pH-responsive hydrogel corresponds to the ionization competition between the network and solvent (water). Counterions developed in the hydrogel during the ion exchange would generate an osmotic pressure. Consequently, volumetric change would be triggered until the elastic forces inside the hydrogel are in equilibrium with the osmotic force.

2.3. *Remotely controlled hydrogel*

Remote-control capabilities could significantly improve hydrogels' merit in actuator applications, since the actuation drive would not be limited to local environment. A variety of approaches have been proposed to impart remote-control ability to hydrogels, mainly including incorporation of magnetic or light sensitive component in hydrogel synthesis.[45] In this chapter, infrared (IR) light responsive hydrogel and magnetic field (MF) responsive hydrogel will be discussed.

2.3.1. *Light-responsive hydrogel*

Light-responsive hydrogels are promising because the activation process via light can be remotely controlled and non-invasive. By converting absorbed light to heat, incorporated materials can turn traditional thermo responsive hydrogels into light responsive hydrogels. Thermo responsive PNIPAAm hydrogels have been shown to possess light responsiveness as actuators, benefitting from incorporated gold nanoparticles (Au NPs) that could convert absorbed light to heat in hydrogel networks.[47] Zeng *et al.* reported an IR responsive microlens based on Au NP incorporated hydrogel (Fig. 5).

Recently, graphene oxide (GO) has been demonstrated to possess extraordinary physical properties when mixed with polymers to form nanocomposites.[39] Thanks to the giant-IR-absorption band in GO and its superior thermal conductivity to trigger the hydration–dehydration transition within the hydrogel networks, glycidyl methacrylate functionalized graphene-oxide (GO–GMA) thermo responsive PNIPAAm hydrogels exhibits superior IR response.[26] Lo *et al.* demonstrated such IR responsive hydrogel and its application in microfuidics, as illustrated in Figs. 6 and 7.

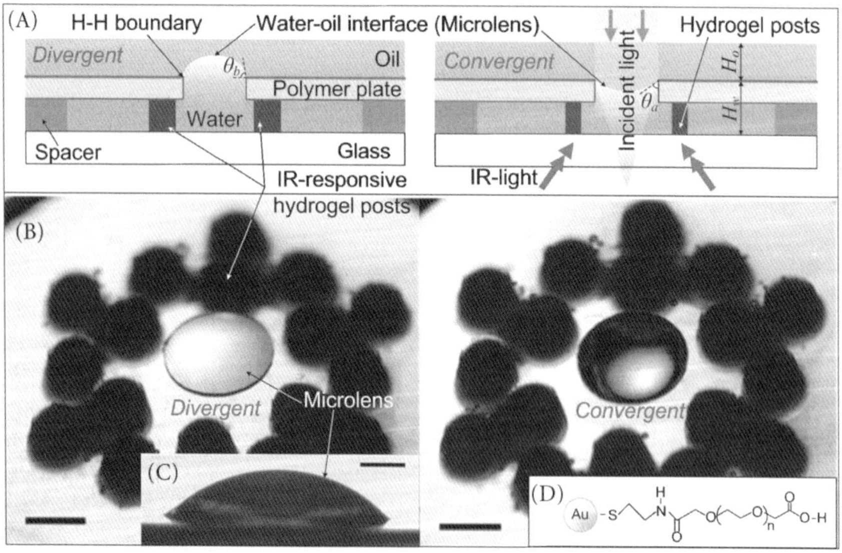

Fig. 5. IR light-actuated tunable microlens. (A) Schematics of the microlens based on hydrogel actuation. (B) A microlens with 18 hydrogel microposts in divergent and convergent states, respectively. (C) Side profile of water meniscus of the microlens in the divergent state at the starting point taken with a goniometer. The scale bar represents 1 mm. (D) Chemical structure of the Au NPs coated with thiolated polyethylene glycol ligands.[47]

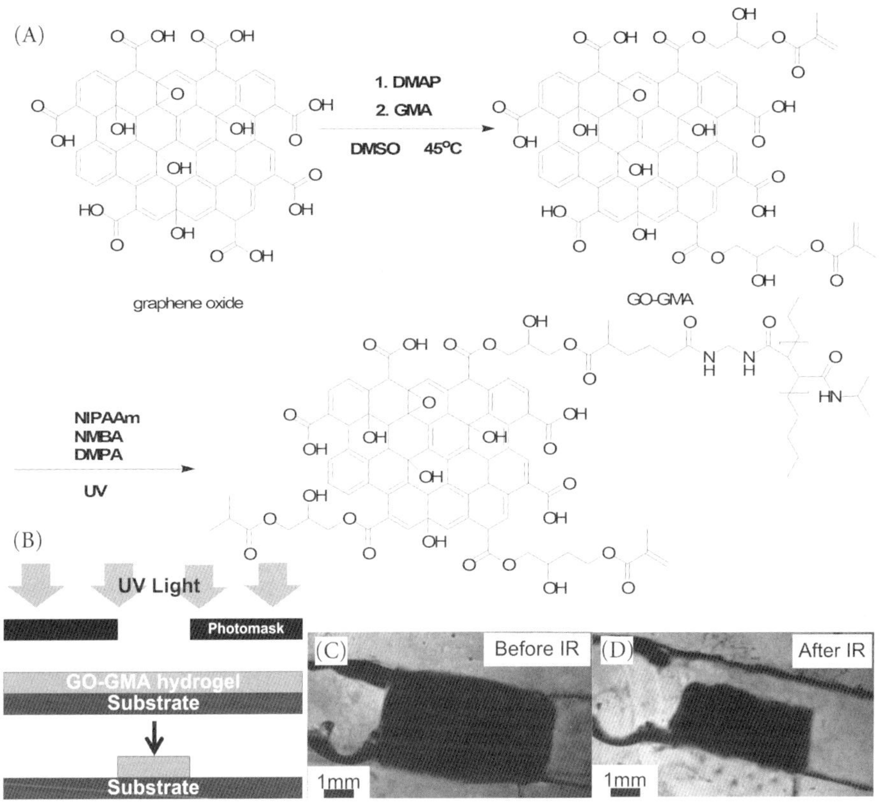

Fig. 6. Synthesis of GO–GMA incorporated PNIPAAm hydrogel. (A) GO and 4-dimethylamino-pyridine (DMAP) are dissolved in dimethyl sulfoxide (DMSO) and GMA is added to the mixture. The mixture is stirred. GO–GMA are purified and redispersed in DMSO solution for hydrogel fabrication. To synthesize GO–GMA hydrogels, NIPAAm, N,N'-methylenebisacrylamide (NMBA), GO–GMA DMSO solution and 2,2-dimethoxy-2-phenylacetophenone (DMPA) in the weight ratio of 4:0.08:4.0:0.154 and photopatterned under UV light. (B) Schematic of the photopatterning process of GO–GMA hydrogel microactuators. IR response of GO–GMA hydrogel (C) before and (D) after.[26]

pH responsive hydrogel also can be integrated into light responsive actuators. Al-Aribe *et al.* reported an ionic hydrogel microactuator driven by an organic photo-electric transducer. The mechanism is illustrated in Fig. 8.[2]

In addition, it has been shown that carbon nanotubes (CNTs) can be selectively heated when exposed to radio frequency (RF) fields or near-IR light.[17,19] Hence, hydrogel mixed with CNTs would also possess light responsiveness.

2.3.2. *Magnetic-responsive hydrogel*

Similar to the actuation mechanism of light sensitive hydrogel, magnetic nanoparticles (MNPs) embedded in a hydrogel polymer matrix could realize magnetic responsivity,

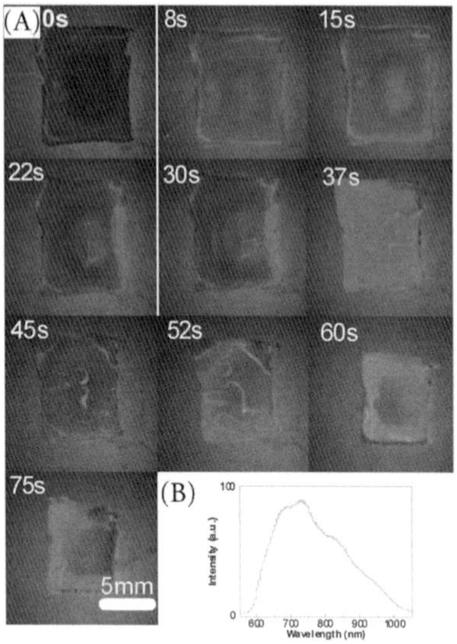

Fig. 7. (A) Dynamic change of a piece of IR-responsive GO–GMA hydrogel film in one illumination cycle as a function of time. The IR is on at 1 s. The intensity of the IR light reaching the hydrogel sample is 1.21 W cm^2. (B) Spectrum of the IR light source.[26]

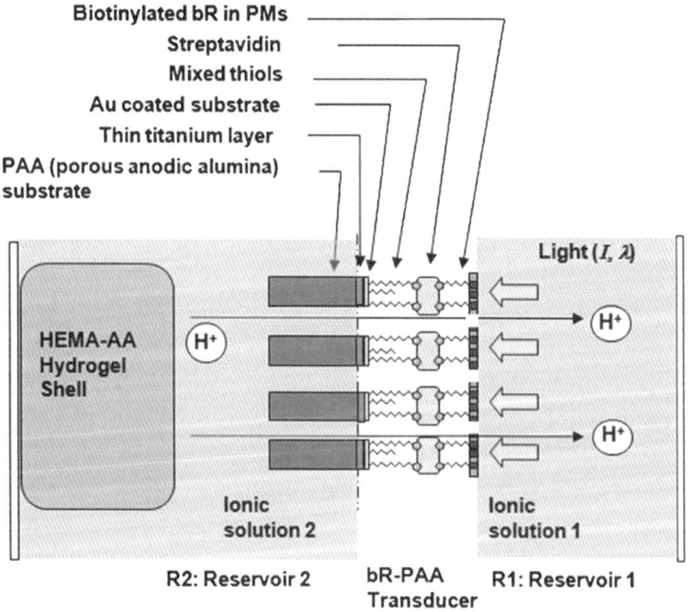

Fig. 8. The design and fabrication of a photoresponsive hydrogel microactuator activated by a bacteriorhodopsin (bR) monolayer. The bR monolayer generates pH gradient when exposed to light beams, which in turn is exploited in triggering actuation of 2-hydroxyethyl methacrylate–acrylic acid (HEMA–AA) pH sensitive hydrogels.[2]

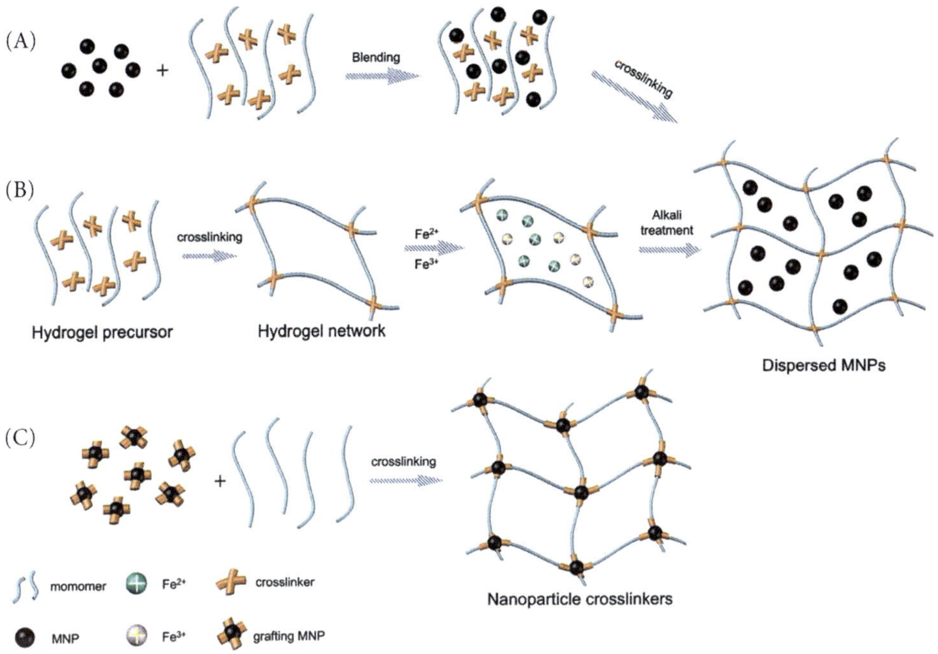

Fig. 9. Schematics of preparation methods of magnetic hydrogels. (A) The blending method: the prepared MNPs mix with a hydrogel precursor solution at a certain molar ratio and crosslink hydrogels to entrap the MNPs. (B) The *in situ* precipitation method: MNPs are fabricated via an *in situ* precipitation reaction in the network of the polymer hydrogels after the crosslinking reaction. (C) The grafting-onto method: grafting several functional groups onto the surface of the MNPs as nano-cross-linkers, leading to the crosslinking reaction.[25]

generally due to their ability to generate heat when exposed to alternating magnetic fields (AMFs) through Neel and Brownian relaxations.[45] DC MFs and low frequency oscillating MFs can also be employed as stimuli to trigger magnetic nanocomposites.[45] The mechanism of actuation in these systems is based on particle–particle or magnet–particle forces. The performance of magnetic hydrogels (e.g., magnetic response) depends on factors including the types of hydrogels and nanoparticles, concentration of embedded nanoparticles, and the size and distribution of the nanoparticles within the hydrogels.[45] Various methods have been developed to fabricate magnetic hydrogels, including a blending method, an *in situ* precipitation method, and a grafting-onto method, as summarized in Fig. 9.[25]

2.4. *Compatibility*

Numerous hydrogel-based applications have emerged in the past half century. Not only does their unique volumetric change satisfy the need for dynamic actuation, the hydrogels' superior compatibility with modern micro-fabrication technology during manufacturing and biological systems and environment during operation excels as well.[14]

2.4.1. *Fabrication compatibility*

With a photo-initiator mixed in the polymer matrix, hydrogel can be patterned under UV light. This would make hydrogel structure with different shapes available, thus feasible in a variety of actuation systems. In Fig. 10,[12] fabrication flow of a hydrogel actuated microlens array is illustrated. Here, hydrogel was photopattened into a ring shape into a cavity by UV light. Compatibility with top–down micro-fabrication flow is one of the most critical features that allow the hydrogel to be widely applied and integrated in miniaturized devices.[14]

Elastic rubbery feature of the hydrogel would also be beneficial in manufacturing. Though not as bouncy as rubber, hydrogels possess sufficient elasticity to function well under certain bending or twisted situations.[52] As demonstrated in Fig. 11, a hydrogel actuated microlens array could be fabricated on a spherical surface without performance degradation. In addition, hydrogel could well adhere to or contact with other structural polymers such as polydimethylsiloxane (PDMS), polyamide, etc. Hence, hydrogel is highly compatible with those elastic polymer structures and is very promising in flexible system applications.[52]

2.4.2. *Operation environment compatibility*

Actuation of hydrogels require aqueous environment, while typical biological systems cannot function without water. Stimuli employed to drive hydrogel

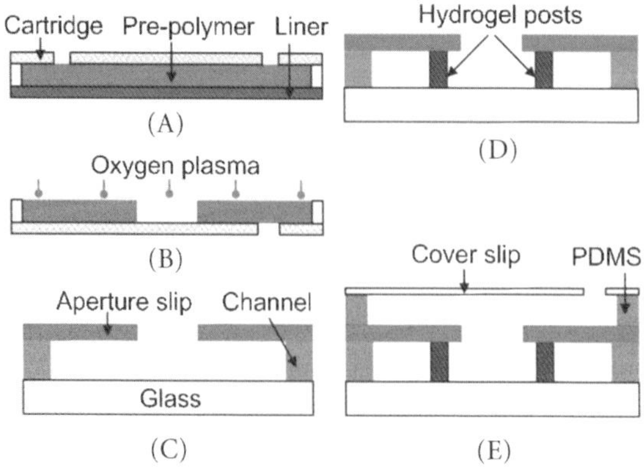

Fig. 10. Fabrication process of the liquid microlens. Hydrogel is photopatterned as actuators.[12]

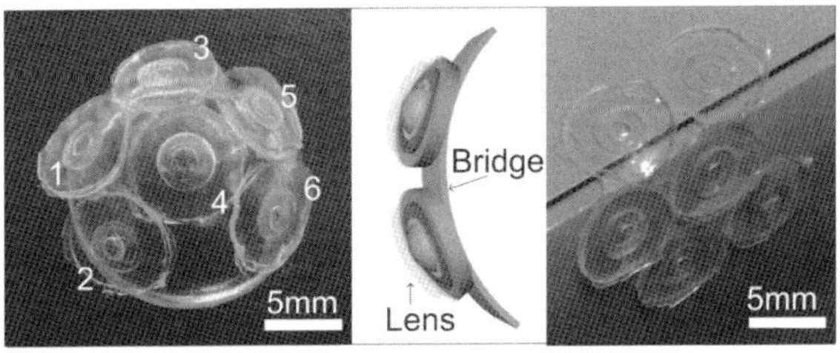

(A) (B) (C)

Fig. 11. (A) Six-element microlens array wrapped onto a hemispherical acrylate shell with a diameter of 20 mm. Each microlens element is connected with the neighboring one by elastic polymer bridges. The scale bar represents 5 mm. (B) Schematic of the elastic polymer bridges, which connect adjacent microlens element. (C) Six-element microlens array folded at the edge of a desk.[52]

actuators such as heat transfer, pH or ion concentration are quite implementable in biological systems. Note that the critical temperature that drives thermo responsive hydrogel is about 32°C, which is close to human body temperature.[35] Therefore, it is convenient to integrate hydrogel structures into bio-related systems such as microfludics labs on chips, and drug delivery. In Fig. 12,[1] a hydrogel-based micro-cooling system in water is illustrated.

2.5. *Characterization of actuation behavior: Size, shape, and response time of hydrogels*

As the field of bio-actuation system grows, precise control in actuator design, which is related to the fundamental understanding of the actuator mechanism, emerges as an important topic in the study of hydrogel.[46] In a hydrogel actuation system, current design and development is partially hindered by limited input from fundamental mechanism study. The patterned shape (size/volume) of hydrogels, as one critical feature in the actuator design, has significant influence on the actuation mechanism including actuation orientation, degree and response time, etc.

The actuation behavior of a hydrogel requires a transport of matter.[32] Initially, a stimulus such as temperature and ion variation diffuses into the hydrogel network, and then osmotic pressure is triggered. Next, by swelling or shrinking of the hydrogel network, equilibrium of the osmotic pressure is achieved.[24,40] This actuation behavior involves continuous mass diffusion, delamination and volume tuning, which is highly associated with characteristics such as shape and response time of hydrogels.[10]

Crone *et al.* performed a comprehensive study on the delamination process during the actuation of hydrogels in various shapes. They tested on a pH sensitive

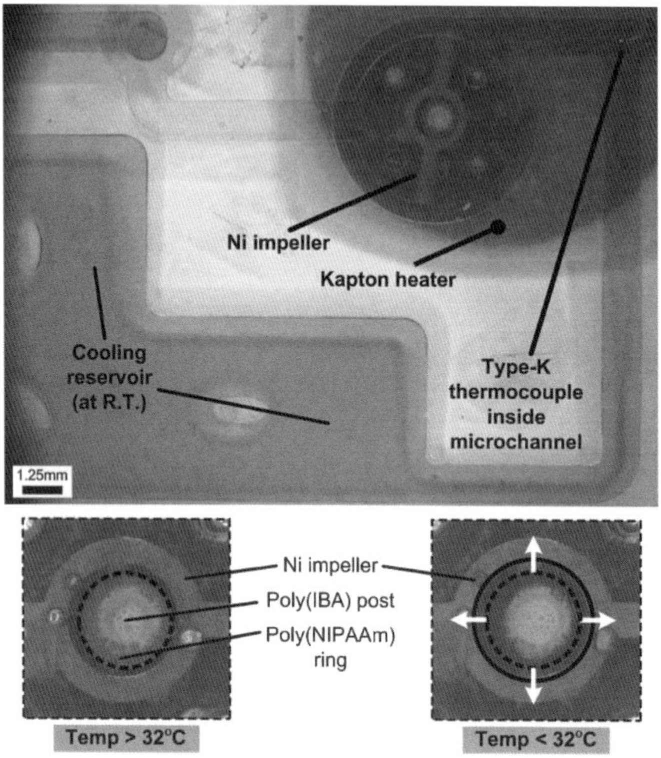

Fig. 12. Experimental setup for an autonomously-triggered on-chip microfluidic cooling pump. A Kapton heater adhered to the underside of the device simulates local heating. A type-K thermocouple inserted inside the microchannel is used to monitor the local fluid temperature. The on-chip cooling reservoir is at room temperature. Here, when local temperatures are above 32°C, the hydrogel contracts; the nickel (Ni) impeller rotates, and pumps recirculates cooler water (see left inset photo; the dashed black circle shows the contracted hydrogel boundary). When local temperatures drop below 32°C, the hydrogel expands (see right inset photo; expansion occurs from dashed black circle to solid circle). Hence, the Ni impeller stops rotating.[1]

hydrogel (hydrogel–liquid solution: 2-(dimethylamino)ethyl methacrylate (DMAEMA), a pH-sensitive monomer, HEMA, an inert monomer, ethylene glycol dimethacrylate (EGDMA), a crosslinker, and 2,2-dimethoxy-2-phenylacetophenone (DMPA), a photo-initiator). The hydrogel was photopolymerized into cylinders and cubes (400–1000 μm in diameter/side, 250–350 μm in height) by UV light.

As the gel swells, the outer ring of the gel delaminates from the substrate first. With further swelling, this ring of delamination propagates radially inward and creates a circular delamination front, as shown in Fig. 13.[10] Regarding shape of cubes or squares, the corners swell faster than the flat edges due to shorter

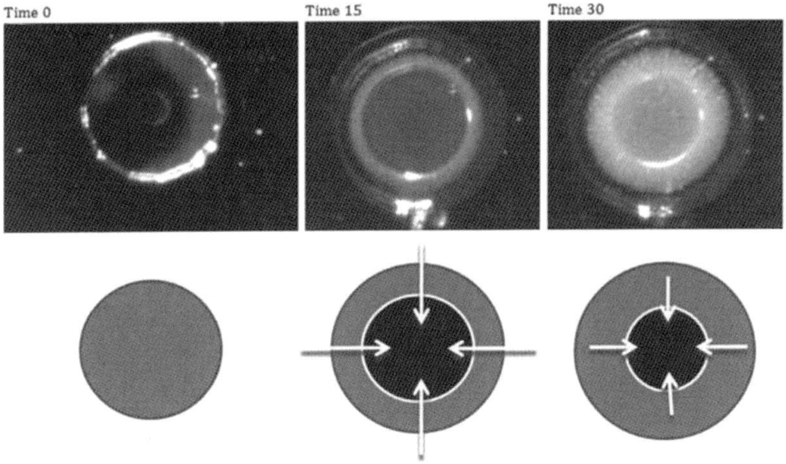

Fig. 13. The delamination process before the addition of a buffer solution, at the 15 m and 30 m marks for a 1,000 μm circular hydrogel post. The delamination proceeds concentrically as illustrated in the schematic drawings.[10]

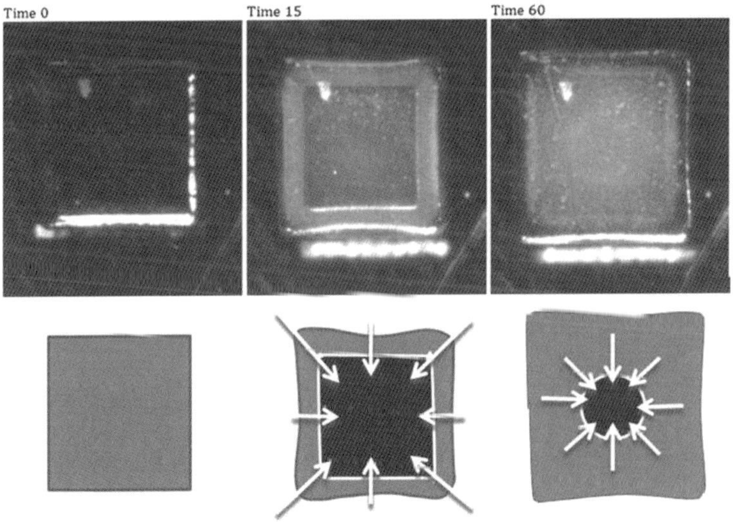

Fig. 14. The delamination process before the addition of a buffer solution, at the 15 m and 60 m marks for a 1,000 μm square hydrogel post. The delamination initially has a square shape, but eventually turns circular as illustrated in the schematic drawings.[10]

diffusion path (Fig. 14). This phenomenon is more prominent in larger square hydrogel posts. Such uneven swelling causes different portions of the delamination front at different rates. Once a sufficient distance from the outer edges of the post has been reached, the delamination front becomes circular.[10]

Zalachas *et al.* studied the swelling and instability of pH-sensitive hydrogels in a ring shape, which is commonly patterned in actuators. A representative setup involves a ring of a pH-sensitive hydrogel bonded between two rigid plates. A creasing phenomenon was observed during the hydrogel ring actuation. It was also found that the height of the ring strongly affects the actuating behavior. When the ring is short, a smooth barrel-shaped swelling develops and the inner surface of the hydrogel ring becomes convex. When the ring is tall, a crease appears and a self-contact develops and propagates, thus the inner surface of the hydrogel ring can even become concave for higher aspect ratio values. Some of the results are shown in Fig. 15.[46]

Though possessing outstanding functionality and compatibility with biological systems, traditional hydrogel actuators suffer from relatively slow response time (several minutes) due to its diffusion-limited mechanism, which restricts its development.[32] Nowadays, with the introduction of nanoparticle mixture, the responsive time of hydrogel can be significantly improved. Lo *et al.* proposed a GO–GMA hydrogel thin film that could realize significant area change within 6 s under thermal variation, as illustrated in Fig. 16.[53]

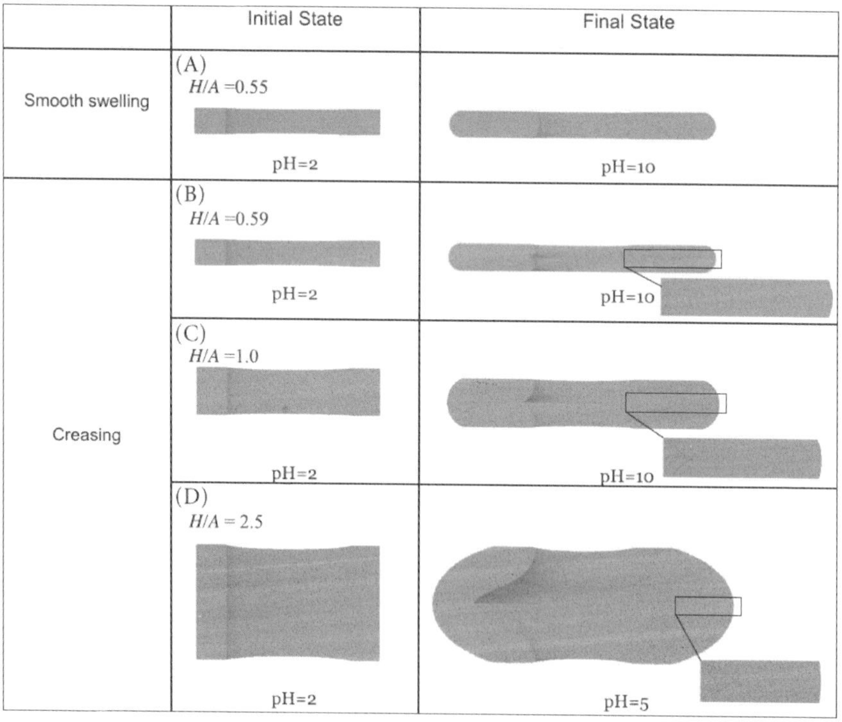

Fig. 15. Initial and final states of the pH sensitive gel ring with different values of aspect ratio H/A. (a) The ring swells smoothly pH = 10. (b) The ring swells smoothly for small pH value, while self-contact begin to propagate at the inner surface to form a crease when pH becomes bigger. (c and d) Concavity is enhanced as aspect ratio increases. The outer surface swells and self contact develops in the center of the inner surface. (d) The ring collapses at a smaller value of pH = 5.[46]

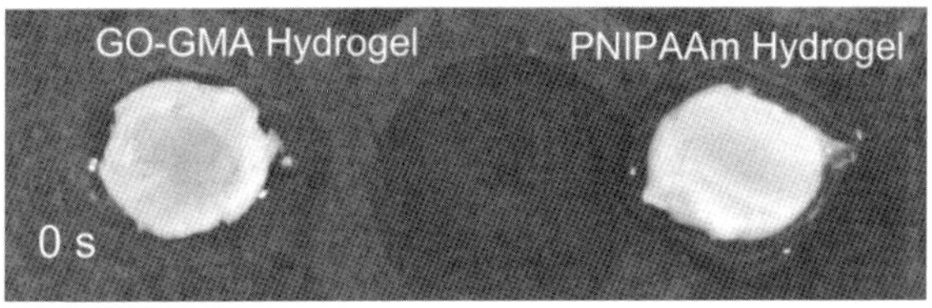

(A)

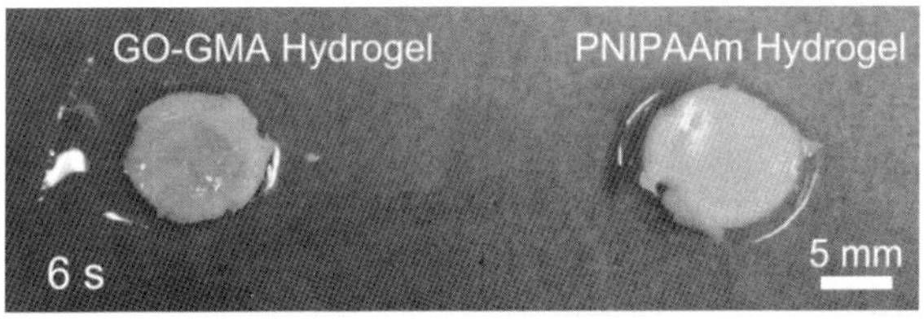

(B)

Fig. 16. Comparison between GO–GMA hydrogel and PNIPAAm hydrogel. (A) Two hydrogel films at initial states. Both are photopatterned into a circular shape with a diameter of 10 mm. Water environment is provided. (B) Both hydrogel films after being heated at 70°C for 6 s. GO–GMA hydrogel becomes smaller than PNIPAAm hydrogel.[53]

3. Hydrogel Actuators in Biological Systems

Hydrogel actuators have been widely explored in various kinds of biological systems. In this section, we will discuss their applications in microfludics, artificial biological devices, and biohybrid systems.

3.1. *Flow control in microfluidics*

Microfluidic technologies, which have broad applications in life science, pharmaceutical, environmental, and agricultural industry, represent one of the fastest expanding fields in the market of commercial instruments. With an estimated world market of about 3 billion dollars in the year 2014, it can be expected that microfluidic devices will continue to find new applications, and to generate interest from a wide range of industries and research fields.[4]

Development of microfluidics depends on a series of multidisciplinary technologies such as materials science, biology and microelectronics engineering. Device miniaturization and exploration of materials with novel functionality and compatibility are the major challenges in microfluidic industry.[14]

Hydrogels as actuators draw considerable attention in the microfluidics field. On one hand, hydrogels exhibit special functionalities as actuators, converting environmental signals into a mechanical drive. On the other hand, hydrogel with photo-initiators embedded could be photopatterned into micro/nano structures

Fig. 17. Schematic of *in situ* photopolymerization of hydrogel posts in a microchannel.[14]

via *in situ* photopolymerization. Multiple components can be realized sequentially or simultaneously, as shown in Fig. 17[14] In addition, the compatibility of hydrogels with fluidics is well aligned with the operation environment of microfluidic systems. Therefore, hydrogel actuators could significantly improve the microfluidic efficiency and reduce the system complexity.

Here, hydrogel-based microfluidic components with various micromechanical properties are introduced, including microvalves, micromixers, and micropumps.

3.1.1. *Micropumps and micromixers*

Micromixers and micropumps are essential components for fluidic control in a microfluidics system. In general, micropumps can be classified as either mechanical or non-mechanical. Mechanical micropumps need physical actuators or mechanisms to function; they include electrostatic, piezoelectric, thermopneumatic and electromagnetic types. Non-mechanical micropumps need further conversion of non-mechanical energy into movement, so that the fluid can be driven. These include magnetohydrodynamic, electrohydrodynamic, electroosmotic, electrowetting, bubble type, electrochemical and evaporation-based micropumps.[4]

Most designs of micropumps require complicated power generation systems and costly fabrication processing. Hydrogel actuators, with their remarkable responsivity and simple manufacturing procedures, have brought in vitality in this area. Hydrogel actuators can play various roles in micropumps and micromixers, including powering source (generating pressure directly or indirectly to drive the pump), control (determining pump characteristics), etc.

Richter *et al.* proposed a diffusion micropump and a displacement pump respectively. Both of them rely on the temperature-sensitive hydrogel PNIPAAm.

In a diffusion pump, when heated, the hydrogel shrinks, causing the solution to be released into the outlet and generate pumping pressure; when the temperature is decreased, the hydrogel swells by absorbing the liquid, leading to the deformation of an elastic membrane which regulates the pressure. In a displacement micropump, unlike in the diffusion type, the polymeric actuator generates pumping pressure when it swells and decreases pumping pressure when it shrinks. These polymeric micropumps can be well controlled by means of local heating elements through a thermo responsive hydrogel.[33]

Agarwal *et al.* demonstrated programmable autonomous micropumps and micromixers based on pH and temperature responsive hydrogel actuators. Timing sequence of fluid pumping can be well determined by external stimuli in these systems. The mechanism is shown in Figs. 18 and 19. Expansion and contraction of a hydrogel clutch is able to stop and restart the Ni rotor, which impels the pump. Operation of a micro-cooling system based on these components was shown in Fig. 12.[1]

To achieve cyclic control, Murase *et al.* developed hydrogel micropumps driven by Belousov–Zhabotinsky (BZ) reaction. The copolymerized temperature-responsive NIPAAm with ruthenium tris (2,20-bipyridine) [Ru(bpy)3] as the catalyst can respond to the BZ reaction, and obtain autonomous peristaltic motion without external stimuli.[31]

3.1.2. *Microvalves*

Microvalve is another crucial component for flow control in microfluidic systems. The opening and closing of valves accomplishes regulation of fluid flow in a channel.

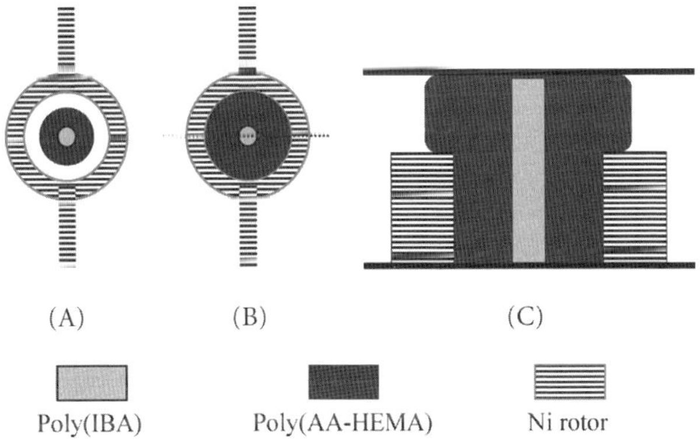

Fig. 18. Conceptual diagram of using poly(HEMA-AA) to control the actuation of a Ni rotor, much like an automotive clutch. (A) Low pH causes the hydrogel to shrink, allowing the Ni rotor to rotate freely. (B) High pH causes the hydrogel ring to expand in volume, constricting the Ni rotor's rotational movement. (C) A cross-sectional image of the expanded hydrogel, taken at the dotted line in (B) that shows the development of an overlaying mushroom cap on top of the Ni rotor.[1]

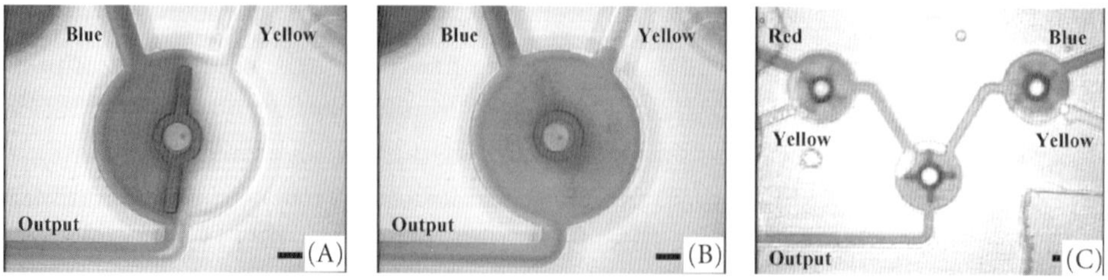

Fig. 19. Two photographs of a single micromixer (Ni rotor measuring 2.0 mm × 0.40 mm × 0:20 mm) and one photograph of an array of three micromixers (Ni rotors measuring 1.5 mm × 0.35 mm × 0.20 mm) with a central core polymer post. Scale bars represent 1.0 mm. (A) Using a syringe pump, the two input dyed water solutions (blue and yellow) flow into the system at an individual flowrate of 5.0 μL/min. Laminar flow results at the output since the external magnetic stirrer is not activated, i.e., no mixing. (B) Upon activation of a magnetic stirrer, the Ni rotor rotates and causes chaotic action inside the mixing chamber, and subsequently mixes the two input fluids into a green output solution. (C) An array of three micromixers mixing dyed water solutions.[1]

Typically, a valve consists of a deformable structure coupled to an actuator that controls the ON/OFF state. The tuning of these systems depends on their intrinsic responsiveness to thermal, electrical or optical stimuli. Thus, hydrogels, whose actuation along with significant volumetric change, possesses great potential in microvalve applications.[14,43] Various hydrogel microvalve geometries and chemistries have been developed to date. Most of them rely on volumetric change of hydrogels under stimuli. Some valves are based on phase/characteristic change of hydrogels.

Beebe *et al.* designed and realized microvalves by either patterning a pH-sensitive hydrogel along the walls of a channel or creating an array of the same hydrogel inside the microfluidic devices. The swelling of these hydrogel posts blocked a channel when a high pH solution flowed into the channel, whereas the contracted state of hydrogel allowed the fluid to pass under low pH condition. A pH-value controlled microvalve is shown in Fig. 20.[14]

Lo *et al.* demonstrated a microvalve based on GO–GMA hydrogel in a microfluidic channel. This valve can be controlled by IR and temperature, thanks to the GO–GMA structure embedded. As shown in Fig. 21, the rectangular hydrogel structure as the microvalve was phototopatterned in the microfluidic channel. Initially at its OFF state, with the IR light turned off, the GO–GMA hydrogel microvalve completely blocked the fluidic flow in the microfluidic channel. When the IR light was turned on, the GO–GMA sheets absorbed IR light, triggered the hydrogel to contract to allow for fluidic flow, thus the ON state. The GO–GMA hydrogels shrunk in every dimension. As the IR light was turned off, the heat dissipated to the surrounding environment and the hydrogel absorbed water and expanded back to its original volume, preventing the fluidic flow (OFF state again).

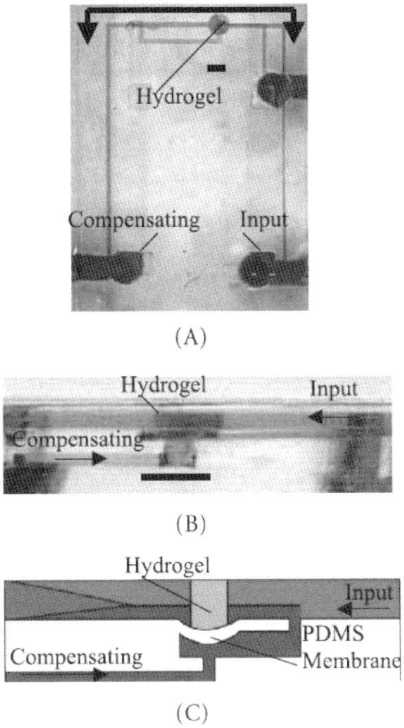

Fig. 20. (A) Top view of a pH regulation device, (B) side view of the regulation device with a slice taken through the segment indicated in (A), and (C) side-view schematic demonstrating the flow conditions. The compensating buffer enters from the bottom left (blue), while the input enters from the top right (red). The two streams meet and then flow past the hydrogel post from right to left and exit the outlet on the top left (purple). The scale bars in (A) and (B) indicate 1 mm.[14]

The function of the GO–GMO hydrogel microvalve here was remotely controlled by IR light (Fig. 21).[26]

Yang *et al.* developed microvalves in microfluidics with a different perspective. They investigated membrane valves with tuning permeability. This type of microvalves is based on a combination of a rigid porous membrane with a soft stimuli-responsive hydrogel. The fabrication is realized by depositing a thin layer on the top surface (ts) of a membrane or by introduction of pores. The position of the hydrogel (on top or within the porous barrier) can be critical in the actuation performance. With different hydrogel integrated, this microvalve can be driven by stimuli such as temperature, pH, light, and MF. The mechanism of this type of microvalves is shown in Fig. 22.[44]

The characteristic wettability of hydrogel along with its phase change can also be exploited to design microvalves. Chunder *et al.* demonstrated hydrophobic/hydrophilic switchable surface driven by thermal responsive hydrogel. This surface

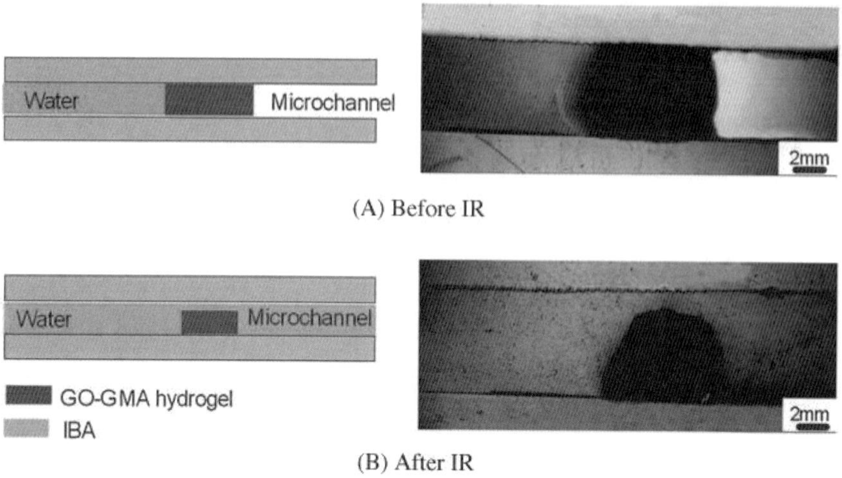

Fig. 21. A microvalve made of GO–GMA IR-responsive hydrogel. Top views of the microvalve are on the left. Respective images taken under a microscope are on the right. (A) GO–GMA hydrogel microvalve before actuated by IR irradiation (OFF state) and (B) after exposure to IR light (ON state). The valve exhibits significant change in dimensions before and after IR actuation. Initially, the fluid flow is blocked by the valve at its OFF state. Then, at its ON state, the flow is able to pass through the microvalve.[26]

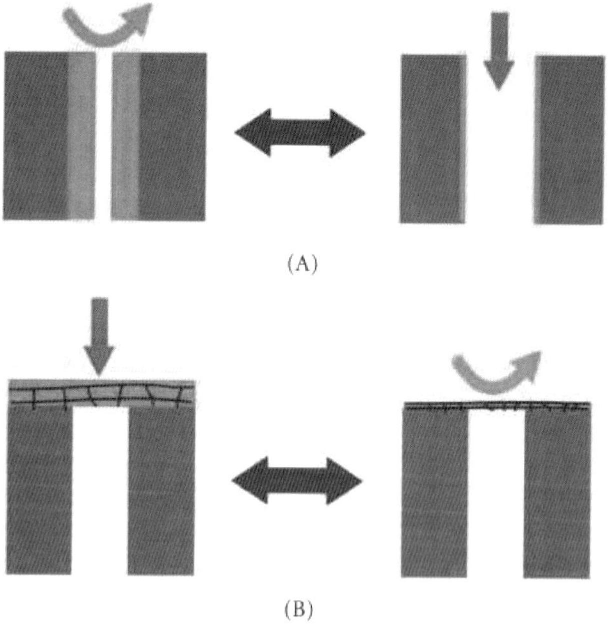

Fig. 22. Schematic overview on two types of membrane valves with stimuli-responsive hydrogel layer at different locations: (A) layer on the pore wall (hydrogel de-swelling leads to increasing permeability); (B) crosslinked layer on the top of membrane covering the pore (hydrogel de-swelling leads to decreasing permeability).[44]

can be fabricated in microfluidic channel and function as a thermo-sensitive valve that regulates the flow.[8]

3.2. *Bionics*

Bionics emerges with more and more frequent interaction of biology and engineering, physics, chemistry, materials science, and electronics. Artificial biological system has evolved from unreachable theory to realistic device at hand. With huge potential market such as artificial intelligence, medical care, environmental science, and agriculture, bionic technology is destined to bloom in academia and industry in near future. Materials with good bio-compatibility like hydrogels inevitably draw great attention in bionic field.

3.2.1. *Artificial muscles: Mechanical actuation of hydrogels*

Hydrogels were referred to as "muscle-like" materials when they were initially developed.[21] The mechanical actuation behavior of hydrogels (deformation or the swelling/shrinking process) under external controllable stimuli shows their potential to function like biological muscle system. Their elastic nature and good bio-compatibility are also a big advantage. There have been many attempts to fabricate artificial muscles sensitive to external stimuli with hydrogels.[29] A couple of typical works are introduced here.

Techawanitchai *et al.* demonstrated reversible bending and straightening of bilayer hydrogels by alternating their exposure to UV light. The hydrogel employed exhibits certain bending feature and controllability, resembling human hand, as shown in Fig. 23.[41] This photo-switchable, reversible self-bending pH responsive hydrogel actuators were realized by integrating a pH-jump reaction. By tuning the nitrobenzaldehyde (NBA) concentration in a solution environment, the gel thickness, bending orientation, kinetics, and reversibility can be well controlled.

Actuation of most light sensitive materials is localized only at the illuminated regions and the responsiveness is discontinuous when the light source is removed. Unlike these polymers, Techawanitchai's hydrogel-based system induces self-bending of inactivated regions via proton diffusion. This ability provides opportunities for the design of systems that exhibit predictive and programmable actuations as well as complex movements as muscle-like materials.[41]

MF is also a convenient source as external stimulus to provide controllable mechanical actuation. MF responsive hydrogel actuators are also very promising acting as artificial muscles.

Zhou *et al.* reported physically crosslinked hierarchical hydrogels through the swelling of a series of polymer hydrogels consisting of the amphiphilic

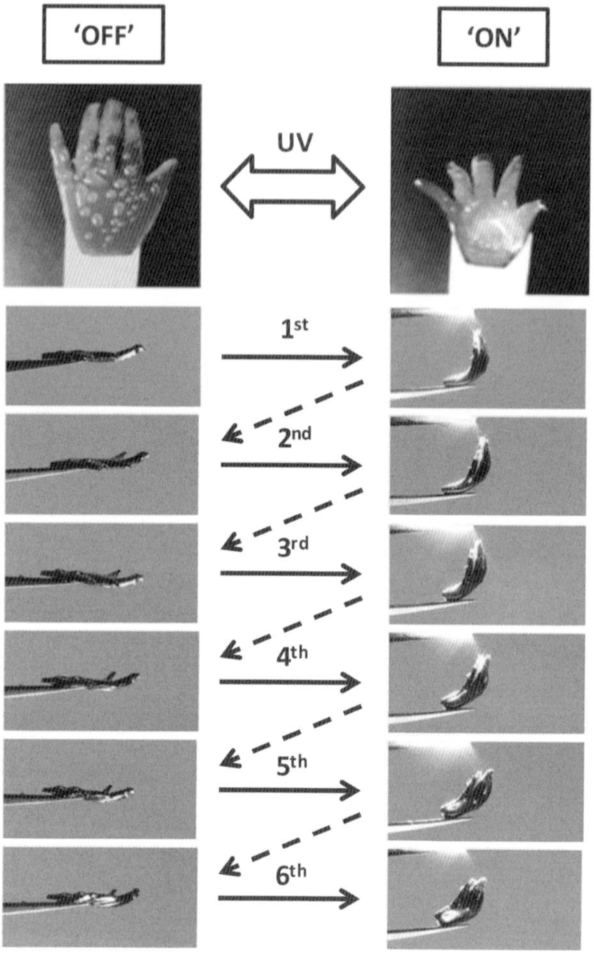

Fig. 23. Photographs show the top view and side view of the reversible bending ("ON" state) and straightening ("OFF" state) behavior of 10 mM NBA-loaded palm-shaped bilayer gels consisting of polyacid P(NIPAAm-co-CIPAAm) and polybase P(NIPAAm-co-DMAPAAm) layers with 0.5 mm of each layer when alternating exposure to UV light on the polyacid side.[41]

liquid-crystalline (LC) pentablock copolymer PB20-C5MA *x*-Aa *y* and Fe_3O_4-based nanoparticles as shown in Fig. 24. These magnetic responsive hydrogels possess significant swelling degrees. They also show good mechanical and super-paramagnetic properties, which make them good potential candidates for artificial muscle materials.

3.2.2. *Artificial skin: Mechanical reaction of hydrogels*

Not only could hydrogels provide mechanical drive corresponding to external stimuli, they also are able to transfer or convert mechanical signals such as strain, stress, etc.

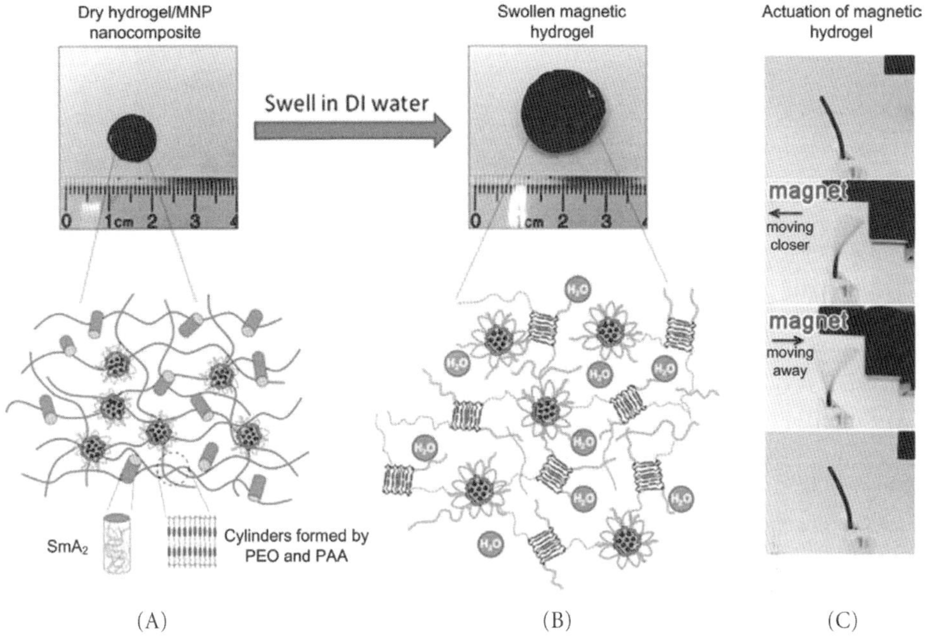

Fig. 24. Magnetic hydrogels for soft actuator application. (A) Schematic of a dry hydrogel/MNP nanocomposite with a hierarchical structure dispersed in a polymer matrix. (B) Schematic of a swollen magnetic hydrogel with a network structure connected by LC domains and MNP clusters. (C) Actuation of magnetic hydrogels using a magnet.[49]

Manandhar *et al.* conducted studies on the reaction of elastomeric ionic hydrogels under external strain and stress. They proposed a strain sensing system on a robot arm. Hydrogel integrated in this system functions as potential artificial skin.[28]

Proprioception, the sense of position of a body part, is a critical challenge for artificial biological system design. In humans and many other animals, proprioceptions are realized through nerve fiber endings in the skin that can respond to skin stretch. The hydrogel-based system reported by Manandhar *et al.* can provide comprehensive mechanical sensing, resembling proprioceptions.[28] As shown in Fig. 25, an array of hydrogel-based sensors was integrated on a robot arm to monitor the strain field at various locations within a low elastic modulus sheet. Unlike other textile "skin", this hydrogel-based system has a unique "memory-less" sensing feature, i.e., the response is fast enough to monitor joint motion in real time. Output of this hydrogel transducer does not depend on previous inputs and can be much more accurate. This work investigated mechanical reaction of hydrogels, and proved their potential as artificial skin structure.

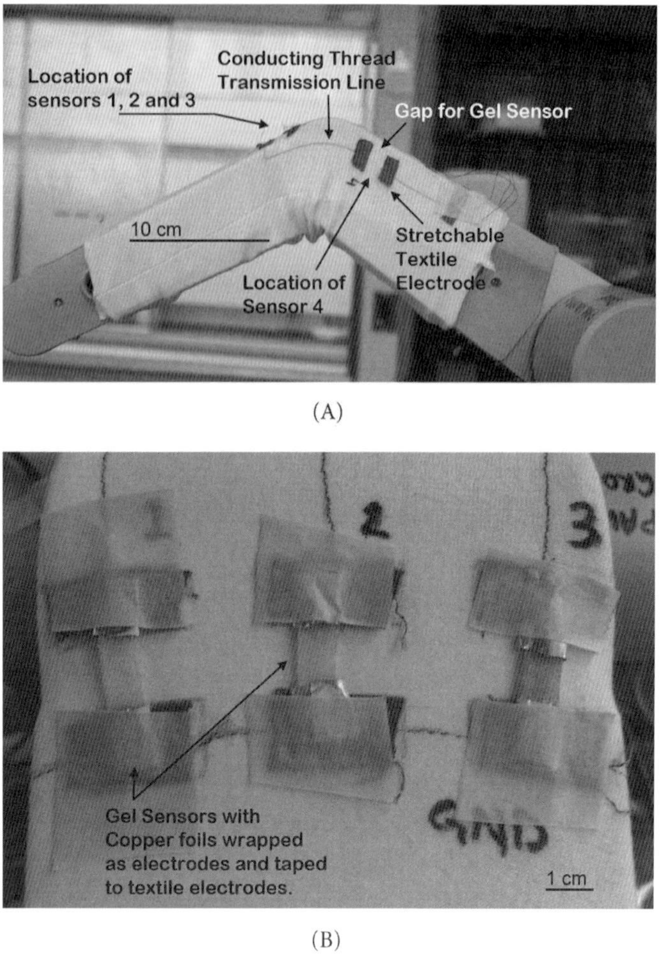

Fig. 25. (A) Sleeve with textile transmission lines and electrodes worn on robotic arm and (B) with hydrogel sensors attached to electrodes.[28]

3.2.3. *Artificial eyes: Hydrogel-based optical components*

3.2.3.1. Imaging

A natural eye is one of the most complicated structures in human beings or animals. Human eye belongs to the category of camera eye, in which focusing at different distances is made possible by altering the tension on the flexible crystalline lens through ciliary muscles. It has a forward-facing horizontal field of view (FOV) less than 180°. Although with high resolution and variable focus, human eyes suffer from limited viewing angles. As a result, the head or body have to move or rotate to capture the complete scene of the surrounding.[50] On the other hand, compound eyes of some insects, which consist of thousands of small eye volumes distributed on a dome structure, have a much larger FOV.[23] Those eye volumes with a simple photoreceptor structure are referred to as ommatidia. The image is obtained by combination of inputs from the numerous ommatidia. However, due to the small size of the ommatidia, diffraction effect limits the compound eye resolution. In summary, camera eyes generally owned by mammals can provide high resolution, focus tunability and high acuity, but are limited in their viewing angles and must be supported by complicated systems and high energy consumption to complete the imaging process. On the other hand, compound eyes preferred by insects could provide large FOV. They have simple structure but inferior optical performance.

Optical systems have been developed to mimic natural eyes in the past decade,[22] in which microlenses play significant roles in the imaging process. A focus-tunable microlens-based system could mimic a camera eye. Hydrogel actuators can realize focus tunability without complicated control system. Dong *et al.* developed liquid microlenses based on thermo responsive hydrogels (Fig. 26).[12,13] Focus of these

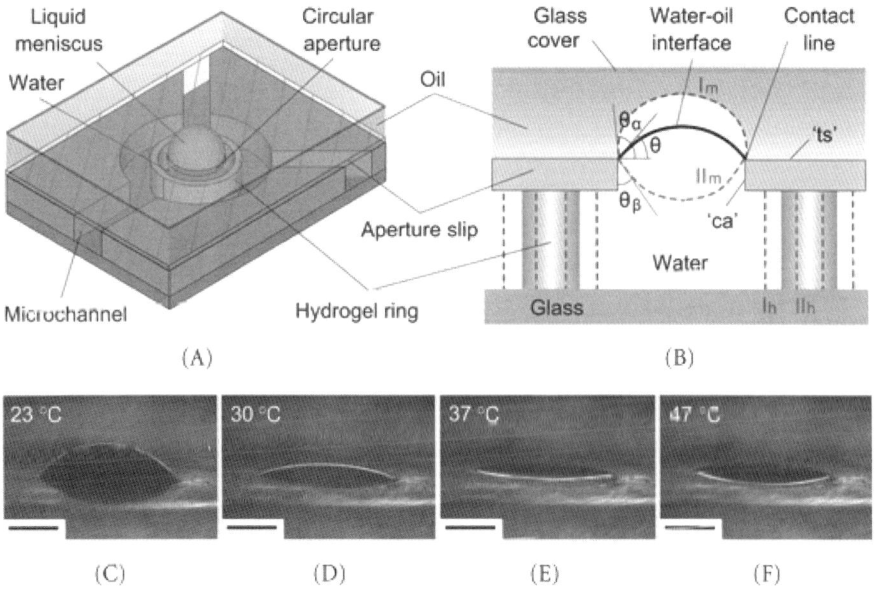

Fig. 26. Smart microlens using a pinned liquid–liquid interface. (A) The water–oil interface forms the liquid microlens. The microchannels allow the flow of fluids to the microlens structure. (B) Smart variable-focus mechanism. The hydrophilic sidewall and bottom surface ('ca') and hydrophobic ts of the aperture pin the water–oil meniscus along the contact line "ca-ts". The expansion and contraction of the hydrogel regulates the shape of the liquid meniscus by changing the angle q of the pinned water–oil interface. The blue dashed lines show the expanded state of the hydrogel ring ('Ih') and the corresponding divergent microlens ('Im') at q_a. The red dashed lines show the contracted state of the hydrogel ring ('IIh') and the corresponding convergent microlens ('IIm') at q_b. C–F, The shape of the liquid microlens varies with local environmental temperature. Scale bars represent 1.0 mm.[12]

microlenses can be tuned by temperature variation through responsive hydrogel structure, mimicking the ciliary muscles in a human eye. Water-oil meniscus pinned at a lens aperture serves as the liquid microlens. The hydrogel actuator resides in a water cavity under the meniscus. The expansion and contraction of the hydrogel regulates the shape of the liquid meniscus, thus realizing focus tunability of this microlens structure.

In addition, flexibility and compatibility of hydrogel could make it possible to realize microlens array on a curved or spherical surface, resembling the compound eyes.

By further exploiting hydrogel's elastic nature, Zhu *et al.* demonstrated thermo responsive hydrogel-based focus-tunable microlens arrays with multiple elements on spherical surfaces, which have the potential to function as artificial eyes combining advantages of both camera eyes and compound eyes as shown in Fig. 27.[51] Liquid-based microlenses were omni-directionally fabricated on transparent hemispherical bases in order to realize large FOV. The microlenses were distributed on a cap subtending an angle of 113° (for a 20-mm diameter base) or 116° (for a 18-mm-diameter base). Thin flexible PDMS bridges connecting the neighboring microlenses were designed to reduce the wrapping stress and deformation of the microlens. Similar to the microlens structure in Dong *et al.* focus of microlens unit in this array could also be tuned by temperature variation. Possessing both focus tunability and large FOV, this hydrogel-based microlens array exhibits great potential in artificial eye development.

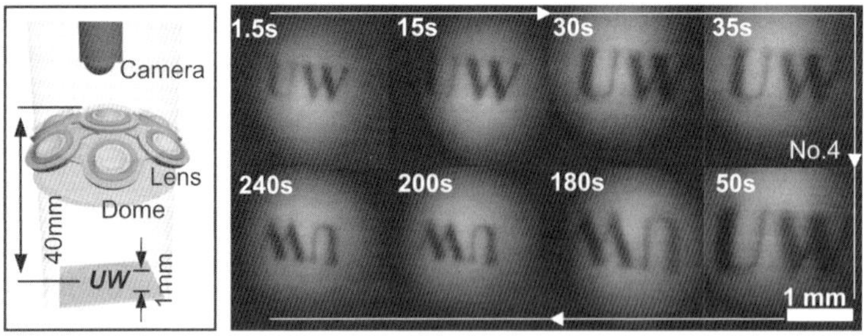

(A) (B)

Fig. 27. (A) 3D schematics of testing the imaging of a microlens array on a glass hemisphere with a diameter of 18 mm. A transparency film with a logo of "UW" (1 mm in size) is placed 40 mm below the microlens array. A CCD-coupled stereoscope records images. (B) Frame sequence of the focused images from one typical microlens in this microlens array in one scan. The microlens is initially convergent, and real images are obtained. Then, the focal length of the microlens gradually increases as the hydrogel actuator expands with thermal dissipation. As a result, the real images are magnified with time. Then, the microlens switches to the divergent status after around 80 s; thus, the images are inverted as virtual images after being refocused. As the temperature further cools down, the negative focal length keeps increasing, resulting in the increase in size of virtual images. The scale bar is 1 mm.[52]

3.2.3.2. Color recognition

Color recognition is also critical in artificial eye imaging system. Hydrogel actuators can also be applied in this field.

Shin *et al.* showed a tunable inverse opal hydrogel colorimetric sensor with fast and reversible color changes. They synthesized a dually tunable inverse opal sensor driven by temperature and pH, using opal-templated photopolymerization of copolymeric hydrogel consisting of HEMA, NIPAAm, AA, and N,N′-Methylenebisacrylamide. Temperature drop or pH increase would induce hydrogel swelling and subsequent color changes in the sensor system, due to critical temperature actuation behavior of NIPAAm and the altered Donnan potential by AA units. The color change driven by hydrogel actuation could be quantitatively recognized by reflectance spectra. The mechanism and results are shown in Figs. 28

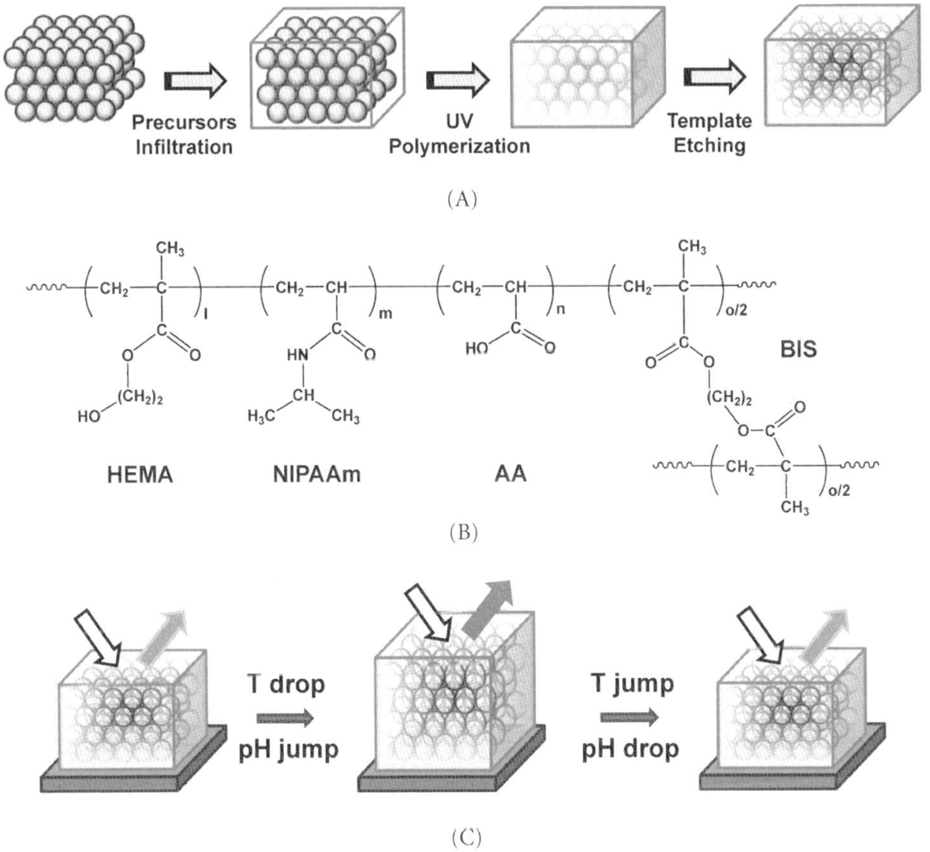

Fig. 28. Schematic illustrations showing a hydrogel colorimetric sensor. (A) Brief procedures for the fabrication of the hydrogel sensor. (B) Chemical structure of poly (HEMA-*co*-NIPAAm-*co*-AA-*co*-BIS) hydrogel. (C) Longitudinal swelling of the dually tunable sensor induced by temperature or pH stimuli, and specific wavelength of visible light is diffracted to exhibit structural color.[37]

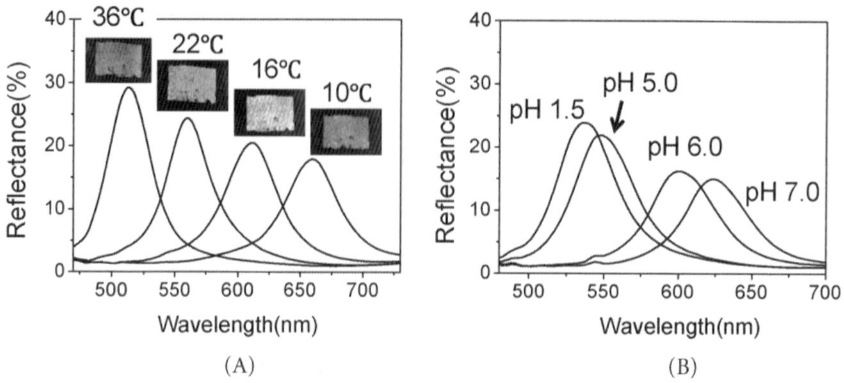

Fig. 29. Reflectance spectra obtained from a dually tunable sensor induced by temperature or pH stimuli. (A) Temperature responses of the sensor at a fixed pH of 1.5, (B) pH responses of the same sensor at an ambient temperature (26°C). The photographs on top of each reflectance spectrum show the corresponding structural colors of the sensor. The size of photograph is 1.5 cm × 2 cm.[37]

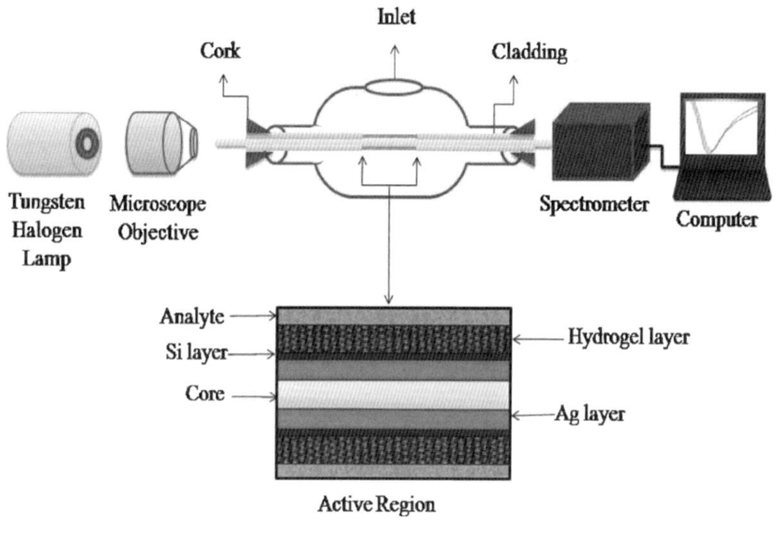

Fig. 30. Schematic diagram of a surface plasmon resonance-based fiber optic sensor.[38]

and 29. This hydrogel colorimetric transducer has a response time of 0.5 s (temperature variation) and 28 s (pH variation).[37]

3.2.3.3. Optical transmission

Transmission of optical signal is another importnat criterion in artifical eye systems. Generally, fiber structures are responsible for delivering optical information. Hydrogel can be integrated in the fiber design as well, in order to achieve variable optical transmission chacteristics.

Singh *et al.* reported the fabrication and characterization of a surface plasmon resonance-based fiber optic sensor to measure the pH of an aqueous fluid as shown in Fig. 30. The sensing probe was fabricated by coating an uncladded core of an optical fiber with three consecutive layers of silver, silicon, and pH-sensitive hydrogel. Variation in pH of the fluid near the probe caused the swelling/shrinking of the hydrogel layer, resulting in the change in its refractive index. This change in turn caused a shift in the resonance wavelength in the transmitted spectrum. Singh *et al.* established the relationship between the pH variation and the spectrum by integrating the hydrogel layer into a fiber structure. In other words, by controlling the pH value, output spectrum could be tuned. This work shows a promising method in optical transmission design in artificial eye design.[38]

3.3. *Advanced medical instrument*

Recently, owing to technical improvements such as the invention of wireless capsule endoscopy and robotic surgery, minimally invasive surgery (Fuchs, 2012) has drawn considerable attention. Hydrogel, whose capability has been proved in micro-optical components, holds promise in the minimally invasive medical instrument design.

Endoscopy, a key technology in advanced minimally invasive procedures, was intended as a methodology to acquire images of the interior surfaces of cavities in human bodies. Current commercially available rigid or flexible endoscopes enable exploration of the gastrointestinal tract, including the esophagus, stomach, colon, the abdominal cavity, thorax, etc. These endoscopes aid in diagnosing pathology and delivering therapy. In addition, endoscopy helps the understanding of interior organ structures and functions.[9]

Zeng *et al.* reported prototype fiber endoscopes with tunable-focus liquid microlenses integrated at their distal ends and actuated through IR light. Hydrogel actuators were integrated in the microlenses to realize focus tunability. This tunability allows for minimal back-and-forth movements of the scopes themselves and different depths of focus, thus having spatial depth perception in the obtained images. The liquid microlens was formed by a water–oil meniscus pinned at a hydrophobic–hydrophilic boundary at a lens aperture. IR light-responsive hydrogel posts were formed by thermo-responsive NIPAAm hydrogel with Au NPs. The volumetric change in the hydrogel posts regulated the pressure difference across the water–oil interface and thus varied its focal length. The operations of the microlenses were realized through light transmitted via optical fibers, and a separate fiber bundle was responsible for the transmission of images obtained from the microlenses. The response times of the microlenses were on the range of 20 s. The range of focal length of one typical microlens was from 52 to 8 mm. The angle of view of an endoscope was from 77° to 128°. The mechanism, structure and images of this endoscope based on hydrogel are shown in Figs. 31–33, respectively.[48]

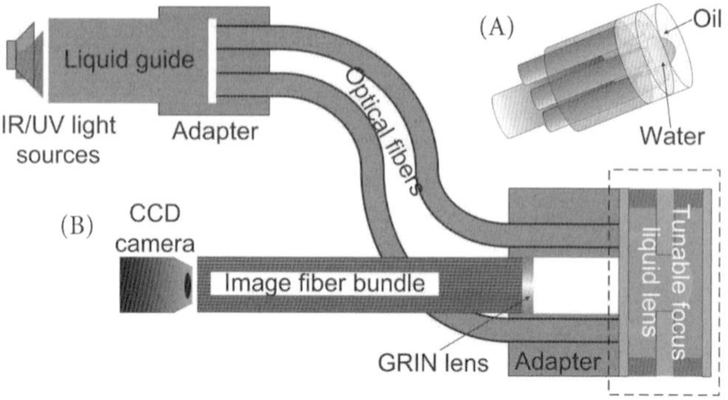

Fig. 31. (A) Schematic and (B) 3D schematic of a prototype endoscope with a liquid tunable-focus microlens integrated at its end and actuated through IR light. Two sets of optical fibers are used: the actuation fibers and the image acquisition fiber bundle. IR light is transmitted via the actuation of optical fibers to tune the focal length of the microlens at the end of the fibers. Images from the microlens are transferred to a camera via the image acquisition fiber bundle.[48]

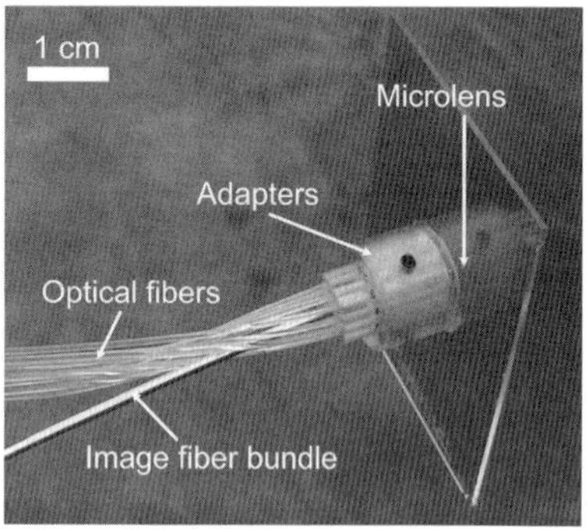

Fig. 32. Photo of a prototype endoscope with a tunable-focus liquid microlens at its end. An image acquisition fiber bundle and 12 optical fibers are bound together and attached to the back side of the microlens by adapters.[48]

Laparoscopy, similar to endoscopy, is another operation performed in the abdomen or pelvis with small incisions. This technology allows doctors to perform both minor and complex surgeries with minimum invasion to the human body. Hydrogel-based micronlens array can also be implemented in the camera system integrated in a laparoscope.

Zhu *et al.* realized an optical imaging system with potential in laparoscope integration.[52] This prototype optical imaging system consists of a hydrogel-based microlens array on a spherical surface, a fiber bundle, a rotational stage, and a CCD

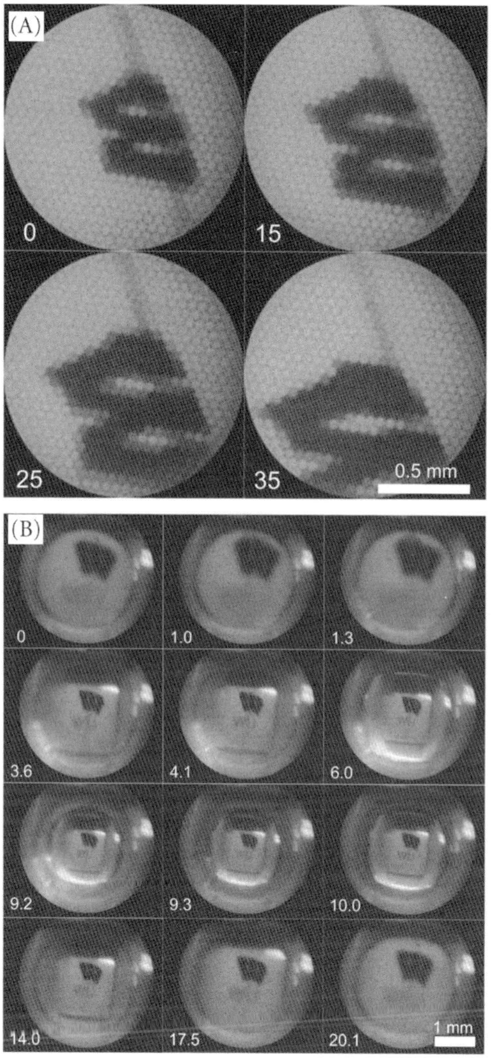

Fig. 33. (A) Frame sequence of focused images obtained from one endoscope in the reverse scanning cycle of the microlens. The microlens is actuated by six hydrogel microstructures formed through optical fibers. Each hexagon pixel stands for one fiber in the image acquisition fiber bundle. At 0 s, when the IR light is switched off, the hydrogels are most contracted. Owing to the dissipation of heat, the hydrogels begin to expand to the original swollen status. The focal length of the microlens increases, and thus the size of the resultant image increases. (B) Frame sequence of the focused images obtained from a second endoscope in one scanning cycle of the microlens. The microlens is actuated by 42 hydrogel microstructures. From 1.3 to 9.3 s, the IR light is on, and the microlens scans the area in a forward cycle. The logo "W" is focused at 3.6 s. The logo "UW" is focused at 6.0 s. After 9.3 s, the IR light is off, and the microlens begins to scan reversely. At 10.0 and 14.0 s, the microlens subsequently scans the two images plates again. Finally, at 20.1 s, the microlenses complete one scanning cycle.[48]

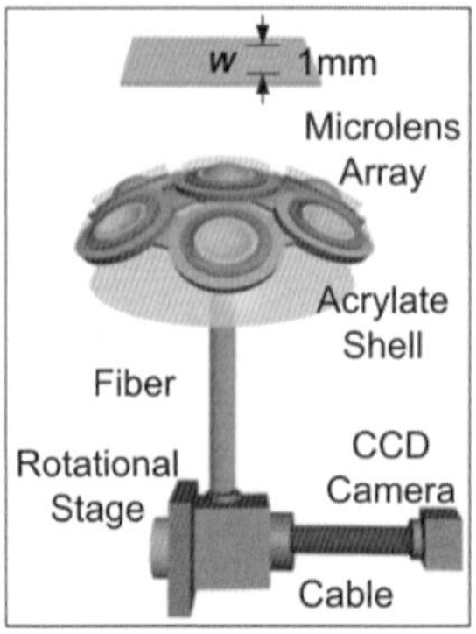

(A)

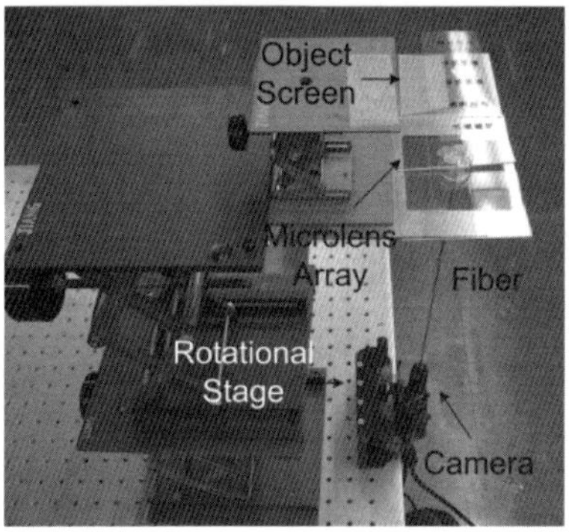

(B)

Fig. 34. Schematic of an optical imaging system demonstrated based on the curvilinear focus-tunable microlens array. This system consists of a six-element microlens array on a hemisphericial acrylate shell with a diameter of 20 mm, an optical fiber bundle, a rotational stage and a CCD camera that connected with a computer. A transparency film with a logo of "W" (1 mm in size) was placed around 20 mm above the microlens array. (B) Photo of this prototype system.[52]

camera. Figure 34 illustrates the schematic of this detection system. A six-element microlens array with focus-tunability was fabricated on a transparenthemispherical shell with a diameter of 20 mm. An optical fiber bundle, connected with a CCD camera, was inserted into the shell to capture images obtained from each microlens. The fiber bundle was attached to a 3D rotational stage, which allowed the fiber bundle to move forward or backward and rotate. As a result, it could receive optical information from different microlenses in this array and record images while the focal length of the target microlens was thermally tuned. An example of frame sequence from the recorded images of one microlens is shown in Fig. 35. The magnification of the focused images indicates the variance in focal length of the microlens.[52]

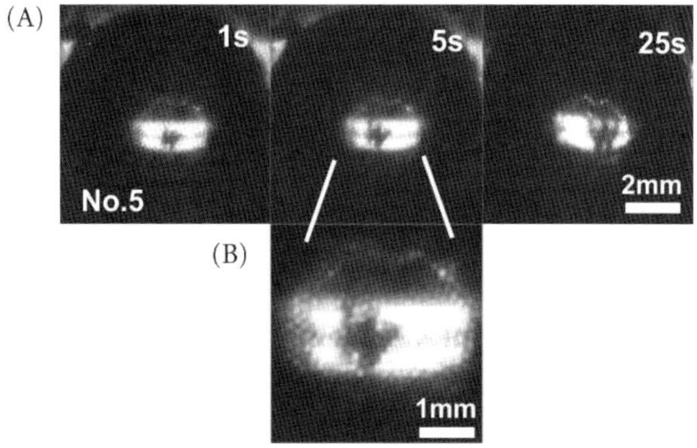

Fig. 35. Frame sequences of focused images captured by simple optical system in Fig. 34. The images are obtained from one microlens: in this microlens array in one scan. Owing to thermal dissipation, the focal length of the microlens gradually increases, and thus, the real images are magnified with time. The scale bar is 2 mm. (B) Magnified photo from one lens at 5 s. The image quality is limited by the relatively low resolution of the optical fiber bundle. The scale bar is 1 mm.[52]

3.4. *Biohybrid actuators*

Cell-based biohybrid actuators are integrated systems that employ biological elements to power synthetic structures. The biological and synthetic components have a dependent relationship: biological information passes through in one or both directions between them. This direct interaction forms a "biohybrid" system. For example, cells are able to sense the mechanical properties of the substrate where they grow and respond through biological functions such as migration, differentiation, or proliferation.[11] Dependent on cell structure and this direct interaction, biohybrid actuators can be autonomous, or controlled chemically or electrically.

Cells can generate power by converting chemical energy source (e.g., glucose) to mechanical energy. Cell-based biohybrid systems could be used to promote the design of more effective and intelligent bio-machines, help us better understand the cellular systems, and realize applications in drug discovery, energy harvesting system, etc.[6]

Significant progress in developing cell-based biohybrid actuators have recently been demonstrated.[20] Employing hydrogel structures to realize the cell energy harvesting is one of them,[6] in which the good elasticity and compatibility hydrogel prove critical.

Chan *et al.* proposed cantilevers based on poly(ethylene glycol) (PEG) hydrogel. PEG is a synthetic polymer that has tissue-like elasticity, which can be fine-tuned by changing its molecular weight or percent composition. PEG-based hydrogels are hydrophilic and can operate by activating with Sulfo-SANPAH

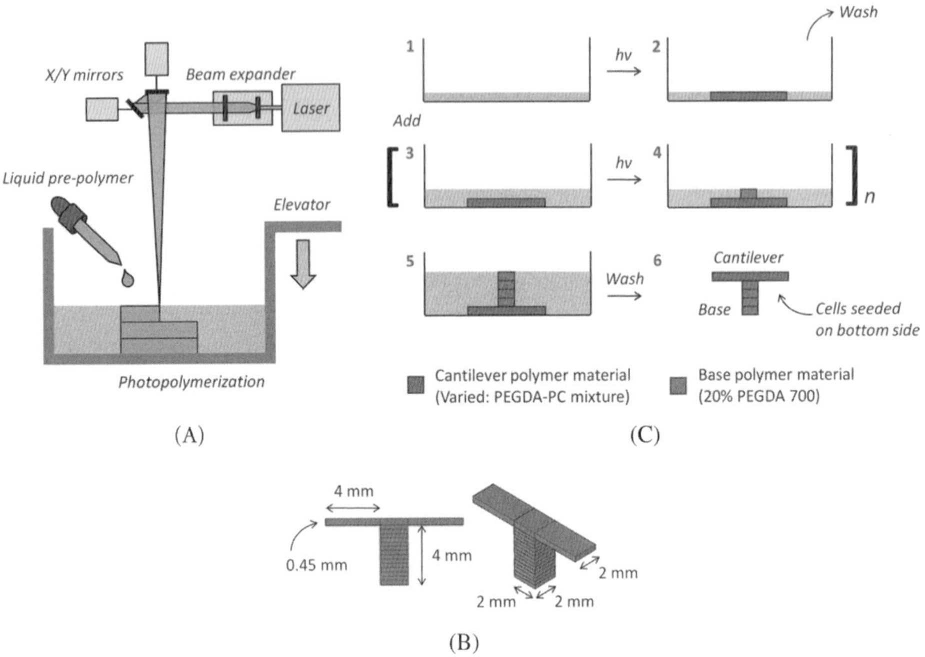

Fig. 36. Fabrication of a multi-material cantilever. (A) The cantilevers were fabricated with a 3D stereolithographic printer, which uses a UV laser to construct layer-by-layer patterns. (B) Two separate cantilevers were built on opposite ends of one base. The molecular weight of the PEG diacrylate acrylic-PEG-collagen (PEGDA–PC) cantilever beam was varied using either PEGDA-PC 700 or 3400, while the base was kept constant using PEGDA-PC 700. (C) A simplified fabrication process flow is shown, which begins with the formation of the cantilever beam before the base.[6]

crosslinker and conjugating with full-length extracellular matrix molecules. As backbone structure, PEG-based hydrogel can react to the actuation from the cells. The mechanism, experimental setup and results of this cell driven hydrogel cantilever are demonstrated in Figs. 36 and 37.[6]

3.5. *Scaffolds for tissue engineering*

In the past two decades, tissue engineering has been successful in handling thin or avascular tissues, such as skin, bladder, and cartilage.[25] As a template for cell growth and tissue formation, a porous scaffold plays important role for tissue engineering. A variety of synthetic polymers [e.g., poly(L-lactide) or PLLA and poly(lactic-*co*-glycolic acid) or PLGA], and natural polymers (e.g., collagen, alginate, agarose) have been tested to fabricate the porous scaffolds. However, due to their limited biocompatibility and functionality, limited available cell density and active controllability of the cell growth were observed.[25]

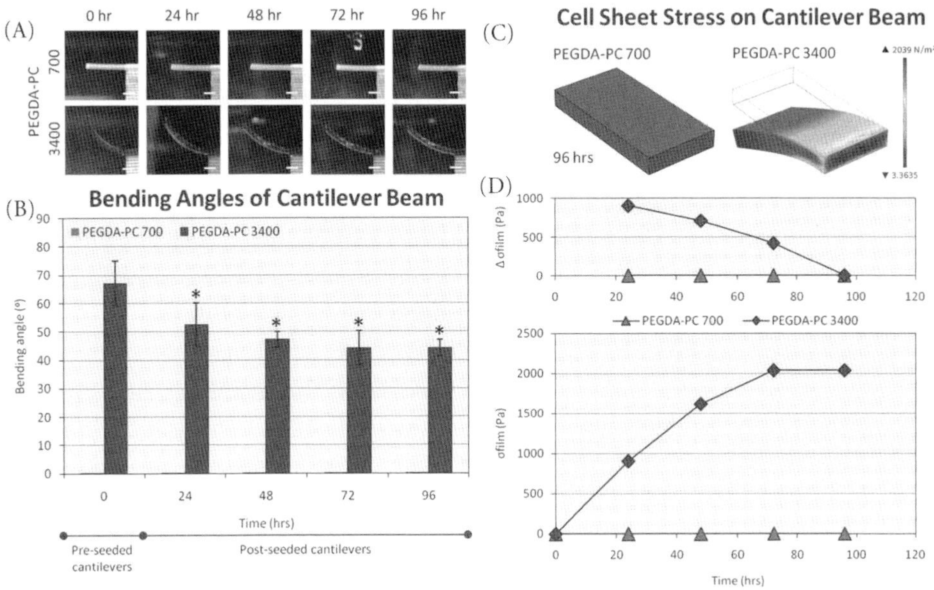

Fig. 37. Cell sheet stress calculations using the cantilever beams. Cells from the ventricles of neonatal rat hearts were seeded on the backside of the cantilever beams. (A) The traction forces of these cells, which are responsible for migration, proliferation, and differentiation, caused the PEGDA-PC 3400 cantilever beams to deflect downward in the Z-direction over time. (B) The average bending angles of the cantilevers were measured over a 96 h period, which was used to calculate the deflection at the tip of the beams. (C) These deflections were simulated using finite element analysis to calculate the stresses exerted by the cell sheets due to traction forces. (D) These stresses exerted by the cell sheet and modeled as a thin film were plotted over time. The change in stress over 24 h time points decreased and reached 0 by 96 h. Scale bars are 1 mm.[6]

Magnetic-hydrogel exhibits the potential to realize scaffolds with high performance. First, magnetic scaffolds are able to stabilize biological agents bound to MNPs through an external MF, and then provide a nutritious environment for initial cell seeding and further cell proliferation. Secondly, magnetic scaffolds can respond to physical cues like mechanical stimulus through interactions between nanoparticles and an AMF, offering controllability on cell biological behavior.[25]

Skaat *et al.* fabricated a series of magnetic fibrin hydrogel scaffolds with fluorescent rhodamine γ-Fe_2O_3 nanoparticles conjugated by various growth factors, including basic fibroblast growth factor (bFGF), beta nerve growth factor (β-NGF), and glial-cell-derived neurotrophic factor (GDNF).[36] They demonstrated that the bFGF-conjugated particles could significantly improve the growth of nasal olfactory mucosa cells seeded in the magnetic scaffold. This type of magnetic hydrogel should be further investigated as a potential nanocomposite implant for tissue engineering. Sapir *et al.* found that aortic endothelial cells in magnetite-impregnated alginate scaffolds exhibited elevated metabolic activity and organized into early capillary-like structures when exposed to an AMF for a total of 14 days *in vitro.*[34]

In conclusion, hydrogel actuators have been widely applied in artificial biological or biomimetic systems. With their unique functionality and characteristics, hydrogels have brought in a new era in the bio-actuation field: their significant volumetric change under a certain stimulus leads to a host of actuator applications in microfluidics, micro-optics and bionics; their simple fabrication and operation allows for effective actuation without complicated control system; their superior compatibility with bio-environment benefits tissue engineering and biohydrid systems; their variety with different nanoparticles mixed offer diversity in the actuator design.

Today, there are still many unsolved issues with the hydrogels such as their responsivity and tuning behavior. In the future, further investigations on the underlying mechanism of hydrogels and their applications may produce more effective and diversified actuators.

Acknowledgments

The materials are drawn from the works of many other researchers. We thank these colleagues for their tremendous effort in advancing the field.

References

1. A. K. Agarwal, L. Dong, D. J. Beebe, and H. Jiang, Autonomously-triggered microfluidic cooling using thermo-responsive hydrogels. *Lab Chip*, **7**(3), 310–315 (2007).
2. K. M. Al-Aribe, G. K. Knopf, and A. S. Bassi. Photo-electric biotransducer for activating ionic hydrogel microactuators. *Sensor Actuat. B-Chem.*, 176, 1056–1064 (2013) .
3. M. Alger, *Polymer Science Dictionary*, 2nd edition, London: Chapman & Hall (1997).
4. S. Argentiere, G. Giuseppe, M. Mariangela, G. Iriniand and B. Laura (2012). *Smart Microfluidics: The Role of Stimuli-Responsive Polymers in Microfluidic Devices, Advances in Microfluidics*, Ed. Dr. R. T. Kelly, Janefa Trdine, 51000 Rijeka, Croatia: In Tech (2012).
5. D. J. Beebe, J. S. Moore, J. M. Bauer, Q. Yu, R. H. Liu, C. Devadoss, and B.-H. Jo. Functional hydrogel structures for autonomous flow control inside microfluidic channels. *Nature*, **404**(6778), 588–590 (2000).
6. V. Chan, J. H. Jeong, P. Bajaj, M. Collens, T. Saif, H. Kong, and R. Bashir. Multi-material bio-fabrication of hydrogel cantilevers and actuators with stereolithography. *Lab Chip*, **12**(1), 88–98 (2012).
7. G. Chen, Z. Guan, C.-T. Chen, L. Fu, V. Sundaresan, and F. H. Arnold. A glucose-sensing polymer. *Nat. Biotechnol.*, **15**(4), 354–357 (1997).
8. A. Chunder, K. Etcheverry, G. Londe, H. J. Cho, and L. Zhai. Conformal switchable superhydrophobic/hydrophilic surfaces for microscale flow control. *Colloid Surface A*, **333**(1), 187–193 (2009).

9. P. B. Cotton and C. B. Williams, Colonoscopy and flexible sigmoidoscopy. *Practical gastrointestinal endoscopy: The fundamentals,* 6th edition, Oxford (UK): Wiley-Blackwell, pp. 87–175 (2008).
10. W. C. Crone, S. Chindhy, J. C. Springmann, and C. Benjamin, Experiments on hydrogels of varying shape. In *Mechanics of Biological Systems and Materials,* New York: Springer, Vol. 5, pp. 217–222 (2013).
11. D. E. Discher, P. Janmey, and Y.-L. Wang, Tissue cells feel and respond to the stiffness of their substrate. *Science,* **310**(5751), 1139–1143 (2005).
12. L. Dong, A. K. Agarwal, D. J. Beebe, and H. Jiang, Adaptive liquid microlenses activated by stimuli-responsive hydrogels. *Nature,* **442**(7102), 551–554 (2006).
13. L. Dong and H. Jiang, pH-adaptive microlenses using pinned liquid–liquid interfaces actuated by pH-responsive hydrogel. *Appl. Phys. Lett.,* **89**(21) 211120–211120 (2006).
14. D. T. Eddington, and D. J. Beebe, Flow control with hydrogels. *Adv. Drug Deliv. Rev.,* **56**(2), 199–210, (2004).
15. P. J. Flory and J. Rehner Jr., Statistical mechanics of cross-linked polymer networks I, rubberlike elasticity. *J. Chem. Phys.,* **11**, 512(1943).
16. K. H. Fuchs, Minimally invasive surgery. *Endoscopy,* **34**(2), 154–159 (2002).
17. C. J. Gannon, P. Cherukuri, B. I. Yakobson, L. Cognet, J. S. Kanzius, C. Kittrell, R. B. Weisman *et al.*, Carbon nanotube-enhanced thermal destruction of cancer cells in a non-invasive radiofrequency field. *Cancer,* **110**(12), 2654–2665 (2007).
18. F. Heath, P. Haria, and C. Alexander, Varying polymer architecture to deliver drugs. *AAPS J.,* **9**(2), E235–E240 (2007).
19. N. W. S. Kam, M. O. Connell, J. A. Wisdom, and H. Dai, Carbon nanotubes as multifunctional biological transporters and near-infrared agents for selective cancer cell destruction. *Proc. Natl. Acad. Sci. USA,* **102**(33), 11600–11605 (2005).
20. J. Kim, J. Park, Y. Sungwook, J. Baek, B. Kim, S. H. Lee, E.-S. Yoon, K. Chun, and S. Park, Establishment of a fabrication method for a long-term actuated hybrid cell robot. *Lab Chip,* **7**(11) 1504–1508 (2007).
21. W. Kuhn, B. Hargitay, A. Katchalsky, and H. Eisenberg, Reversible dilation and contraction by changing the state of ionization of high-polymer acid networks. *Nature,* **165**, 514–516 (1950).
22. M. F. Land and R. D. Fernald, The evolution of eyes. *Annu. Rev. Neurosci.,* **15**(1), 1–29 (1992).
23. L. P. Lee and R. Szema. Inspirations from biological optics for advanced photonic systems. *Science,* **310**(5751), 1148–1150 (2005).
24. Y. Li and T. Tanaka. Phase transitions of gels, *Ann. Rev. Mater. Sci.,* **22**(1), 243–277 (1992).
25. Y. Li, G. Huang, X. Zhang, B. Li, Y. Chen, T. Lu, T. J. Lu, and F. Xu. Magnetic hydrogels and their potential biomedical applications, *Adv. Funct. Mater.* (2012).
26. C.-W. Lo, D. Zhu, and H. Jiang. Microfluidic actuators based on infrared-light responsive PNIPAAm hydrogel nanocomposite incorporating graphene-oxide. In *Solid-State Sensors, Actuators and Microsystems Conference (TRANSDUCERS), 2011 16th International,* IEEE, pp. 2430–2433 (2011).

27. C. W. Lo, D. Zhu, and H. Jiang, An infrared-light responsive graphene-oxide incorporated poly (N-isopropylacrylamide) hydrogel nanocomposite. *Soft Matter*, **7**(12), 5604–5609 (2011).
28. P. Manandhar, P. D. Calvert, and J. R. Buck, Elastomeric ionic hydrogel sensor for large strains, *Sensors Journal, IEEE*, **12**(6), 2052–2061 (2012).
29. L. Mao, Y. Hu, Y. Piao, X. Chen, W. Xian, and D. Piao, Structure and character of artificial muscle model constructed from fibrous hydrogel. *Curr. Appl. Phys.*, **5**(5), 426–428 (2005).
30. T. Miyata, N. Asami, and T. Uragami, A reversibly antigen-responsive hydrogel. *Nature*, **399**(6738), 766–769 (1999).
31. Y. Murase, S. Maeda, S. Hashimoto, and R. Yoshida, Design of a mass transport surface utilizing peristaltic motion of a self-oscillating gel. *Langmuir*, **25**(1), 483–489 (2008).
32. A. Richter, G. Paschew, S. Klatt, J. Lienig, K.-F. Arndt, and H.-J. P. Adler, Review on hydrogel-based pH sensors and microsensors. *Sensors*, **8**(1), 561–581 (2008).
33. A. Richter, S. Klatt, G. Paschew, and C. Klenke, Micropumps operated by swelling and shrinking of temperature-sensitive hydrogels. *Lab Chip*, **9**(4), 613–618 (2009).
34. Y. Sapir, S. Cohen, G. Friedman, and B. Polyak, The promotion of *in vitro* vessel-like organization of endothelial cells in magnetically responsive alginate scaffolds. *Biomaterials* **33**(16), 4100–4109 (2012).
35. H. G. Schild, Poly(N-isopropylacrylamide): Experiment, theory and application. *Prog. Polym. Sci.*, **17**(2), 163–249 (1992).
36. H. Skaat, O. Ziv-Polat, A. Shahar, D. Last, Y. Mardor, and S. Margel, Magnetic scaffolds enriched with bioactive nanoparticles for tissue engineering. *Adv. Healthcare Mater.*, **1**(2), 168–171 (2012).
37. J. Shin, S. G. Han, and W. Lee, Dually tunable inverse opal hydrogel colorimetric sensor with fast & reversible color changes. *Sensor Actuat. B: Chem.* **168**, 20–26 (2012).
38. S. Singh and B. D. Gupta, Fabrication and characterization of a highly sensitive surface plasmon resonance-based fiber optic pH sensor utilizing high index layer and smart hydrogel. *Sensor Actuat. B: Chem.* **173**, 268–273 (2012).
39. S. Stankovich, D. A. Dikin, G. H. B. Dommett, K. M. Kohlhaas, E. J. Zimney, E. A. Stach, R. D. Piner, S. T. Nguyen, and R. S. Ruoff, Graphene-based composite materials. *Nature*, **442**(7100) 282–286 (2006).
40. T. Tanaka and D. J. Fillmore, Kinetics of swelling of gels. *J. Chem. Phys.*, **70**, 1214 (1979).
41. P. Techawanitchai, M. Ebara, N. Idota, T.-A. Asoh, A. Kikuchi, and T. Aoyagi, Photo-switchable control of pH-responsive actuators via pH jump reaction. *Soft Matter*, **8**(10), 2844–2851 (2012).
42. H. van der Linden, H. Sebastiaan, O. Wouter and B. Piet, Stimulus-sensitive hydrogels and their applications in chemical (micro) analysis. *Analyst*, **128**(4), 325–331 (2003).
43. J. Wang, Z. Chen, M. Mauk, K. -S. Hong, M. Li, S. Yang, and H. H. Bau, Self-actuated, thermo-responsive hydrogel valves for lab on a chip. *Biomed. Microdevices*, **7**(4), 313–322 (2005).

44. Q. Yang, N. Adrus, F. Tomicki, and M. Ulbricht, Composites of functional polymeric hydrogels and porous membranes. *J. Mater. Chem.*, **21**(9), 2783–2811 (2011).
45. J. Z. Hilt, Hydrogel nanocomposites: A review of applications as remote controlled biomaterials. *Soft Matter*, **6**(11), 2364–2371 (2010).
46. N. Zalachas, S. Cai, Z. Suo, and Y. Lapusta, Crease in a ring of a pH-sensitive hydrogel swelling under constraint. *Int. J. Solid. Struct.*, **50**(6), 920–927, (2013).
47. X. Zeng and H. Jiang, Tunable liquid microlens actuated by infrared light-responsive hydrogel. *Appl. Phys. Lett.*, **93**(15), 151101–151101 (2008).
48. X. Zeng, C. T. Smith, J. C. Gould, C. P. Heise, and H. Jiang, Fiber endoscopes utilizing liquid tunable-focus microlenses actuated through infrared light. *J. Microelectromech. S.*, **20**(3), 583–593 (2011).
49. Y. Zhou, N. Sharma, P. Deshmukh, R. K. Lakhman, M. Jain, and R. M. Kasi, Hierarchically structured free-standing hydrogels with liquid crystalline domains and magnetic nanoparticles as dual physical cross-linkers. *J. Am. Chem. Soc.*, **134**(3), 1630–1641 (2012).
50. D. Zhu, C. Li, X. Zeng, and H. Jiang, Hydrogel-actuated tunable-focus microlens arrays mimicking compound eyes. In *Solid-State Sensors, Actuators and Microsystems Conference. TRANSDUCERS* 2000, *International*, pp. 2302–2305, (2009), IEEE.
51. D. Zhu, C. Li, X. Zeng, and H. Jiang, Tunable-focus microlens arrays on curved surfaces. *Appl. Phys. Lett.*, **96**(8), 081111 (2010).
52. D. Zhu, X. Zeng, C. Li, and H. Jiang, Focus-tunable microlens arrays fabricated on spherical surfaces. *J. Microelectromech. S.*, **20**(2), 389–395 (2011).
53. D. Zhu, C.-W. Lo, C. Li, and H. Jiang, Hydrogel-based tunable-focus liquid microlens array with fast response time. *J. Microelectromech. S, Journal of* **21**(5), 1146–1155 (2012) 1–10 (2012).

Chapter 8

Advances in Smart Hydrogels for Biosensing Applications

Wei Wang, Jimmy Amin, and Zhiqiang Cao*
Chemical Engineering and Materials Science
Wayne State University, Detroit, MI 48202, USA
**zcao@wayne.edu*

Smart hydrogels are widely used for biosensing applications due to their stimuli-responsive, analyte diffusive and potentially biocompatible features. In this chapter, recent advances in smart biosensing hydrogels will be discussed, with a focus on three categories of classification: traditional smart hydrogels, conductive hydrogels, and new smart materials. Unique properties of state-of-the-art hydrogels, that are critical in enabling robust, user compliant, and highly sensitive biosensors for implantable applications, will also be highlighted.

1. Introduction

Smart hydrogels are able to alter their properties in response to environmental changes (i.e., variations in temperature, pH, ionic strength, light, glucose, antigens, redox, etc.);[1–8] this stimuli-responsive nature makes them promising materials for biosensing applications. A typical biosensor is composed of a biological recognition device, a transducer to transform the recognition signal to another easily measured one, and a reader device to interpret the results for an end user. A smart hydrogel usually acts as a recognition device and is typically located at an environment/sensor interface (e.g., the tissue/sensor or biological fluid/sensor interface of an implantable sensor). A hydrogel, a three-dimensional (3D) matrix holding a lot of water, is considered to be potentially biocompatible due to its resemblance to tissue.[9] This feature makes a hydrogel less likely to be rejected by the body and maintains the regular working condition of a sensor. In addition, the porous structure of the hydrogel readily allows diffusion of analytes (mostly water soluble) and charge neutralizing ions; this further strengthens the ability of the hydrogel device to recognize biological signals. These two important features, along with smart or stimuli-responsive properties, provide a rationale for the wide use of hydrogels in biosensing applications.

In this chapter, recent advances of smart hydrogels for biosensors will be discussed with a focus on three categories of smart hydrogels: (i) Traditional smart hydrogels, (ii) Conductive hydrogels, and (iii) New smart materials.

(i) **Traditional smart hydrogels** — These hydrogels alter their dimensions (i.e., through swelling or shrinking) in response to small changes in environmental conditions such as temperature, pH, light, ionic strength, salt type, solvent, etc. Accordingly, these hydrogels can be used as component materials in constructing sensors.

(ii) **Conductive hydrogels** — These are a unique class of materials that synergize the advantageous features of hydrogels and organic conductors. They have been used in many high-performance electrochemical devices such as implantable biosensors. Several notable examples are given due to their immediate relevance to common electro-chemical-based biosensing applications, such as glucose detection.

(iii) **New smart materials** — Antifouling properties and biocompatibility are highly desirable for a sensor to reduce its background signal/noise (i.e., by resisting non-specific binding with an analyte) and to enable its implantation into the body (i.e., by resisting foreign body reactions, and consequent fibrotic tissue formation). Functionalization capability is also needed in order to immobilize biosensing elements on the sensors (e.g., biomolecules that specifically bind/recognize a targeted analyte). A recent advance shows new smart materials having switchable antifouling, biocompatible, and functionalization properties. The new material is ideal for constructing biosensors.

2. Recent Advances of Smart Hydrogels

2.1. *Traditional stimuli-responsive smart hydrogels*

Typical smart hydrogels are those that respond to heat and pH (such as hydrogels made from poly(N-isopropylacrylamide) (PNIPAAm) and poly(acrylic acid), respectively), and change their dimensions (i.e., either swell or shrink) based on a hydrophobic change in polymer matrix chains.[6,7,10,11] Other stimuli include light, glucose, antigens, redox, etc. and these stimuli can be further converted to a change in heat/pH, or directly break up/join together a molecular network to affect hydrogel properties.[6,12–14] These traditional smart hydrogels (materials) have been well covered by several review papers.[1,2,4–8] Under this section, a few examples of recent advances are given with notable properties highlighted.

One significant advance in thermo-responsive hydrogels was reported by Chu *et al*,[15] where they described a simple method to synthesize smart hydrogels that are rapidly responsive, highly swellable (e.g., 8,000% swelling ratio) and stretchable (e.g., 1,800% of initial length) (Fig. 1). The basic material used to make the hydrogels was still PNIPAAm, but the researchers created a nano-structured architecture with controllable activated nanogels (ANGs) as nano-crosslinkers. The

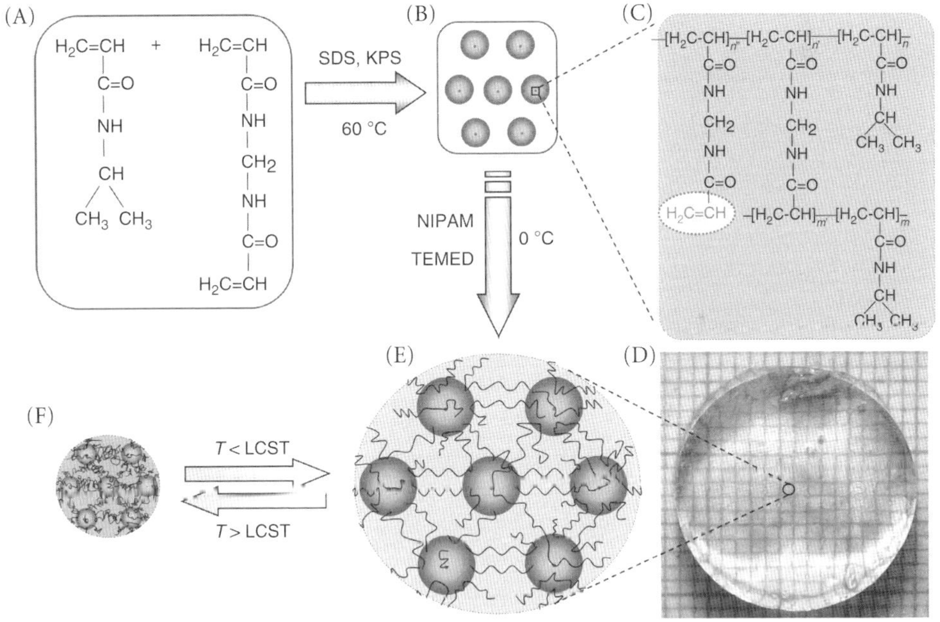

Fig. 1. (A) Chemical structures for NIPAAm and MBA as monomers and cross-linkers, respectively. (B) ANGs were prepared through by precipitation polymerization of NIPAAm and MBA, with KPS as the initiator and SDS as the emulsifier. The resulting ANGs have unsaturated double bonds for further reaction. (C) Chemical structure of the ANGs. (D) Optical image of NSG. NSG was obtained from ANGs (as crosslinker) and NIPAAm (monomer) in presence of TEMED (accelerator) at 0°C. (E) NSG was in the swollen state at temperature below the lower critical solution temperature (LCST) and (F) in the shrunken state at temperature above the LCST. Ref. 15.

Source: Copyright 2013 Nature Publishing Group.

ANGs are responsive nanoparticles with hydrodynamic diameter <60 nm, and have double bonds available for further polymerization or gelation. The resulting nano-structured smart hydrogels (NSGs) show extraordinarily large responsive swelling ratio, e.g., a temperature change from 45°C to 15°C led to a swelling ratio as high as 8,000%. It took only 6 min for the NSGs to shrink from fully swollen state (at 25°C) to fully collapsed state (at 55°C). The NSGs are highly elastic and able to withstand high compressions, slicing, and high levels of deformation (e.g., bending, twisting, knotting, and extensive stretching). They can be stretched beyond 1,800% of their original length at room temperature.

Importantly, the large responsive swelling ratio, rapid response rate, and high elasticity of the NSGs can be integrated simultaneously, which is rarely found within the category of smart hydrogels. The mechanism for obtaining both stimuli-responsive and elastic properties comes from the unique design of responsive polymer chains crosslinked by responsive nanogels. A change in polymerization time, monomer or crosslinker (i.e., nanogel) content will finely tune responsive and elastic properties. Being able to combine rapid response rate, extremely large responsive swelling ratio, high elasticity, and ease of fabrication, the NSGs are highly promising for potential sensing applications (e.g., as coatings or component materials in constructing sensors). The existing practice of using PNIPAAm hydrogels on cells or tissues, along with their demonstrated cytocompatiblity, makes NSGs particularly desirable for bio-related sensing applications.

For recent advances in pH-sensitive hydrogels, Liu *et al.* reported a facile approach to fabricating a pH-sensitive high-strength hydrogel containing two types of hydrogen bonding motifs: 4,6-diaminotriazine (DAT) and phenylboronic acid (PBA) (Fig. 2).[16] Specifically, two hydrogen bonding motif-containing monomers, 3-acrylamidophenyl-boronic acid and 2-vinyl-4,6-diamino-1,3,5-triazine, a co-monomer of oligo(ethylene glycol) methacrylate, along with a crosslinker *N,N*-methylenebisacrylamide were co-polymerized to form the hydrogel. Since DAT forms hydrogen bonds at neutral and basic pH, and PBA at neutral and acidic pH, the hydrogel can maintain at least one type of hydrogen bonding (i.e., crosslinking) over a broad pH range. Note that both DAT and PBA based intermolecular hydrogen bonding can significantly enhance the tensile and compressive strengths, and the resulting hydrogels show high mechanical strength over a wide range of pH. With an optimized monomer/crosslinker ratio, the tensile and compressive strength reached 0.8 and 23 MPa, respectively, at neutral pH (two types of hydrogen bonding maintained), and 0.33 and 12 MPa, respectively, at acidic or basic pH. Two types of hydrogen bonding are maintained in a neutral environment but one type of hydrogen bonding is available in an acidic or basic environment. The swelling behavior is thus influenced by the environmental pH, with a higher swelling ratio of the hydrogel at acidic or basic pH. The ability to maintain both high tensile

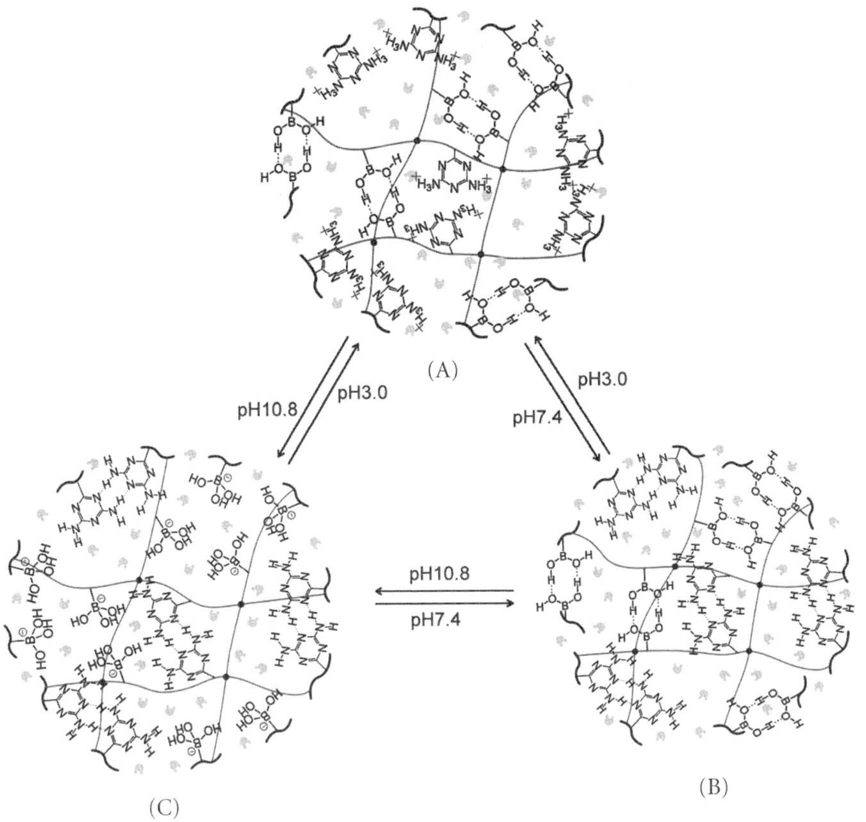

Fig. 2. Illustration of a pH-responsive hydrogel. Both DAT and PBA hydrogen bonding exist at neutral pH. In acidic conditions, only PBA hydrogen bonding is maintained. In basic conditions, only DAT hydrogen bonding is available. Ref. 16.

Source: Copyright 2013 WILEY-VCH Verlag GmbH & Co. KGaA, Weinheim.

and compressive strengths within a wide pH range and respond to pH changes makes hydrogels useful for biosensor systems, especially in the waste water treatment industry, where the pH variation is frequently wide.

As to photosensitive smart hydrogels, Chen and Yu *et al.* reported a straightforward preparation of near-infrared (NIR) sensitive PNIPAAm/graphene oxide (GO) nanocomposite hydrogels (Fig. 3).[17] Typically, a light sensitive hydrogel involves a chromophore which absorbs light and dissipates it locally as heat.[6] A thermo-sensitive hydrogel such as PNIPAAm will respond to local heat and further convert the light signal to a swelling behavioral change.[6] Unlike visible light chromophores, GO has high optical absorbance in the NIR region and shows effective photothermal heating in *in vivo* conditions.[18] It should be noted that other materials, such as gold nanorods and carbon nanotubes, also demonstrate high NIR adsorption, but GO, made from graphite, is more abundant and lower in cost. The reported NIR responsiveness is an advantageous feature in a hydrogel used for biological sensing applications. Note that NIR light penetrates through skin and soft tissues up to four inches.[19]

The reported NIR responsive hydrogel was synthesized from NIPAAm monomer and GO in an aqueous solution with γ-irradiation to initiate the polymerization. The phase transition of the resulting hydrogel can be remotely manipulated through laser exposure. The hydrogel temperature induced by a NIR laser exposure was controlled by the GO content and irradiation time. As shown in Fig. 3B, the temperature of a higher GO loaded hydrogel increases most rapidly. To reach a certain gel temperature, the irradiation time can also be finely tuned. Since the phase transition (i.e., swelling/deswelling transition) is directly driven by thermo-responsive PNIPAAm (indirectly by NIR), it is completely reversible as shown in Fig. 3C.

Recently Harada *et al.* reported a redox-responsive hydrogel with self-healing properties (Fig. 4).[13] The hydrogel contains two polymer components: a poly(acrylic

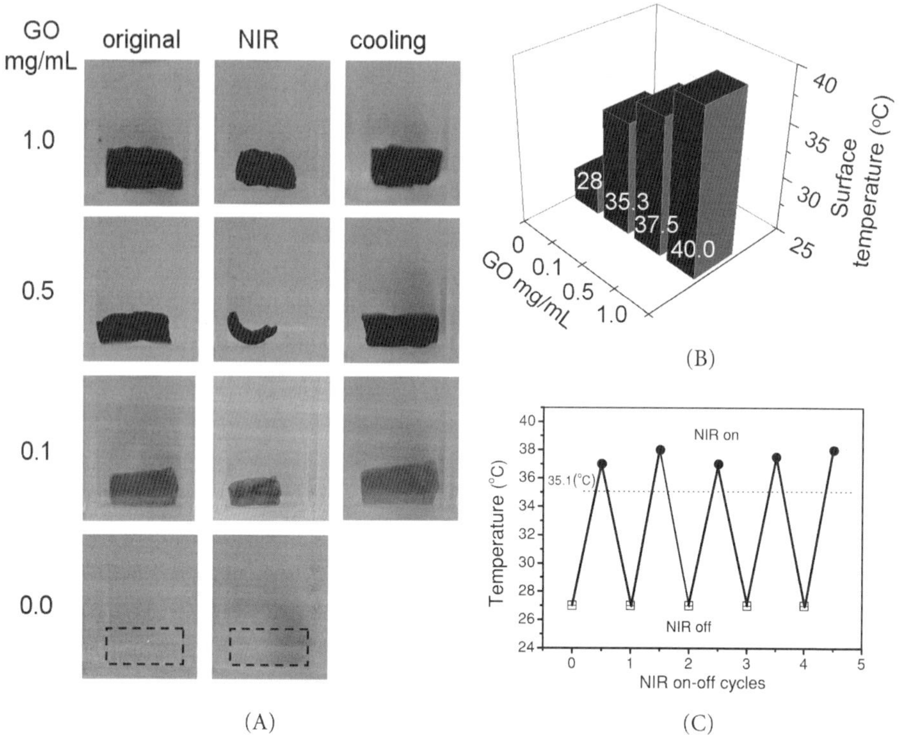

Fig. 3. (A) Photographs of the NIR responsive hydrogels prepared with varied GO content (0.0/0.1/0.5/1.0 mg·mL^{-1}). The original hydrogels were shown in the left column. Hydrogels after 2 min NIR laser irradiation were shown in the middle column. Hydrogels after cooling were shown in the right column. (B) Temperatures of NIR responsive hydrogels after 2 min NIR laserirradiation. GO content in the hydrogels varied among 0.0/0.1/0.5/1.0 mg·mL^{-1}. (C) Transition temperatures of NIR responsive hydrogels after each of the heating-cooling cycle. Ref. 17.

Source: Copyright 2012 WILEY-VCH Verlag GmbH & Co. KGaA, Weinheim.

acid) (pAA) possessing β-cyclodextrin (CD) as the host (pAA– βCD), and a pAA possessing ferrocene as a guest (pAA–Fc). Once the two polymers were mixed, a transparent supermolecular hydrogel quickly formed due to host-guest interactions. When the ferrocene was oxidized, host-guest interaction was not preferred, and the hydrogel turned to a sol phase. Figure 4 shows the redox-responsive reversible sol-gel phase switching through either chemical oxidation or electrochemical reactions. Sodium hypochlorite (NaClO) as a chemical oxidant induced the hydrogel phase transition to a sol state. Glutathione (GSH) as a reductant turned the sol back to the gel and recovered the elasticity. Electrochemical oxidation (+1.0 V Versus Ag/AgCl) also turned the gel into a sol, while reduction by heating at 50°C reversed this process and recovered the elasticity.

Based on the host-guest interactions, the redox responsive hydrogel also has unique self-healing properties such as re-adhesion between cut surfaces. As shown

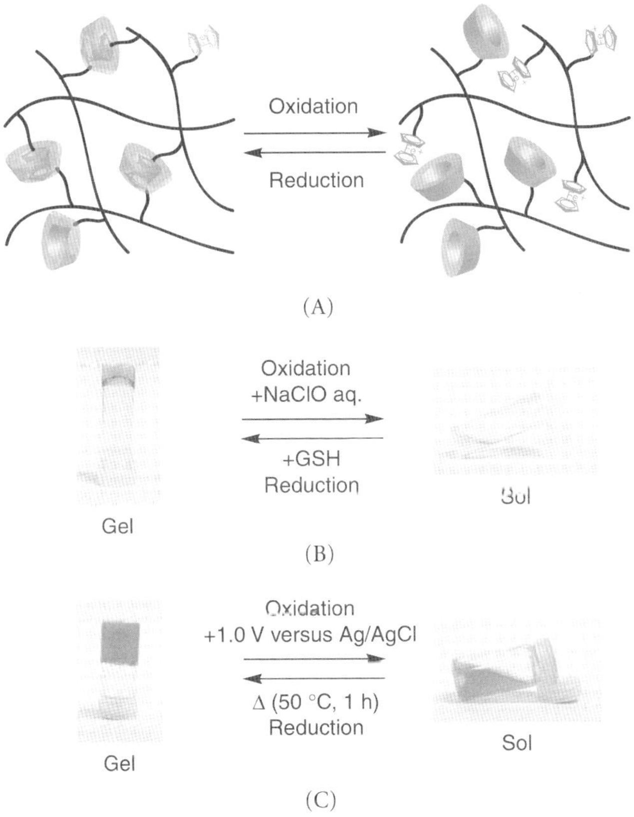

Fig. 4. Redox-responsive sol-gel transition of the pAA–βCD/pAA–Fc hydrogel. (A) Sol–gel transition based on host (βCD)–guest (Fc) interactions. Fc at a reduced state favors host–guest interactions, while its oxidized state disfavors such interactions. (B) Chemical redox agents induced sol–gel transition. (C) Electrochemical oxidation and reductive heating induced sol–gel transition. Ref. 13. *Source*: Copyright 2011 Nature Publishing Group.

in Fig. 5, a pAA–βCD/pAA–Fc hydrogel (2 wt.%) was cut into halves. When reductant (20 μL of 20 mM GSH aqueous solution) was applied on the cut surface, and after two pieces were reattached for 24 h, self-healing was achieved. The two pieces re-adhered into one piece. An oxidant such as NaClO did not provide the right conditions for self-healing due to the disfavored βCD–Fc (oxidative state) interactions. With redox-induced dimensional change (or sol–gel transition) properties, the smart hydrogel can potentially be used to sense the oxidative/reductive levels of an environment. The unique self-healing properties may provide opportunities to construct novel sensor systems for biological applications. Note that the intracellular environment and most tumor sites are reductive.

2.2. *Conductive hydrogels*

Electrochemical biosensors are one of the most popular applications of biosensing technology. They are useful for detecting a broad range of substances such as: glucose, peroxide, cholesterol, deoxyribonucleic acid (DNA), or immune-analytes.[20,21] In a typical electrochemical sensor, electrons will be transferred from the biochemical reaction site (when "sensing" reactions or interactions happen) through a conductive network, and picked up/measured by an electrode.[20] Conductive hydrogels play an important role in constructing such conductive networks, and thus are addressed

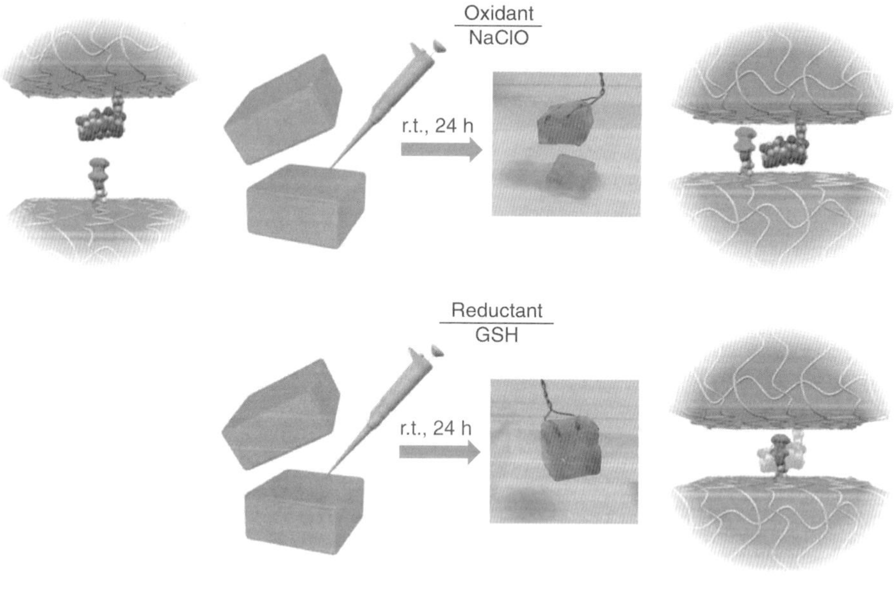

Fig. 5. Self-healing of a redox responsive pAA–βCD/pAA–Fc hydrogel. The self-healing happened when reductants, but not oxidants, were applied on the cut surface. Ref. 13.

Source: Copyright 2011 Nature Publishing Group.

under this section, separately from the previous discussion of traditional smart hydrogels. Conductive hydrogels are a unique class of materials showing advantages of both hydrogels and organic conductors. They are easy to synthesize and process, with electrical and optical properties matching those of metals and inorganic semiconductors. The most important feature of a conductive hydrogel is the ability to facilitate rapid electron transfer, which dramatically enhances the performance of electronic sensors. To construct a conductive hydrogel, either a conductive polymer (e.g., polyaniline (PANi), polypyrrole (PPy))[20–23] or conductive element (e.g., carbon nanotubes, graphene oxide)[24,25] could be used as the gel component. Recent advances in super conductive hydrogels and how they enabled highly sensitive glucose or H_2O_2 sensors are given in the following paragraph.

Bao *et al.* reported a synthesis of PANi-based conductive hydrogel with excellent electronic conductivity and electrochemical properties.[26] Unlike previous conductive hydrogels synthesized from a conducting polymer monomer and a non-conducting hydrogel matrix whose electrical properties are compromised,[27,28] the reported hydrogel is free of insulating polymers and achieves a new record conductivity of 0.11 S·cm^{-1}. The trick involves phytic acid as the gelator and dopant, which reacts with PANi by protonating its nitrogen group, and crosslinks PANi to form a 3D porous nanostructure (Fig. 6). Because of the simplicity of the

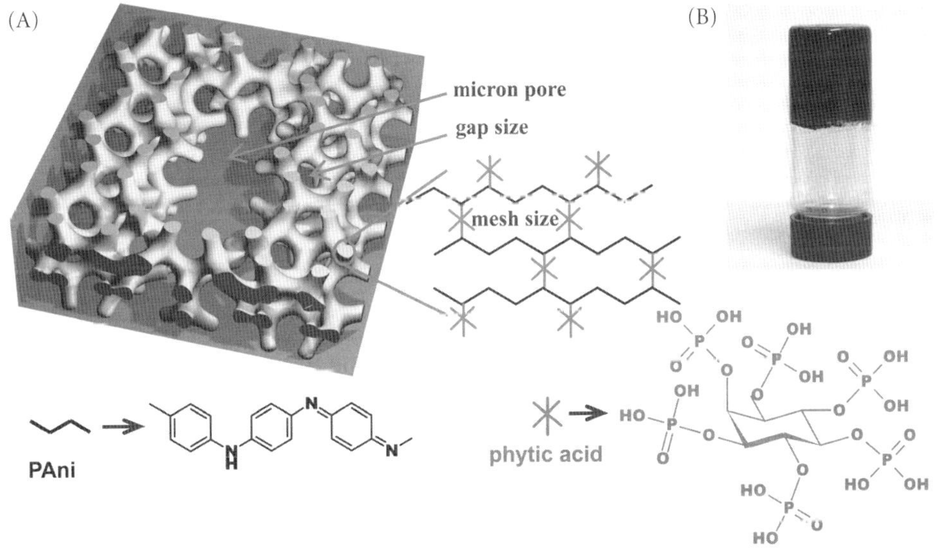

Fig. 6. (A) Chemical structure of phytic acid gelated and doped PANi hydrogel. (B) PANihydrogel inside a glass vial. Ref. 26.

Source: Copyright 2012 National Academy of Sciences, USA.

gelation, the hydrogel is highly processible and can be ink-jet printed or spray coated into a variety of micropatterns.

The high surface area and interconnected super conductive network are attractive features for high-performance chemical and biological sensing. Glucose sensing normally involves glucose oxidase (GO*x*), which oxidizes glucose to gluconic acid; this process can be monitored through electrochemical measurement. When GOx was fabricated with the reported PANi hydrogel along with an electrode, the resulting glucose sensor had a fast response time (~0.3 s), superior sensitivity (~16.7 $\mu A \cdot mM^{-1}$), and high linearity with glucose concentration (up to 2.5 mM).[26] It is worth mentioning that the excessive phosphate groups (on phytic acid) interact with the amine groups on the GOx. Such interactions, along with the hierarchical 3D network structure, promote the GOx immobilization on the hydrogel support. With the stable immobilization of the enzyme, the glucose sensor showed no significant decrease in signal after immersion in phosphate buffered saline (PBS) at 4°C for 7d.

Pan and Yu *et al.* reported an ultra-sensitive glucose sensor based on Pt nanoparticles (PtNP)-PANi hydrogel heterostructured electrodes with GO*x* immobilized (Fig. 7).[29] The conductive hydrogel was made from aniline monomers, crosslinked by phytic acid and initiated by ammonium persulfate. Note that both Bao group and Pan and Yu group used phytic acid as a crosslinker to prepare the super conductive PANi hydrogel. To further improve the sensor performance, monodispersed PtNPs were densely loaded onto the hydrogel matrix. PtNPs acted as highly active catalysts for the electro-oxidation of hydrogen peroxide, a by-product from GO*x* catalyzed reaction, and thus worked with

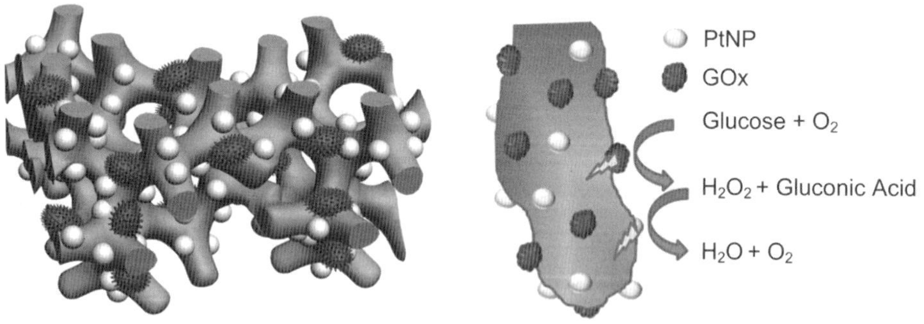

(A) (B)

Fig. 7. (A) 3D heterostructure of the PtNP/PANi hydrogel. PtNPs and GO*x* are well dispersed and immobilized on the hydrogel matrix. (B) Reaction mechanism of the glucose sensor. GO*x* and PtNP co-catalyze the oxidation of the glucose. Reprinted (adapted) with permission from Zhai *et al.* Ref. 29. *Source*: Copyright 2013, American Chemical Society.

GO*x* to efficiently catalyze the oxidation of glucose. Furthermore, the 3D-microstructured hydrogel well dispersed the PtNPs, immobilized the GOx, and provided a conductive 3D network for efficient charge collection. The reported glucose sensor showed unprecedented sensitivity (96.1 $\mu A \cdot mM^{-1} \cdot cm^{-2}$), a response time of 3 s, and a linearity with glucose from 0.01 to 8 mM, with a low detection limit of 0.7 μM.

Besides conducting polymers, conductive elements have also been used to make conductive hydrogels. Elvassore *et al.* developed a GOx loaded polyacrylamide hydrogel modified with conductive single-walled carbon nanotubes (SWNTs) (up to 0.85 wt.%) (Fig. 8).[30] The doping of SWNT is critical to render electroconductivity to a soft hydrogel, and it is believed that the SWNT network allows electric percolation, from an enzymatic reaction site to a collecting electrode, through the biomaterial. The reported glucose sensor had a sensitivity of 0.63 $\mu A\ mM^{-1}$ and a linear glucose range between 0.1 mM and 1.6 mM, with low detection limit at

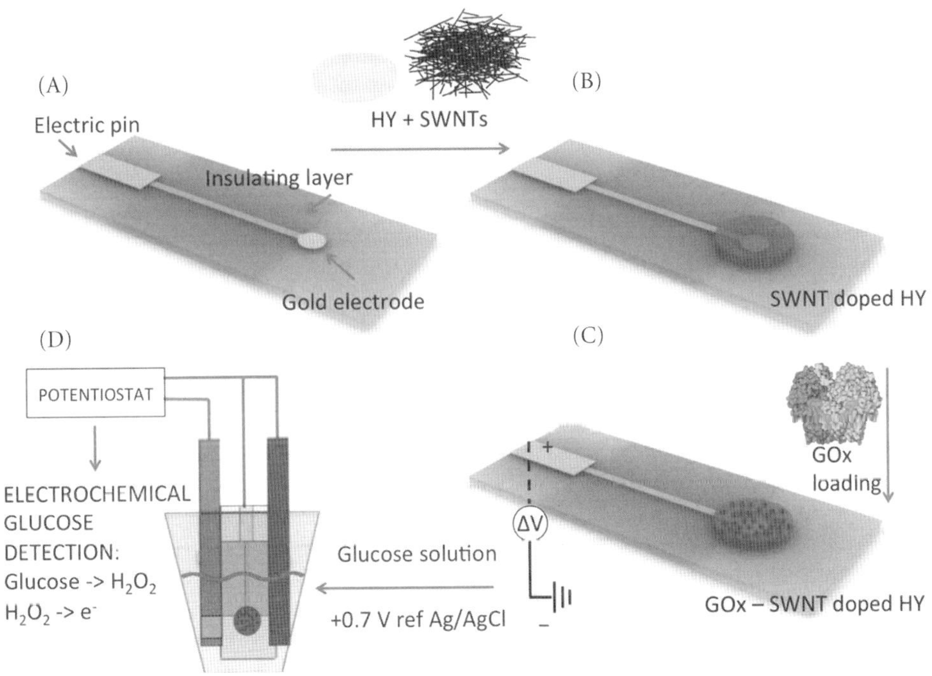

Fig. 8. Preparation of the SWNT modified hydrogel biosensor. A clean, insulated gold electrode (A) is modified with a hydrogel (HY) doped with SWNTs (B). GO*x* is loaded into the HY by applying a positive voltage on the electrode (C). Glucose sensing through the potentiostatic measurement (D) Ref. 30.

Source: Copyright 2013, Royal Society of Chemistry.

15 μM. Because of the low concentration of SWNTs used for the modification, the biophysical properties of the hydrogel did not change.

Li and Wang *et al.* synthesized bovine serum albumin (BSA) crosslinked hydrogel, which adsorbed and reduced GO nanosheets *in situ* during irradiation (Fig. 9).[31] Specifically, BSA was reacted with N-acryloxysuccinimide to render vinyl groups. Such BSA crosslinkers, together with N,N-dimethylacrylamide (DMAA), and GO, formed a hydrogel through photopolymerization with 2,2-diethoxyacetophenone as photo-initiator. GO was adsorbed on BSA through hydrogel bonding and the tyrosine groups of BSA were responsible for the reduction of GO during the UV irradiation. It should be noted that a reduced form of GO (rGO) has better electronic performance than GO. By modifying an electrode with the reported hydrogel, loaded with hemin chloride (electron transport mediator), a H_2O_2 sensor with sensitivity up to 32.90 mA·M^{-1} and detecting limit down to 0.61 μmolL^{-1} was achieved.

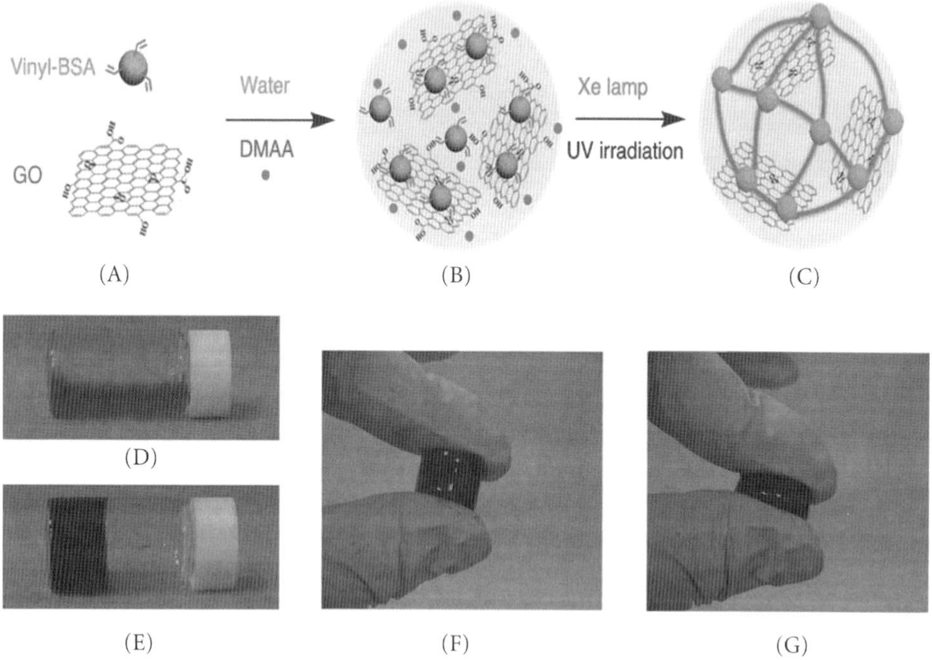

Fig. 9. (A) GO sheet, DMAA monomer, and Vinyl-BSA were mixed in water. (B) Vinyl-BSA was adsorbed onto the surface of GO sheets. (C) Upon UV irradiation, a hydrogel formed. (D) Photo of the reaction mixture in a glass vial before the UV irradiation. (E) Photo of the reaction mixture before the UV irradiation. (F) Photo of the resulting hydrogel. (G) The resulting hydrogel with pressure applied. Ref. 31.

Source: Copyright 2013, Royal Society of Chemistry.

It is worth mentioning that the loading of highly conductive elements into a soft hydrogel is an easy and straightforward strategy to form electroconductive percolative patterns for potential biosensing applications. The overall network conductivity may not be as high as a conducting polymer hydrogel, but there are many more choices for the polymer matrix. This caters to different application purposes, such as good biocompatibility or sufficient mechanical properties to match the strength and elasticity of a tissue environment. Conductive hydrogels integrated with a variety of smart/responsive mechanisms will open up opportunities for new sensing applications.

2.3. *New smart materials for hydrogel biosensors*

Biosensors have been widely used in environmental and industrial applications. However, their use in clinical monitoring is mostly restricted to academic studies. One of the critical issues preventing long-term sensor implantation is biofouling.[32] Most existing materials, upon implantation, will be recognized by the body's immune system as foreign, and become encapsulated within a dense collagen capsule.[33,34] Such foreign-body reactions occur on almost any type of implant, including: teflon, polyurethane, silicone rubber, polyethylene, poly(methyl methacrylate), poly(2-hydroxyethyl methacrylate) (PHEMA), poly(ethylene glycol) (PEG), Dacron, gold, titanium, and alumina. Whether materials are hydrophilic, hydrophobic, hard, soft, polymeric, metallic, or ceramic, they all tend to elicit a foreign body reaction.[34,35] The resulting capsules are impermeable to most molecules in the biological environment, cause the failure of sensors due to limited signal exchange, and can produce tissue distortion and pain.[36,37]

It is hypothesized that if a material can truly eliminate non-specific protein adsorption (i.e., the first step of the foreign-body reaction), or is truly non-fouling, it may be able to prevent capsule formation.[38–41] PEG and PHEMA have been widely considered as good non-fouling materials,[34,42,43] but were still encapsulated upon implantation. In recent years, zwitterionic materials, including carboxybetaine, sulfobetaine and phosphorylcholine, have emerged as an important class of non-fouling materials.[40,44,45] Among them, carboxybetaine (e.g., poly(carboxybetaine methacrylate) (PCBMA)) has ultra-low fouling properties and resists protein adsorption down to <0.3 ng·cm^{-1}, even from 100% blood plasma or serum. Note that a monolayer of protein coverage on a surface is 100–500 ng·cm^{-1}.[40,46] In a recent report, hydrogels made from PCBMA showed the ability to resist capsule formation for up to 3 months when they were subcutaneously implanted in mice.[47] It was also found that PCBMA hydrogels promote angiogenesis in surrounding tissue, which might be due to the presence of macrophages exhibiting anti-inflammatory and pro-healing phenotypes.[47] This remarkable demonstration of biocompatibility implies that zwitterionic hydrogels have an advantage over most existing materials when preparing tissue scaffolds and implantable medical devices (including implantable biosensors).

The use of a PCBMA hydrogel in biosensing applications has been demonstrated on a glucose sensor (Fig. 10).[48] The hydrogel was coated on an electrode with GOx immobilized via photopolymerization. In a so-called "physical adsorption method," a solution containing CBMA monomers, ethyleneglycol dimethacrylate (EGDMA) crosslinker and 2-hydroxy-2-methylpropiophenone photoinitiator was applied on the electrode followed by UV exposure for 30 min (Fig. 10 (A)). In a chemical attachment method, the electrode surface was prefunctionalized with (trimethoxysilyl)propyl methacrylate followed by the physical adsorption procedure mentioned-above (Fig. 10(B)). The chemical attachment strategy outperformed the physical adsorption one; its resulting glucose sensor achieved high sensitivity and good linearity in PBS and human blood serum (dilute as well as 100% concentration).

A unique property of poly(carboxybetaine) (e.g., PCBMA) is its abundance of carboxylate functional groups for the convenient immobilization of biomolecules containing amine groups (i.e., carboxylic acids conjugate with amines via typical coupling chemistry, e.g., 1-ethyl-3-(3-dimethylaminopropyl)carbodiimide and N-hydroxysuccinimide (EDC/NHS)).[40,49] Thus, GOx enzyme can also be immobilized on PCBMA hydrogel to construct a glucose sensor.[50] As shown in Fig. 11, the above mentioned chemical attachment strategy was modified by conjugating GO*x*

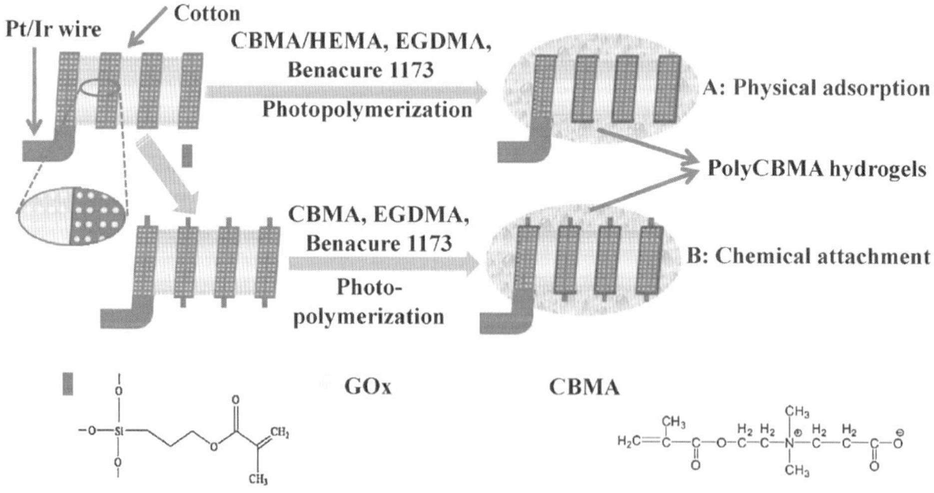

Fig. 10. Zwitterionic PCBMA hydrogel-based glucose sensor. The hydrogel was coated on an electrode through either physical adsorption (A) or chemical attachment (B). Ref. 48.
Source: Copyright 2011, Elsevier.

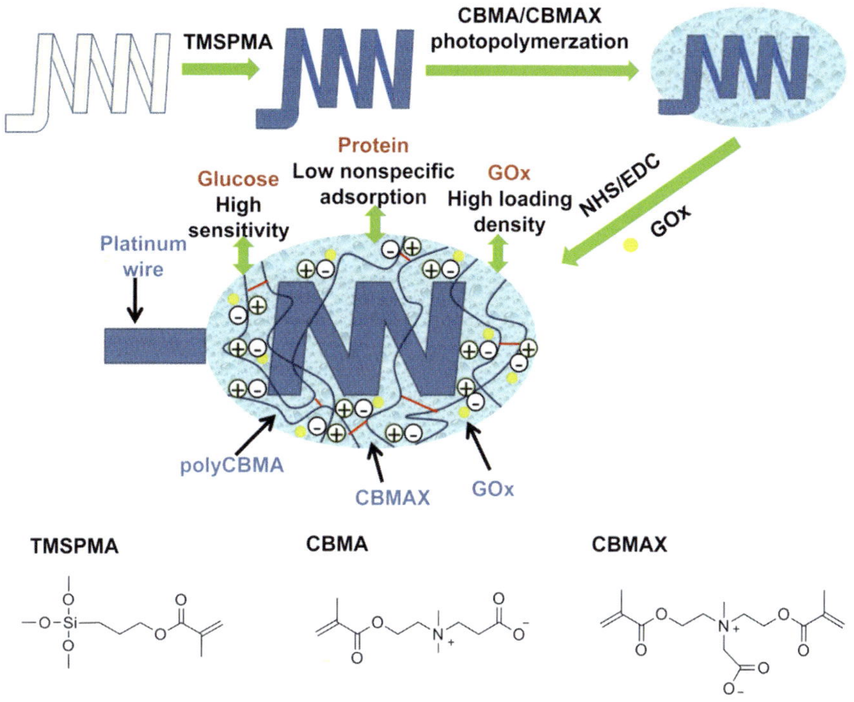

Fig. 11. Zwitterionic PCBMA hydrogel-based glucose sensor. The hydrogel was chemically coated on an electrode. GOx was further covalently immobilized on the hydrogel via NHS/EDC chemistry. CBMAX is a crosslinker with a carboxybetaine group. The constructed PCBMA hydrogel contained solely carboxybetaine side chains. Ref. 50.

Source: Copyright 2012, Elsevier.

into the PCBMA hydrogel matrix rather than physical adsorption on the electrode. A new crosslinker (CBMAX) was also used to form the hydrogel, which has a carboxybetaine group.[51] The solely CBMA-containing hydrogel sensor showed extraordinary performance in whole blood for 42 days, as opposed to 33.3 h for commercial sensors.

Based on carboxybetaine, a smart material that can switch reversibly between non-sticky (resistant to protein adsorption from 100% blood plasma) and sticky (permanent covalent coupling), two incompatible properties, has been proposed.[52,53] As shown in Fig. 12, the new material, driven by either acidic or basic conditions, can alternate between two equilibrium states (CB–OH and CB-Ring, respectively). The CB–OH is ultralow fouling (non-sticky) due to its zwitterionic structure. The CB-Ring is reactive (sticky) towards nucleophiles (e.g., amine moieties) because of the lactone structure.

A simple manipulation strategy to take advantage of this smart material was illustrated, wherein the switchable non-sticky and sticky properties could be used

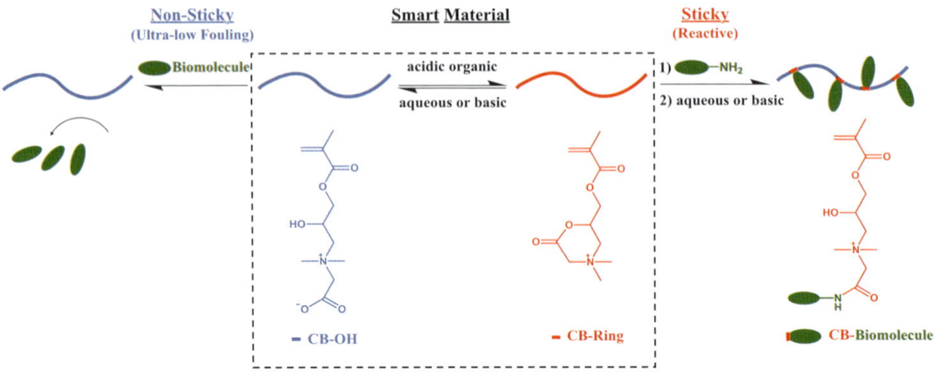

Fig. 12. The switchable properties of a material based on unique equilibrium states, CB-OH and CB-Ring, respectively. Ref. 52.

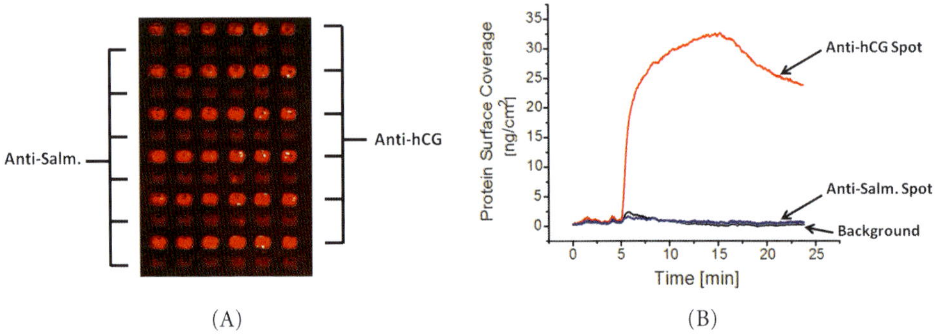

Fig. 13. (A) A 6 × 12 protein array formed on a substrate coated with the new smart material. Anti-hCG and anti-Salm (both red) were printed by reacting with CB-Ring. The background (black) was CB–OH which is ultralow fouling. (B) The specific detection of hCG was observed for only anti-hCG spots (red); there was no sensor response for either the background (black) or the anti-Salm spots (blue). Ref. 52.

to create a protein array.[52] A CB-OH polymer brush was formed on a gold substrate through surface initiated atom transfer radical polymerization (SI-ATRP). The resulting CB-OH film was then converted to CB-Ring via brief acid treatment. Two antibodies (anti-hCG and anti-Salm) were then contact-printed onto the CB-Ring surface (i.e., amine-reactive) in pH 10 buffer, forming a 6 × 12 array (Fig. 13A). Unreacted CB-Ring surface was hydrolyzed using pH 9 buffer; this converted the unspotted background from sticky into ultralow-fouling. The resulting chip with a protein array on the surface was mounted on a surface plasmon resonance (SPR) imaging biosensor. hCG could be sensitively detected by anti-hCG spots. For both the control antibody (anti-Salm) and the background, there was zero response (Fig. 13(B)).

This work provides a new smart material (or monomer) that can potentially solve the biofouling (or biocompatibility) and functionalization issues affecting current biosensors. Its application in the form of a smart hydrogel, though not explored in this work, could be promising.

3. Conclusion and Perspective

This chapter discussed the recent advances in smart hydrogels, specifically for biosensing applications. Traditional smart or stimuli-response hydrogels were covered with unique properties highlighted (e.g., unusually high swelling ratio, high mechanical strength at a wide pH range, NIR responsiveness, and redox induced self-healing features). Because of their immediate relevance to common electro-chemical based biosensors, conductive hydrogels that could be prepared using either conductive polymers or other component elements were also discussed. Lastly, zwitterionic polymer (i.e., PCBMA) and its derived smart material, with the potential to solve the biocompatibility issues and functionalization demands of implantable biosensors, were highlighted. It is envisioned that future smart hydrogels designed with state-of-the-art features will enable robust, user compliant and highly sensitive biosensors for *in vivo* and even clinical applications.

Acknowledgment

This work has been supported by Wayne State University Faculty Start Up Funds.

References

1. M. A. C. Stuart *et al.*, *Nat. Mater.*, 101–113 (2010).
2. T. Miyata, T. Uragami, and K. Nakamae, *Adv. Drug. Deliv. Rev.*, 79–98 (2002).
3. C. D. H. Alarcon, S. Pennadam, and C. Alexander, *Chem. Soc. Rev.*, 276–285 (2005).
4. A. S. Hoffman, *Adv. Drug. Deliv. Rev.*, 10–16 (2013).
5. G. R. Hendrickson and L.A. Lyon, *Soft. Mat.*, 29–35 (2009).
6. Y. Qiu and K. Park, *Adv. Drug. Deliv. Rev.*, 321–339 (2001).
7. J. Kopecek and J. Y. Yang, *Polymer. Int.*, 1078–1098 (2007).
8. J. Kopecek, *Biomater.*, 5185–5192 (2007).
9. A. S. Hoffman, *Adv. Drug. Deliv. Rev.*, 3–12 (2002).
10. G. H. Chen and A. S. Hoffman, *Nature*, 472–474 (1995).
11. A. S. Hoffman, *J. Control. Release.*, 297–305 (1987).
12. T. Miyata, N. Asami, and T. Uragami, *Nature*, 766–769 (1999).

13. M. Nakahata *et al.*, *Nat. Commun.*, 511–517 (2011).
14. S. Singh *et al.*, *Angew. Chemie. Int. Ed.*, 3000–3003 (2013).
15. L. Xia *et al.*, *Nat. Commun.*, 2226–2237 (2013).
16. H. Gao *et al.*, *Macromol. Rapid. Commun.*, 63–68 (2013).
17. C. H. Zhu *et al.*, *Adv. Funct. Mater.*, 4017–4022 (2012).
18. M. Acik *et al.*, *Nat. Mater.*, 840–845 (2010).
19. N. Fomina *et al.*, *Macromol.*, 8590–8597 (2011).
20. C. Dhand *et al.*, *Biosens. Bioelectron.*, 2811–2821 (2011).
21. S. Nambiar and J. T. W. Yeow, *Biosens. Bioelectron.*, 1825–1832 (2011).
22. C. Bayer, Conductive and recognitive hydrogels for biosensing applications. *Biomed. Eng.*, (2009).
23. Y. M. Uang and T. C. Chou, *Biosens. Bioelectron.*, 141–147 (2003).
24. Q. W. Tang, J. M. Lin, and J. H. Wu, *J. Appl. Polym. Sci* ., 1490–1495 (2008).
25. H. J. Dai, *Acc. Chem. Res.*, 1035–1044 (2002).
26. L. J. Pan *et al.*, *PNAS*, 9287–9292 (2012).
27. R. A. Green *et al.*, *Sci. Tech. Adv. Mater.*, 014107–14120 (2010).
28. A. Guiseppi-Elie, *Biomater.*, 2701–2716 (2010).
29. D. Y. Zhai *et al.*, *ACS Nano*, 3456–3540 (2013).
30. F. Lamberti *et al.*, *J. Mater. Chem. B*, 5083–5091 (2013).
31. Z. Tang *et al.*, *J. Mater. Chem. B*, 5393–5397 (2013).
32. S. Brahim, D. Narinesingh, and A. Guiseppi-Elie, *Biosens. Bioelectron.*, 973–981 (2002).
33. R. Langer, *Adv. Mater.*, 3235–3236 (2009).
34. B. D. Ratner, *J. Control. Release.*, 211–218 (2002).
35. W. K. Ward, *J. Diabetes Sci. Technol.*, 768–777 (2008).
36. D. F. Williams., *Biomater.*, 2941–2953 (2008).
37. E. M. Hetrick *et al.*, *Biomater.*, 4571–4580 (2007).
38. B. D. Ratner *et al.*, *Biomaterials Science, an Introduction to Materials in Medicine* (*Second Edition*). Amsterdam: Elsevier (2004).
39. E. Ostuni *et al.*, *Langmuir*, 5605–5620 (2001).
40. S. Jiang and Z. Q. Cao, *Adv. Mater.*, 920–932 (2010).
41. B. D. Ratner and S. J. Bryant, *Annu. Rev. Biomed. Eng.*, 41–75 (2004).
42. A. D. Lynn, T. R. Kyriakides, and S. J. Bryant, *J. Biomed. Mater. Res. Part A*, 941–953 (2010).
43. E. G. Campioni, J. N. Nobrega, and M. V. Sefton, *Biomater.*, 829–837 (1998).
44. K. Ishihara *et al.*, *J. Biomed. Mater. Res.*, 1397–1407 (1991).
45. R. S. Smith *et al.*, *Sci. Transl. Med.*, 153ra132 (2012).
46. J. Ladd *et al.*, *Biomacromol.*, 1357–1361 (2008).
47. L. Zhang *et al.*, *Nat. Biotech.*, 553–556 (2013).
48. W. Yang *et al.*, *Biosens. Bioelectron.*, 2454–2459 (2011).
49. Z. Q. Cao *et al.*, *Angew. Chemie. Int. Ed.*, 3771–3864 (2010).
50. W. Yang *et al.*, *Biomater.*, 7945–7951 (2012).
51. L. R. Carr, H. Xue, and S. Y. Jiang, *Biomater.*, 961–968 (2011).
52. Z. Q. Cao *et al.*, *Angew. Chemie. Int. Ed.*, 6102–6104 (2011).
53. Z. Q. Cao *et al.*, *Angew. Chemie. Int. Ed.*, 2602–2605 (2012).

Chapter 9

3D Cancer Models on Hydrogels

Imran Rizvi*, Heather Gudejko*, Emma Briars*, Shazia Khan*, Chun-Te Chiang*, Hamid El-Hamidi†, Jonathan P. Celli†, and Tayyaba Hasan*,‡

**Wellman Center for Photomedicine, Massachusetts General Hospital and Harvard Medical School, 40 Blossom Street, Boston, MA 02114, USA*

†*Department of Physics, University of Massachusetts Boston, 100 Morrissey Boulevard, Boston, MA 02125, USA*

‡*thasan@mgh.harvard.edu*

Improving the translational efficiency of anti-cancer agents requires advancements on many fronts, including the continued development of three-dimensional (3D) tumor models that capture complex physiological cues *in vitro*. Significant progress has been made to improve the sophistication of 3D models, but many critical issues remain under consideration to realize the value, and importantly the limitations, of using bioengineered 3D models in the drug discovery process. Key challenges to implementing 3D models to inform crucial go/no-go decisions in drug development include creating biologically relevant systems that are cost-effective; making complex models and related analysis routines meaningfully compatible with rapid throughput screening; and defining and validating useful metrics of treatment response. This chapter describes some of the major efforts to develop more predictive 3D models that restore tumor architecture, incorporate heterotypic signaling with stromal partners, and integrate physical cues such as mechanical and fluidic stress. Advancements in creating a high-content analysis framework to study treatment response in complex systems are also discussed. With continued development, 3D models could play an increasingly important role in drug discovery to complement both established and emerging preclinical research tools in order to focus valuable resources on the most promising candidate agents.

1. Introduction

The success rate for the translation of oncology-related pharmaceuticals remains dismal at ~10%, leading to significant costs (more than US$1 billion) and protracted timelines (typically more than 10 years) that impact the availability and accessibility of promising new therapeutics.[1–6] Improving these statistics remains a major challenge and will require advancements on many fronts to realize substantive change.[1,6–8] An improved understanding of the cellular and molecular circuitry that is reprogrammed in tumors will be the cornerstone of progress, and will likely inform the design of patient-tailored combination therapies[9,10] based on the genetic and molecular signatures of a tumor rather than its site of origin.[9–15]

The prioritization and target-informed stratification of the most promising treatment regimens from the vast library of candidate agents will require a multifaceted approach,[1,4,16] including the development of sophisticated preclinical models that capture key aspects of complex human physiology *in vitro* and *in vivo*.[1,17] The high costs and failure rates of drug development require the exclusion of failed agents as early as possible in the evaluation phase.[1–5] This need for early exclusion highlights a particularly important role for cell-based screening systems that meaningfully inform crucial go/no-go decisions in the development process.[2,3,17] Three-dimensional (3D) cell cultures are becoming increasingly sophisticated, and with continued development could emerge as important tools to inform the decision to focus valuable resources on the most promising candidates.[2,3,17–33] Several key ideas in the development of 3D models remain open for consideration and frame the issue of how these systems can be optimized to add value to the discovery process and to identify their limitations as research tools. These include important trade-offs between the biological relevance of the models, costs involved to develop and scale up the models for industrial applications, as well as the throughput and ultimate value of the results provided by these systems. Addressing these questions requires a better understanding of how much complexity needs to be engineered into 3D models by integrating physiologically-informed architectural and microenvironmental cues, including communication with the stroma, signaling with heterotypic cellular partners and the influence of physical forces such as flow and mechanical stress.[17,20,22,24,30–38] Additionally the analysis framework that is necessary to extract critical information, such as viability and mechanisms of death, from the models remains an important consideration to successfully implement these systems.[26,27,39–42]

This chapter provides an overview of some of the efforts being made to address the need for more predictive *in vitro* models and is organized into five major sections. Section 2 summarizes the approaches to generate both non-adherent spheroids and adherent 3D monocellular tumor models, and discusses the relevance of these systems with respect to architecture and gene and molecular expression profiles. The approaches highlighted in this section include the use of rotary cultures, the hanging drop method and non-adherent substrates to create tumor spheroids, compared to naturally-derived and synthetic matrices used for adherent 3D cultures. The importance of communication with the extracellular matrix (ECM) is discussed in Section 3 with a particular emphasis on the communication that is restored between cancer cells and the surrounding stroma by some of the matrices that are being used to generate 3D tumor models. Section 4 focuses on the relevance of signaling with heterotypic stromal partners such as tumor endothelial cells (TECs), fibroblasts and macrophages, and highlights the need to integrate these cell types into tumor-specific 3D cultures. The challenges and opportunities associated with imaging and analyzing treatment response in 3D cultures is the topic of Section 5 with an emphasis on quantitative optical imaging and the hurdles to achieving more rapid throughput. Section 6 discusses the relevance of physical cues, such as matrix rigidity and fluidic stress, in 3D systems as determinants of tumor biology.

2. 3D Monocellular Tumor Models

A few commonly used methods to generate non-adherent and adherent monocellular 3D tumor models and the relevance of these systems with respect to tumor architecture and gene and molecular expression profiles are presented in this section.

2.1. *Non-adherent spheroids*

2.1.1. *Spinner flasks and gyratory rotation*

A spinner flask culture system has traditionally been used to grow non-adherent cells in suspension; however it can also be used for growing 3D spheroids of adherent cell types. This model is composed of a flask containing a stirring mechanism, which maintains the cells in suspension. Gyratory rotation of Erlenmeyer flasks or the rotation of cylindrical tubes may also be used. Since the cells are not able to interact with the surface of the flask, the adhesion forces between the cells are greater than that of the cell to the flask, leading to the formation of 3D spheroids. The primary advantage of this system is the formation of high numbers of spheroids per culture vessel, giving the researcher a large sample size for each experiment. In addition, there are multiple commercially available spinner flask systems available, making this method easy to implement and use. The disadvantages of these systems include shear forces that break apart spheroids over time, as well as the inability to observe treatment response on a single-nodule basis. Sutherland *et al.*[43] pioneered the use of the spinner flasks in order to investigate the growth characteristics of Chinese hamster V79 lung cells. They found that V79 nodules had a similar morphology to transplanted mammary carcinomas in mice, containing a mitotic outer zone, an intermediate zone with a lower proliferation rate, and a necrotic inner core.[43] In addition, the growth rate of these spheroids was initially exponential, but over time plateaued, as observed in *in vivo* tumors. Since the 1970's, the use of spinner flasks and gyratory rotation have become commonplace for the study of multiple tumor types, especially when large numbers of spheroids are required.

The most common method for the growth of 3D tumor models in rotary flasks is the use of a Rotary Cell Culture System (RCCS). Originally called Rotating-Wall Vessels (RWV), RCCS was developed by National Aeronautics and Space Administration (NASA) in the early 1990s to investigate the effect of microgravity and shear stress on cells.[44] This system involves the use of a sterile, sealed dish or tube filled with media, which is kept under constant rotation, preventing the cells from coming into contact with, and thus adhering to, the sides of the dish or tube.[44,45] This constant rotation leads to the formation of multiple spheroids per dish, allowing for a large sample size for each experiment. The size of the resulting multicellular spheroids is dependent on the cell type, but ranges from 1–9 mm.[46,47] The main limitation of this culture system is the fragmentation of nodules due to stress forces in the moving media. In order to overcome this issue, microcarrier beads may be used in conjunction with both spinner flask and rotary cultures in order to enhance cell aggregation. These beads are normally made of dextran and give cells added surface area on which to adhere in suspension.[48] Multiple cell-bead aggregates can combine to create larger nodules.[49]

Grun *et al.*[46] used the RCCS system to develop and classify 3D tumor models derived from ovarian and endometrial tumor cell lines These classifications revealed that 3D spheroids were similar to primary tumors in both morphology (necrosis or apoptosis, nuclear organization, and mitotic activity) and epithelial biomarker expression. Smith *et al.*[47] also observed similar morphology between primary tumors and 3D nodules of a glioma-derived cell line using an RCCS system. In this system, spheroids are organized into an outer region of viable mitotic cells, surrounding a necrotic center. These spatial patterns in cell viability and phenotype with respect to 3D architecture are inherently absent in two-dimensional (2D) monolayer culture, and are significant in their correspondence to tumor physiology *in vivo*. In addition, the proliferation rates of the cell lines investigated were significantly lower in 3D cultures than when compared to 2D cultures.

2.1.2. *Hanging drop*

The hanging drop method involves plating suspended droplets containing cells often on the lid of a petri dish or in a specialized plate so the cell suspension is allowed to hang freely below the plate, forming a hanging droplet.[50] The freely hanging cells in suspension gather at the bottom of the droplet, and without the surface of a well to adhere to, the cells are attracted and attach to one another, forming a nodule. This method was originally used in the early 1900s for the study of embryoid bodies[51,52] and has since been adapted for use in 3D tumor modeling.[50,53] The main advantage of this protocol is the consistency of producing one spheroid in each hanging droplet, allowing for the direct comparison and observation of individual nodules. In addition, hanging droplet plates are commercially available in 96 and 384 well formats, making them a useful tool for high-throughput screening systems. Zietarska *et al.* used the hanging drop method to grow 3D spheroids of multiple ovarian cancer cell lines.[54] These spheroids were compact, and comprised of actively proliferating cells throughout the nodule. These spheroids were highly consistent between drop cultures of the same cell line and were used to perform xenograft and gene expression experiments.[54]

2.1.3. *PolyHEMA*

Since early experiments by Costachel *et al.*[55] and Sutherland *et al.*[43] demonstrating the growth of multicellular spheroids on non-adherent substrates, a variety of approaches and materials have been used to generate reproducible suspension cultures. Among these, poly(2-hydroxyethyl methacrylate) (polyHEMA) is a hydrophilic hydrogel commonly used to make soft contact lenses, however its biocompatibility makes it ideal for use in tissue culture methods as well.[56,57] While polyHEMA was initially described in the late 1950s, it was not used for the production of 3D spheroids until the late 1970s,[58] where it was shown that cells plated on this substance exhibited a rounded morphology compared to plating on uncoated surfaces. The researchers also showed that rounded cells had less deoxyribonucleic acid (DNA) synthesis than flattened cells, suggesting that there may be a link between cell shape and the rate of cell division. Additionally, cell suspensions plated on polyHEMA-coated plates spontaneously form 3D spheroids. This technique has its advantages over other hydrogels because it is a synthetic plastic and not naturally-derived, avoiding the possibility of inadvertently exposing spheroids to external growth factors.

2.1.4. *Agar*

A less expensive alternative to polyHEMA is coating plate wells with agar, a gelatinous polysaccharide derived from red algae. When solidified, agar becomes a porous, inert semi-solid substance. Agar is easy to use and readily available in most laboratories, and alterations in pore size and solidity can be made by varying the percentage of agar used. In this method, a culture dish or well is coated with a low percentage agar solution, atop which a cell suspension is plated. The agar coating prevents the cells from adhering to and spreading across the dish surface, causing cell–cell interactions to become dominant, resulting in the formation of 3D spheroids. Agar-coated plates were shown to be successful for the growth of multiple mammalian cell lines as unattached multicellular spheroids.[59] While the morphology varied between cell types in these agar-coated plates, most spheroids had a similar structure to the three zones mentioned above. It was also found that these spheroids grew at a rate proportional to their *in vivo* counterparts, while the growth rates of monolayer cultures did not correlate with either the spheroids or *in vivo* tumors.[59] In addition to growing spheroids on top of agar-coated plates, Folkman and Hochberg[60] showed that multicellular spheroids of several cancer cell types could be grown suspended in an agar gelatin. These spheroids exhibited the classic organization of a mitotic outer layer, an intermediate zone, and a necrotic center.[60]

2.2. *Adherent nodules*

2.2.1. *Naturally-derived substrates*

By far the most common adherent substrate for the growth of 3D nodules is the laminin-rich basement membrane (lrBM) derived from murine Engelbreth-Holm-Swarm sarcomas, sold commercially as Matrigel and Cultrex. Matrigel is ECM, composed of laminin, type IV collagen, heparan sulfate proteoglycan, and entactin, isolated from the Engelbreth-Holm-Swarm murine sarcoma tumor.[61] After its discovery in the mid-1980s, it was soon observed that the epithelial cells grown on Matrigel form 3D tubular structures.[62–64] Human umbilical vein endothelial cells (HUVEC) cells plated on lrBM form long, branched tubular structures similar to capillaries, and the formation of these tubes is dependent on the presence of laminin in the lrBM.[62] This method has also been used to study breast cancer. Normal mammary cells form acini surrounding a lumen, the formation of which is regulated by apoptosis.[63,64] On lrBM, mammary epithelial cells form spheroid acini, similar to the *in vivo* observations.[63] These spheroids are formed by both normal and carcinoma primary cell lines. While normal mammary cell spheroids grow rapidly at first (faster than 2D cultures), the growth arrests after 7 days, and a layer of basement membrane is secreted by the outer layer of cells. Cancer-derived cultures do not show this arrest, resulting in larger spheroids with no basement membrane and incorrect cell polarization.[63]

Not all components of the ECM matrix are necessary to create a sufficient surface for the formation of 3D nodules *in vitro*. Collagen is a main component of the ECM and the most abundant protein in the mammalian body. Due to the fact that most cells readily adhere to collagen through the binding of integrins, collagen has been used to grow adherent 3D nodules since the mid-1980s. Collagen can be used to coat the bottom of dishes or wells, creating an artificial ECM-like environment. Alternatively, the cells can be suspended in a collagen matrix for the formation of 3D nodules.[65] Since collagen is naturally derived, it contains inherent signaling properties that may affect the growth and gene expression profiles of the cells in culture.

An interesting alternative to mammalian-derived scaffolds, such as collagen, is silk, a biocompatible, highly adaptable material that has been used in medicine for many years. The biocompatibility and mechanical strength of the silk is due to the core silk fibroin fibers, isolated after the primary silk is rid of its sericin (glue-like proteins).[66] For example, the silk from the silkworm *Bombyx mori* (*B. mori*) has been used for decades as a material for sutures.[66] The use of silk as a substrate for 3D cell culture became accepted in the 1990s and continues to become more popular today in the field of tissue scaffolds. The silk fibroin of *B. mori* is composed of the amino acids glycine, alanine, serine and tyrosine, resulting in a homogeneous network of β-sheets with important mechanical properties.[66] The growth of multiple carcinoma cell lines on silk fibroin, including human colon adenocarcinoma (SW116, Colo201), human mouth epidermoid carcinoma (KB), and human lung carcinoma (QG56), was similar to that observed on a collagen matrix, suggesting that silk is an acceptable substrate for growing mammalian cells in culture.[67] In addition, silk fibroins have been used as scaffolds for the growth of replacement anterior cruciate ligaments.[66,68]

2.2.2. *Synthetic substrates*

PuraMatrix, a chain of hydrophobic peptides that form a β-sheet conformation creating an insoluble membrane,[69] was developed in 1993 by Zhang *et al.* and now has the convenience of being commercially available. In contrast to Matrigel but similar to silk, PuraMatrix is not derived from mammalian cells, therefore reducing the possibility of introducing extraneous growth factors to the cell culture. Ovarian cancer cells encapsulated in PuraMatrix form 3D acinar structures, similar to morphologies observed in *in vivo* models.[29,70] Yang *et al.*[70] also found that there was no difference in the attachment potential of cells to PuraMatrix compared to type I collagen. Additionally, ovarian cancer cells were able to migrate within PuraMatrix similar to when in a collagen matrix.[70]

2.3. *Gene expression and signaling in monocellular 3D cultures*

The validation of 3D models as an intermediate between 2D monolayer cultures and *in vivo* models is ongoing. Kenny *et al.*[71] performed a comprehensive study of 25 epithelial cell lines and observed their 3D morphologies. These cell lines were then categorized into four groups: round, mass, grape-like, and stellate. While there was no difference in cell proliferation seen between the nodule types, significant differences in gene expression were found. For example, glycosylated β1-integrin was expressed more in the invasive stellate cell lines compared to other morphologies. Stellate cell lines did not express E-cadherin, consistent with their elongated morphology and invasive *in vivo* phenotype. Between 2D and 3D cultures, genes involved in signal transduction were significantly overexpressed in 3D nodules.

Genes encoding matrix metalloproteinases (MMPs), TGF-β, integrins, and cell adhesion molecules are upregulated in 3D brain cancer cultures when compared to 2D.[47] In PC3 cells, the invasive prostate cancer cell line increases expression of β1 integrin and vimentin when grown in 3D when compared to 2D.[72] These cells also show a decrease of α6 integrin when compared to 2D cells.[72] Interestingly, nonmetastatic and metastatic cell lines, LNCaP and PC3 respectively, both expressed active CXCR4, a marker for increased metastasis in 2D cultures; however, when grown in 3D, PC3 showed increased expression compared to LNCaP, similar to what is seen in patient samples.[72] The genes encoding the ECM proteins laminin and hyaluronic acid (HA) were also found to be upregulated in melanoma cells grown as 3D spheroids when compared to 2D monolayers.[73] The chemokines CXCL1 and IL-8, which have been shown previously to be involved in melanoma progression, were significantly upregulated in 3D spheroids compared to 2D cultures of melanoma cells.[73] In addition, pro-angiogenic genes also showed increased expression.

It has become clear, however, that no *in vitro* model is an exact match to *in vivo* tumors. Dangles *et al.*[74] showed that cell lines derived from patient samples had gene expression profiles distinct from the parent tumors, and that the *in vitro* model that most closely resembled the *in vivo* tumor varied from sample to sample.

Further proteomic analysis has shown that there are significant changes in protein expression when cells are cultures in 3D relative to 2D in multiple cancer cell lines.[46,47,73] Many of these genes are involved in stem cell renewal and differentiation, as well as ECM interactions and cytokines.[47,73] Comparing 2D and 3D cell cultures, Windus *et al.*[72] found that 3D models of prostate cancer more accurately represent *in vivo* protein expression patterns. Invasive PC3 cells lost E-cadherin and increased N-cadherin when grown as 3D spheroids, similar to expression profiles found in primary tumor samples. Vimentin and laminin have increased expression in 3D spheroids compared to 2D monolayers of human ovarian surface epithelial cells, possibly for the structural integrity of the spheroid.[75] Yang *et al.*[76] investigated the expression of the stem cell marker CD133 in SW620 colorectal cancer cells in 2D and 3D model systems. Using the hanging drop method to obtain 3D spheroids, the researchers found that CD133 expression was increased in 3D cultures compared to 2D cultures, and that this increase appeared to be caused by cell–cell adherence.

The change in cell shape that accompanies the formation of 3D nodules has also been shown to cause changes in chromatin organization. Mammary epithelial cells grown as 3D nodules had lower levels of acetylated histones H3 and H4 compared to cells grown in 2D monolayer.[77] This deacetylation correlated with a decrease in gene expression and promoter repression in 3D nodules.[77]

3. Communication with the ECM in 3D Cultures

ECM consists of two major compartments, the interstitial matrix and basement membranes (BM), both of which provide structural and biochemical support to surrounding cells. Though previously believed to be a largely static and mainly structural component, it is increasingly evident that the ECM is a dynamic and key contributor to the tissue microenvironment as discussed in this section.

3.1. *Composition and role of the ECM*

The ECM is composed of a diverse set of biomolecules. The components of the interstitial matrix as well as BM are synthesized inside cells and secreted into interstitial spaces.[78] Collagens are the most abundant biomolecules of the ECM, found in both the interstitial matrix and BMs.[78,79] The BM interacts with the epithelium and endothelium, separating these from the stroma. The components of the BM are produced by epithelial, endothelial, and stromal cells.[80] The major BM components include collagen IV, laminins, linker proteins like entacten/nidogen and heparin sulphate proteoglycans.[79,81] The interstitial matrix is made up of fibrillar collagens like collagen I and III, glycoproteins like fibronectin and tenascin, and proteoglycans.[79–81] Different organs and tissues, however, have different expression levels of these basic ECM components, as well as specialized factors exclusive to a tissue type, which confer unique properties.[78,80] Integrins are the most prominent ECM receptors, and belong to the class of cell surface adhesion molecules.[82,83] Other cell surface receptors such as syndecans and dystroglycan are also found in the ECM.

The role of the ECM is two-fold: structural and biochemical. The ECM provides structural support for tissues, acting as the scaffold for tissues, as wells as a compression buffer that provides differential elasticity. Additionally, the ECM is involved in signaling and may provide docking sites for cytokines, growth factors, and other paracrine regulatory molecules. Biochemical as well as biomechanical changes to the ECM can be both the cause and effect of signaling networks in tissues.[78,80,81] The physical properties like rigidity, porosity, spatial arrangement, as well as biochemical properties may be altered following changes in composition and/or by the action of remodeling factors. Importantly, the ECM is a highly dynamic structure that undergoes continuous remodeling; the different properties of the ECM are inter-connected and there is significant cross-talk between the ECM and cellular components.[80]

3.2. *The ECM and cancer*

The ECM is continuously remodeled, either to maintain the homeostasis observed in adult healthy tissues, or may undergo more drastic changes such as those that occur in normal development, wound healing, inflammation, and disease states. ECM remodeling is dynamic, and involves at least one, and often both of the following seemingly antagonistic processes: synthesis and assembly of ECM components, and proteolysis of ECM components, respectively.[84] In healthy tissues, these opposing processes balance for self-renewal, however in disease states this balance is disrupted and there is subsequent dysregulation of ECM remodeling.

Tumor formation is influenced by the 3D context of the tissue, with the ECM being an integral part of this 3D microenvironment. The ECM is directly involved with tumorigenesis at both the early stage — cell invasion — and the late stage — metastasis. Modifications in the BM such as a loss in structural integrity, thickening or loss of cell attachments are early events in cancer progression.[85] Uncontrolled cell growth and loss of cell-to-cell contact, the latter a hallmark of epithelial-to-mesenchymal transition (EMT), occur in conjunction with altered BM.[86,87] Additionally, the BM underlies the endothelium in blood vessels and in a healthy state acts a barrier between the blood and tissues. However in cancer metastasis, the BM can be compromised, mitigating this physical barrier and facilitating extravasation and invasion.[88,89]

The ECM in tumors is known to be distinct from normal ECM. In different cancers, both the ECM components and the ECM remodeling enzymes may be upregulated.[81,90–93] Matrix stiffening due to increased deposition and crosslinking of collagen is an important event in tumourigenesis. A notable aspect is that hypoxia, an important driver of cellular transformation, can also support increased matrix deposition, crosslinking, and remodeling events.[81,94–98] Hypoxic tumors also show increased expression of matrix remodeling enzymes.[99,100]

The classes of ECM remodeling enzymes include lysyl oxidases (LOX) and MMPs. Notably, both classes of these enzymes are upregulated in cancers.[78] LOX and the LOX family of amine oxidases catalyze collagen crosslinking.[101,102] It has been shown that an increase in collagen production leads to an increase in LOX is production, leading to increased collagen crosslinking and subsequent tissue stiffness that can support cells' progression to malignant tissue, invasion, and metastases.[78,81,91] Essentially, the collagen fibers bundle and align radially with respect to tumors to facilitate cancer cell invasion.[103–105] These invading cancer cells usually produce MMPs.[106–109] MMPs are proteolytic enzymes that degrade the ECM, with important roles in normal tissue remodeling, bone development, wound healing, inflammation, and angiogenesis.[84] The action of MMPs also contributes to re-organization of collagen bundles to create a 'microtrack' to facilitate cell migration.[110] Different MMPs may be elevated in cancers and contribute to the replacement of the normal ECM with tumor cell secreted ECM, and are key players in signaling networks that drive tumor progression and metastases.[78,84]

The ECM-cellular cross-talk, and resultant changes in tumors, are as important as changes to the ECM structure itself. Integrins serve as mechanoreceptors and are important players in tumor biology.[111] Changes in integrin expression and localization promote cell migration, which is important for tumor initiation, growth, and metastasis. These integrin based signal transduction processes are mediated mainly via pathways involving Rho GTPases.[35,78,112] Rho can also promote EMT.[113] Matrix stiffening and transduction through integrins also affects numerous other signaling pathways including the ERK, MAPK, calcium, and Src pathways.[85,114]

In summary, the ECM is an integral and complex component of tissues, both contributing to and modulated by tumorigenesis through intertwined mechanisms and processes. The 3D microenvironment of the tumor, including the ECM, affects different properties of the tumor such as invasiveness, metastatic capability, and drug sensitivity.

3.3. *Matrices in 3D tumor models*

As discussed in a previous section, 3D cancer models present features closer to *in vivo* cancers compared to 2D models.[115,116] In 3D models, cancer cells show a rounded morphology and tend to grow in clusters, and also show a different growth rate compared to the monolayer and spread out growth pattern seen in 2D.[116–119] The cancer cells in 3D also show different metabolism, and expression profiles for key signaling factors including those involved in angiogenesis and cell migration.[120–125] Differences in anti-cancer drug resistance are also noted in 3D models.[118,126–128] Along with these factors, the 3D models provide a platform to study tumor cells in conjunction with other cell types and non-cellular components.

The ECM is an integral part of the 3D microenvironment and thus an indispensable part of complete 3D cancer models. 3D matrices for tumor models can be hydrogels or pre-fabricated scaffolds. Hydrogels refer to biocompatible polymers that have high water content and resemble the gel-like properties of natural ECM.[129] Both hydrogels and scaffolds may be derived from natural substances or made synthetically, and can be engineered for properties like pore size.[130–132]

The most commonly used matrices in 3D models of cancer currently are prepared from biological ECM. These natural ECMs help to provide the appropriate biomechanical properties, as well as molecular signaling to support 3D tumor architecture.[133]

3.3.1. *LrBM/matrigel*

LrBM derived from murine Engelbreth-Holm-Swarm (EHS sarcomasis the most popular matrix material for 3D cancer models.[134] This lrBM, commercially available as Matrigel or Cultrex, is derived from mouse tumors that contain abundant ECM proteins,[134] and is best suited for the growth of glandular epithelial cancers.[133] A liquid at lower temperatures, lrBM solidifies at 37°C, and can support cell morphogenesis, differentiation, and growth of tumors.[134] LrBM is composed of collagen IV, laminins, perlecan (a proteoglycan) nidogen/entactin, and is proficient in providing ECM related signaling cues for cell motility, angiogenesis and drug resistance.[134–139] Fragments of laminin-1, collagen IV and other matrix proteins can provide signaling for angiogenic and growth promoting processes,[135] and these matrix components can be anti-apoptotic through integrin mediated signaling.[134]

Variants of the standard lrBM that contain no growth factors, higher levels of collagen or mimic a more 'aged' environment (e.g., matrigel derived from aged mice) have been used for different studies,[134] showing that lrBM can be customized for specific applications or studies. Besides supporting tumor cell growth and proliferation, lrBM is utilized for studying tumor cell invasion and angiogenesis.[134,140–142] Notably, Matrigel can also support the growth of embryonic stem cells.[134,143] Although widely used for multiple applications, a limitation of lrBM is that it is not well-defined and there can be batch-to-batch variability, introducing variability across experiments.[144]

3.3.2. *Collagen and gelatin*

Collagen I (type I collagen) or gelatin are the preferred matrices for 3D cancer models of stromal or bone metastatic cancers.[145] Collagen I, a fibrillar collagen, is a major component of the stroma and bone. Gelatin is denatured collagen I. Both collagen I and gelatin may be isolated from biological sources such as bovine skin, rat tail or human placenta, and are commercially available.[145,146] Collagen I and gelatin can form gels in which matrix pore size and fiber size can be modulated depending on pH and the concentration used.[147–149] They may also be electrospun to form membranes.[150,151]

Collagen I is an important signaling molecule by virtue being a key ligand for integrin receptors, in particular integrins $\alpha(2)\beta(1)$ and $\alpha(3)\beta(1)$.[152] MMP production, cell proliferation, and migration (through E-cadherin or Rho GTPases) as well as sensitivity to drugs may be affected through such signaling.[145,153–155] Also, collagen matrices may be derived from aged mice, to provide a very specific 3D tumor microenvironment.[156] An important application of collagen is that microtracks for cancer cell invasion can be generated and studied in this matrix.[157,158]

3.3.3. *HA*

Hyaluronic acid or hyaluronan is another biologically derived hydrogel matrix.[117,126] HA is a ubiquitous glycosaminoglycan biomolecule in the ECM and thus can provide a relevant 3D microenvironment for cancers.[145] HA mediates signaling through its CD44 receptor and its receptor for hyaluronan mediated motility (RHAMM), where subsequent activation can affect different processes including apoptosis and cell proliferation.[159–161] HA lends itself to chemical modification and is more commonly utilized in a semi-synthetic form. HA chains can be modified with acrylate, methylacrylate, or to present reactive thiols, thus altering crosslinking and providing binding sites for additional biomolecules.[162,163]

3.3.4. *Silk*

Silk is produced by silkworms and spiders and silk matrices are used widely used in tissue engineering.[66,164] It is emerging as a biomaterial for use as the matrix in 3D cancer models, among other applications, due to its biocompatibility, tensile properties, and relative stability.[165–168] Furthermore, amino acid side chains of silk may be functionalized to attach growth factors or adhesion factors; a prominent example being modification of silk fibers or films with immobilized Arg-Gly-Asp (RGD) peptides, which can act as an integrin recognition sequence and facilitate cell attachment.[66,164] Modifications such as these can allow for a more robust 3D environment by stimulating cell growth and proliferation.

Silk may be used as a film, membrane, powder, sponge, scaffold or gel, and can support growth of different cell types as a matrix for 3D cancer models.[164,169–171] Studies have used silk fibronin protein as a porus polymer scaffold or a porus silk sponge, mainly in 3D tumor modeling.[172–174]

3.3.5. *Composite and other biologically-derived matrices*

Recent studies have demonstrated the feasibility of using hydrogels derived from biological materials containing more than one type of material (composite hydrogels). Preparation of such matrices may involve a semi-synthetic process, such as modification of hyaluronan chains to present reactive groups, as mentioned in a previous section. For example, hydrogels that are composites of hyaluronan and collagen may be used. Such hydrogels have been found to be suitable as 3D brain mimetic models, as hyaluronan is a major component of brain ECM and collagen in blood vessels plays a prominent role in glioblastoma migration respectively.[175] Another example of a suitable glioblastoma microenvironment is HA-chitosan composites that can form a porus scaffold.[176] A recent study reports gelatin-methacrylamide based hydrogels as a 3D matrix for ovarian cancer models.[177] This hydrogel can be modified further to incorporate laminin and HA, which was shown to enhance cancer spheroid growth in the composite matrix.

In addition to the materials described earlier, hydrogels such as fibrin, chitosan, and alginate may be used for 3D cancer modeling.[129,178–180] Recently, amniochrionic membranes have also been used as a scaffold for 3D cancer models.[181]

3.3.6. *Synthetic matrices*

Synthetic matrices are commonly employed as scaffolds in tissue engineering. These matrices can also be used for 3D cancer modeling, though they are less popular than biologically derived matrices. Polyacrylamide synthetic hydrogels have been widely used for 2D studies and for fabricated 'microtracks', but these are not the most suitable choice for 3D models due to their inability to provide biological cues and toxicity.[104,182–184] Polyethylene glycol (PEG) hydrogels have been used as 3D matrices in tissue engineering.[185] PEG is biocompatible, relatively inexpensive and can be chemically modified for desired properties, including the modulation of mechanical properties.[186,187] Ligands such as growth factors or oligopeptides can also be incorporated in PEG hydrogels to engineer bioactivity.[184] PEG hydrogels have been successfully utilized as a matrix for modeling 3D tumors.[188] Poly(D,L-lactic-co-glycolic acid) (PLGA or PLG), polylactic acid (PLA), poly(E-caprolactone) and polyvinyl alcohol (PVA) are other synthetic substances that can be used as scaffolds for 3D tumor models.[145,184]

Though chemically pure and relatively simple to engineer, synthetic matrices are not as biologically relevant, and are unable to provide signaling cues essential to the ECM microenvironment.[145] Inclusion of biomolecules such as growth factors is possible in these matrices to provide a more natural microenvioronment for tumor modeling. Also, hybrid matrices, discussed in the next section, that integrate both synthetic materials like PEG, and biologically derived materials, can offer alternatives to purely synthetic matrices.

3.3.7. *Semi-synthetic matrices (biological-synthetic hybrids)*

Scaffolds composed of peptides designed for self-assembly have emerged as 'designer' 3D matrices for tissue modeling.[189,190] The peptide solutions are capable of forming scaffolds upon exposure to physiological media or salt solutions, making it possible to have a 3D microenvironment for encapsulated cells.[190] These scaffolds have been shown to be able to support cell growth and differentiation for a variety of cell types.

Due to high water content of these biomaterials, they are also called peptide hydrogels. PuraMatrix is a commercially available peptide hydrogel composed of peptide RAD16-I.[29,191] Interestingly, peptide hydrogels such as PuraMatrix can be combined with other materials like PolyHIPE polymer to successfully yield a composite that is a variation of the peptide hydrogel.[191] Peptide amphiphiles or beta-hairpin peptides are special types of peptides that may also form scaffolds.[145,192,193] Also, biologically active motifs can be included in the self-assembling peptide design to stimulate and promote cell growth and differentiation of specific cell types.[194]

PEG is a popular base for hybrid hydrogels. Extracel is a semi-synthetic hydrogel that contains chemically modified hyaluronan and gelatin (to present thiol-groups), that are both crosslinked polyethylene glycol diacrylate.[195,196] An elastin-mimetic polymer consisting of PEG and alanine-rich lysine containing peptides and a collagen mimetic polymer consisting of PEG diacrylate and collagen mimetic peptides are other examples.[197,198]

Such hybrid hydrogels containing multiple components can be variously prepared and fine-tuned for desired properties, and also functionalized to provide relevant signaling cues.[105,199]

Different matrices are available to support and form a complete 3D tumor microenvironment to facilitate the development of robust and biologically relevant cancer models. Appropriate matrices may be selected based on the cell types used and the specific application desired.

4. Heterocellular Tumor Models

4.1. *Heterotypic communication in tumor ecosystems*

Healthy tissues and organs exist as ecosystems: heterotypic environments that support the needs of their constituents. As part of being a robust system, normal tissue, and cellular microenvironments resist tumor cell proliferation and cancer formation.[200] Although these ecosystems are robust, they are not impervious to disruption. In normal tissue, the stroma plays a critical role in maintaining the balance of homeostasis. However, this balance can be tilted due to the expression of growth factors that promote angiogenesis, altered ECM expression, accelerated fibroblast proliferation, and increased inflammatory recruitment.[201] This shift in balance and associated changes in cellular phenotypes lead to the ecosystem we call a tumor.

It is well established that tumors are complex environments of cooperating cell types, supported by a diverse matrix and vasculature. The stroma itself is not monotypic, but is composed of a variety of cell types such as immune cells (for example, lymphocytes and dendritic cells), inflammatory cells (for example, monocytes and granulocytes), muscle and myofibroblasts, and vascular cells (for example, lymph-endothelial, vascular-endothelial cells, and pericytes).[202] The multitude of molecular and structural components of the tumor ecosystem point to the heterogeneity of this system. In order to be successful in understanding, and ultimately treating tumors, model systems must reflect these heterogeneities and relate these components as an ecosystem.

As described in the previous sections, the relationship between the tumor and its supporting matrix is a critical component of the tumor microenvironment. The ECM supports fibroblasts that produce abundant stromal proteins. The basement membrane, which separates the epithelial and endothelial cells from the mesenchymal cells of the stroma, supports basal epithelial cells, myoepithelial cells, and fibroblasts that express major structural proteins such as collagen VI, laminin, entactin, and heparan-sulphate proteoglycans.[203] In normal tissue, the basement membrane functions as a deliberate barrier for the passage of growth factors and signaling molecules between the epithelial and stromal cells.[204] However in tumorigenesis, extracellular proteases can degrade the basement membrane in order to break down this barrier, and mobilize the growth factors and signaling molecules of the stroma.[205]

Materials such as Matrigel, collagen, and PuraMatrix have been utilized to model both the structure and signaling of the ECM and the basement membrane. These materials usually contain a set of purified proteins or a complex mixture from a host extract.[206] Methods for incorporating cells into these structures include cell-matrix mixtures, or seeding cells on top of existing matrix structures including the seminal work of Bissel and colleagues. For example, HMT-3522 human breast cancer cells cultured within a reconstituted basement membrane, formed large, heterogeneous structures[207] and ovarian cancer cells seeded on top of Matrigel formed 3D nodules with a hypoxic core leading to decreased diffusion and perfusion of drugs.[208] Thus, using such materials to model the critical ECM microenvironment yields tumor models with features similar to an *in vivo* tumor.

4.2. *Heterocellular partners*

4.2.1. *Fibroblasts*

(a) The Biology of Tumor Associated Fibroblasts

The anchor of the tumor stroma and the tumor microenvironment are the tumor-associated fibroblasts. Normal fibroblasts in healthy tissue play a critical role in maintaining the 3D structure that supports the epithelia through the ECM. Impaired interactions between the epithelial cells and the ECM, however, can result in transformation of the epithelia.[201] Once the stromal-epithelial relationship is dysregulated, the stroma is activated and fibroblasts are no longer background supporters of tumor growth, but instead are inducers of tumor progression.

Normal stromal fibroblasts can mediate the regulatory activities of estrogen, epidermal growth factor, transforming growth factor α, and transforming growth factor β-1.[207] Notably, it has been shown in numerous tumor systems, such as breast and skin, that stromal fibroblasts immediately adjacent to transformed epithelial cells are modified in both their phenotypic and genotypic patterns.[209–212] The activation of the stroma is identified as expression of α-smooth muscle actin, ECM proteins, and growth factors acting in an autocrine and paracrine fashion such that they potentiate and support the survival of the tumor. Fibroblasts play a key role in the activation of the stroma; they are responsible for the synthesis, deposition, and remodeling of the ECM in the tumor, and are also a source of paracrine growth factors.[201] At the site of a tumor, fibroblasts remain in a perpetually activated state, continuously expressing these growth factors and inducing proliferative signaling.[213]

(b) Fibroblasts in Heterocellular Tumor Models

Activation of the stroma is a crucial component in carcinogenesis. *In vitro* models have demonstrated success in recapitulating the tumorigenic properties of activated fibroblasts. An *in vitro* model of prostate cancer demonstrated that tumor fibroblasts, but not normal prostatic fibroblasts, stimulated epithelial proliferation and malignant transformation, and that tumor fibroblasts do not stimulate the growth of normal primary epithelial cells under identical conditions.[209] *In vitro* models have shown that not only do normal fibroblasts and tumor fibroblasts differ in their histology, but also in their dynamics, as tumor fibroblasts exhibit enhanced proliferation and migration.[214,215] In a 3D model using collagen, fibroblasts cultured with an intracellular fluid-like buffer were shown to maintain their cytoskeleton and migrate through the ECM.[216]

The incorporation of fibroblasts into *in vitro* tumor models is not only important for recapitulating the signaling and cellular behavior of the activated stroma, but also leads to an observable effect on treatment response. In a heterocellular pancreatic cancer *in vitro* model, human pancreatic stellate cells increased pancreatic tumor cell proliferation, migration, invasion, and colony formation, and reduced the effectiveness of gemcitabine and radiation therapy.[217]

4.2.2. *Endothelial cells*

(a) The Biology of Tumor Endothelial Cells

The growth and nourishment of tumors is made possible in part by endothelial cells in the tumor microenvironment. Endothelial cells are ubiquitous within tumors because tumors are vascular.[218] Without the formation of new blood vessels, carcinomas will not grow beyond a very small size or metastasize and reform in distant organs.[219,220] Yet, it is important to note that TECs are genetically the same and homogenous with healthy endothelial cells. However, TECs have a more pro-angiogenic phenotype than normal endothelial cells.[221] The altered phenotype of TECs may be due to their differences in responsiveness to epidermal growth factor, adrenomedullin, and vascular endothelial growth factor.[222]

Through paracrine regulation, endothelial cells modulate a diverse spectrum of pathophysiologic processes in normal and hyperplastic tissues.[218] Complementarily, factors secreted by tumors may alter the characteristics of healthy endothelial cells to a more pro-angiogenic phenotype. Highly angiogenic tumor cells secrete soluble factors, which are in turn received by TECs, resulting in the pro-angiogenic characteristics of endothelial cells.[222] Therefore, the heterogeneous behavior of TECs is a direct result of the factors expressed by the heterogeneous tumor cells in the microenvironment.

(b) Endothelial Cells in Heterocellular Tumor Models

Endothelial cells provide important signaling cues and structural support for tumor angiogenesis. An *in vitro* co-culture of normal endothelial cells and highly metastatic tumor cells led to an increase in expression of pro-angiogenic genes and phenotypic changes in the normal endothelial cells.[222] Endothelial cell-conditioned media has been shown to reduce the growth of both breast and lung carcinoma cells by 40% *in vitro*, and significantly reduced the invasiveness of both cancer cell lines. In this same study, fibroblast-conditioned media was used as a control, and had no effect on cell proliferation or invasiveness.[218] In contrast, a murine 3D co-culture model demonstrated that endothelial-tumor heterocellular spheroids had a higher growth rate than tumor-only spheroids.[22] The contrasting results of these two models generate two important points: the behavior of TECs is heterogeneous and tissue-dependent, and different types of heterocellular models provide different insights. Endothelial cells play an important role in heterocellular tumor models as partners with tumor cells in the heterogeneous cross-talk that influences the phenotype of both the tumor cells and the endothelial cells.

In vitro heterocellular models of tumor cells and endothelial cells have suggested that endothelial cells play a protective role in a tumor's response to treatment. In a 3D co-culture model of murine endothelial cells (2H11) and murine mammary tumor cells (4T1-GFP), the endothelial cells sensitized the tumor cells to taxol chemotherapy while also protecting the tumor cells from ionizing radiation. Molecular analysis of these co-cultures revealed up-regulation of pro-angiogenic factors in the heterocellular spheroids compared to the tumor-only spheroids.[22]

4.2.3. *Macrophages*

(a) The Biology of Tumor Associated Macrophages

Tumor associated macrophages (TAMs) play a supportive role in developing tissues via matrix remodeling capacities, synthesis and growth of angiogenesis factors, and engulfment of apoptotic cells. Not by coincidence, TAMs are recruited by tumors to the tumor microenvironment. It is in this microenvironment that the intrinsic roles of the TAMs are exploited for the benefit of the tumor: chronic inflammation, matrix remodeling, tumor cell division, intravasation, angiogenesis, and seeding at distant sites.[223] The supportive roles that TAMs are recruited for align with evidence that tumor-associated macrophages are abundant in most solid tumors.[224]

One notable role of TAMs is their role in chronic inflammation. In fact, cancers frequently arise in areas of chronic inflammation, and infections that cause chronic inflammation are responsible for more than 15% of cancers worldwide.[225] Mechanisms whereby inflammatory cells influence cancer initiation and promotion likely involve secretion of cytokines, growth factors, and chemokines by inflammatory cells that stimulate proliferation of epithelia, as well as the generation of reactive oxygen species that can cause DNA damage.[225] Clinically, it has been found that there is a correlation between an abundance of TAMs and poor prognosis in breast, prostate, ovarian, and cervical cancers.[226–228]

(b) Macrophages in Heterocellular Tumor Models

The incorporation of macrophages into tumor models is still an emerging practice. An early model probed the generation of TAM *in vitro* using co-cultures of monocytes, monocyte-derived macrophages, and urothelial carcinoma cells in 3D spheroids. As mentioned earlier, macrophages can be activated to both inhibit tumor growth, but also to support tumor growth and metastasis, presenting the need for a model that can demonstrate this function *in vitro*. Even without the presence of any other stromal cells, the tumor cells in the 3D environment directly influence the naive macrophages, yielding TAMs with an infiltrating and pro tumor development phenotype.[229] In a 3D *in vitro* model of mesothelioma, tumor fragment tissue was grown in culture and formed tumor spheroids. These spheroids maintained many characteristics of the original mesothelioma tumor, including the presence of viable macrophages and a collagen-rich stroma, even months after plating. When comparing this 3D system to monolayer cultures of mesothelioma, it was found that the 3D cultures were more resistant to apoptosis.[230] In another system, a co-culture model of macrophages and human tumor spheroids were used to study breast cancer *in vivo.* In this model, the macrophages were shown to infiltrate deep into the central, hypoxic cores of the tumor spheroids.[231] These models serve as examples of the importance and effect of macrophages on tumor cells *in vitro*, and their ability to influence the biology of the tumor microenvironment.

4.2.4. *Adipocytes*

(a) Adipocytes: An Energy Source for Tumors that Promote Growth

It has been shown in breast and prostate cancers that adipocytes promote the growth of cancer cells.[232,233] The omentum, a large fat pad that extends from the stomach and covers the bowel, is primarily composed of these trigclyeride-storing adipocytes.[234] Importantly, in serous ovarian cancer, 80% of cases present themselves with metastases on the omentum. Most remarkably, ovarian cancer metastasis transforms the omentum from a primarily adipocyte organ, to a solid tumor devoid of adipocytes.[235] These observations promote adipocytes as an important cell type in the metastatic niche of intra-abdominal tumors.

(b) Adipocytes in Heterocellular 3D Models

In vitro co-cultures have been used to demonstrate that the communication of soluble factors between tumor cells and adipocytes leads to an invasion of ovarian, gastric, breast and colon cancer cells.[235] Two-dimensional co-culture models of omental or peritoneal adipocytes with ovarian cancer cells demonstrated that adipocytes provide a proliferative advantage and transfer fatty acids to the tumor cells.[235] Adding another dimension to this relationship, a 3D *in vitro* co-culture of primary human mammary epithelial cells and adipocytes resulted in epithelial structures embedded in the adipocyte layers, resembling the human breast *in vivo.*[236]

4.2.5. *Mesothelial cells*

The mesothelium is a crucial layer that lines the body's serous cavities and internal organs. This monolayer of mesothelial cells provides a non-adhesive and protective surface with well-developed cell-cell junctions.[237] A unique characteristic of mesothelial cells is their ability to express both mesenchymal filaments, vimentin and desmin, but also epithelial filaments, cytokeratins.[238] The ability of mesothelial cells to undergo EMT has implications for their role in tumorigenesis as it implies the mesothelium is a probable source of fibrogenic cells during inflammation of the serous cavities.[237,239] The mesothelium is also actively involved in fluid and cell transport through the serosal cavities.[240] In ovarian cancer, a disease that relies on the dissemination of cancer cells in large part through the fluidic currents of the peritoneal cavity, attachment to and breaching of the mesothelial layer plays an important role in metastatic progression.[241–245] In a 3D model of the omentum to study ovarian cancer, Kenny *et al.* incorporated an ECM mixed with fibroblasts from normal human omentum, covered by a layer of primary mesothelial cells from normal human omentum.[30] It is becoming increasingly evident from these and other efforts that a critical component of the microenvironment that influences the biology of ovarian cancer, and other peritoneal tumors, is the layer of mesothelial cells that line the peritoneal cavity.[30,237]

5. Optical Imaging of 3D Tumor Models: Challenges and Opportunities for Quantitative Assessment of Growth and Treatment Response

The restoration of 3D tumor architecture, ECM interactions, and heterotypic cross-talk in the *in vitro* model systems described in this chapter presents both challenges and opportunities pertinent to the interrogation of these systems by imaging. In a variety of applications, the utility of gleaning the high density of information embedded in image data sets has become increasingly recognized.[40,41] For the same reasons that high content screening (HCS) approaches have become mainstays for preclinical oncology screening,[246,247] it makes sense that more biologically sophisticated 3D cancer models should be adapted for this application with even greater success. However, to be utilized to their full potential these models must be accompanied by appropriate and practical analysis tools that provide robust quantification of the structural complexity and heterogeneity that these systems restore.

The desirability of depth-resolved optical sectioning, whether by confocal or multi-photon, is a natural consideration in imaging any biological specimen with significant 3D structure. In 3D cancer models, confocal microscopy has played a key role in elucidating some of the earliest and most profound insights into specific signaling events and tissue architecture in carcinogenesis.[248–254] At the same time, confocal imaging has many limitations, particularly in any efforts that require speed and throughput, not to mention the high cost of these instruments, which often have limited availability in core facilities. For example, a typical confocal z-stack with sectioning of a few microns at 512 by 512 pixel resolution spanning the depth of a field containing 3D tumor nodules on the order of a couple of hundred micrometer thickness might require a few minutes to acquire. Obtaining statistically significant datasets from hundreds of such fields, subject to different treatments and/or growth conditions presents a major pragmatic challenge. Furthermore, unless reliable transfections with a fluorescence reporter (e.g., GFP) are available for a particular cell line being studied, confocal fluorescence microscopy images are typically non-longitudinal, acquired in terminal fixed cultures, immunostained with labeled conjugates. So, while depth-resolved sectioning is clearly essential to reveal structural or functional differences in the cells in the interior Versus the outer layers of a multicellular 3D structure, it is not necessarily a practical treatment screening or monitoring tool. Similarly, multi-photon excited fluorescence, while increasingly widely integrated into commercially available confocal systems and offering the benefit of excellent tissue penetration is, with few exceptions a laser scanning approach which suffers from the same challenges of imaging speed.

On the other hand, the most basic widefield contrast enhancement techniques, phase contrast, darkfield, and differential interference contrast (DIC), taken with reasonable approximations of 3D geometries can be powerful tools in imaging the growth of *in vitro* 3D cancer models. For example, an approach relying on darkfield imaging of large datasets in combination with time-lapse phase contrast microscopy was used to elucidate time dependent evolution of the size distribution and aggregation processes of 3D ovarian nodules.[26]

One recently published implementation of 3D tumor models in image-based therapeutic screening is shown in Fig. 1.[27] This approach, using quantitative analysis of wide-field fluorescence images of the vital dyes calcein acetoxymethyl ester and ethidium bromide, makes the deliberate trade-off of depth resolved 3D imaging for low magnification, but highly statistically powered fields of thousands of 3D nodules. This study, using a combination of 3D pancreatic and ovarian tumor models demonstrates the ability of pseudo-high throughput screening to quantify differential effects of treatment on tumor architecture that may account for previously observed synergism and contrasting growth inhibitory Versus cytotoxic effects. The method works in the limit of small nodules (less than approximately 200 μm in diameter) for which the optical attenuation over the thickness of each nodule is small relative to the total integrated fluorescence signal from each channel (linear dependence of fluorescence signal Versus nodule size).

Advanced optical microscopy techniques have also been shown to be capable of providing label-free longitudinal imaging of 3D growth behavior and treatment response in 3D tumor models. Studies using optical coherence tomography (OCT) have revealed contrasting dynamic structural changes in chemotherapy and PDT

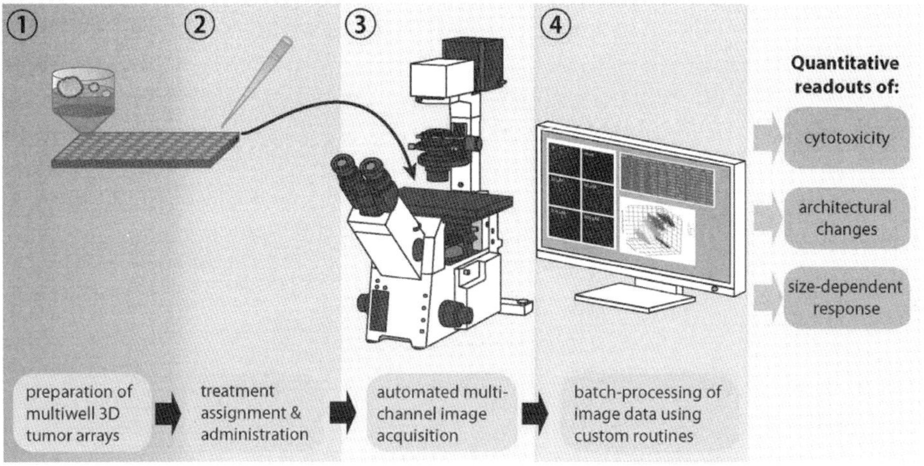

Fig. 1. Representative workflow for image-based therapeutic assessment in 3D tumor models. In this approach, 3D tumor models are plated in multi-well basement membrane overlays on glass bottom imaging plates, imaged via automated microscopy and batch processed off-line to obtain quantification of nodule sizes and dose-dependent cytotoxicity and architectural changes induced by treatments. Figure 1 is reproduced from Celli *et al.*[27] as permitted under the terms of the open access license (http://creativecommons.org/licenses/by-nc-nd/3.0/legalcode).

response in Matrigel-based multicellular 3D ovarian cancer overlays.[42] Digital Holographic Microscopy (DHM) has been demonstrated to provide longitudinal readouts of 3D tumor volume during normal growth and chemotherapy treatment in a similar cancer model.[255] These methods, while clearly powerful tools, are yet to be established as mainstream techniques in cell biology laboratories. Nevertheless, it is hoped that such methods will be increasingly simplified for integration into routine implementation by non-expert users for 3D cancer models and other experimental systems.

6. Physical Forces and 3D Tumor Models

6.1. *Cancer mechanobiology: Studying the impact of matrix mechanics on tumor growth, invasion, and treatment response*

The increasingly prominent role of hydrogel-based 3D tumor models has spurred the further development of these systems as a means to study mechanobiological effects in cancer. Indeed, the mechanical compliance, along with other biophysical properties of the reconstituted basement membrane materials used in 3D cell culture, is a key property of these materials that promotes *in vitro* 3D nodular growth of epithelial cancer cells that is not observed on a rigid plastic substrate. Probing the role of scaffold mechanics further, using customized basement membrane derivatives, mixtures and synthetic materials as model extracellular microenvironments with tunable rheological properties (typically expressed in terms of Young's Modulus, *E*, and/or shear modulus, *G*), a number of studies have provided a foundation in which new insight into the role of interactions with the mechanical microenvironment in regulating tumor growth, morphogenesis, invasiveness as well as response to treatment can be systematically examined. In any study intended to specifically examine the role of extracellular mechanics on tumor growth, the isolation of mechanical effects from biological effects of varying the concentration of native ECM components, is a key experimental consideration. This is particularly important as the integrins, the family of transmembrane heterodimers that act as the crucial link in mechanotransduction from extracellular force to intracellular cytoskeletal tension,[256] are ligated by ECM proteins, as discussed earlier. Yet integrin signaling also plays multiple complex roles in regulating cancer progression and survival.[257,258] Thus, simply studying mechanobiology in cancer models by varying the concentration of reconstituted ECM hydrogels is confounded by the multiple levels of biological effects, in addition to the practical consideration that this approach probes an inherently narrow range of extracellular rigidity. A variety of experimental means of achieving independently tunable ECM mechanics for evaluation of mechanobiology in 3D tumor models have been explored, with varying tradeoffs and limits of applicability. Specific strategies using commercial basement membrane, collagen, and a variety of biofunctionalized synthetic materials as well as the relative advantages and the related evaluation of influence on 3D tumor growth behavior, as depicted schematically in Fig. 2, are discussed in this section.

6.1.1. *Modulation of scaffold mechanics in biologically-derived ECM*

The crucial roles of matrix mechanics in regulating tumor growth via integrin signaling was elucidated in particular by pioneering studies conducted by the Weaver laboratory examining the role of tissue stiffness on malignant behavior.[35,114] In an elegant and comprehensive study, Paszek *et al.* systematically revealed how matrix rigidity regulates integrin signaling, induces focal adhesions, and modulates cell behavior in 3D mammary epithelial cell (MEC) cultures.[35] This work, as in a number of subsequent studies described herein, begins with investigation of hydrogels consisting of mixtures of basement membrane and collagen I that were prepared with equal proportion of the former component (to promote acinar adhesion) and varying concentrations of the latter to modulate rigidity. In this manner, elastic modulus was varied over a relatively narrow range from 175 to 1200 Pa to reveal a profound impact on MEC colony size, EGF-dependent growth, and $\beta 1$ integrin adhesions. These observations indicate that substrate compliance not only impacts morphology

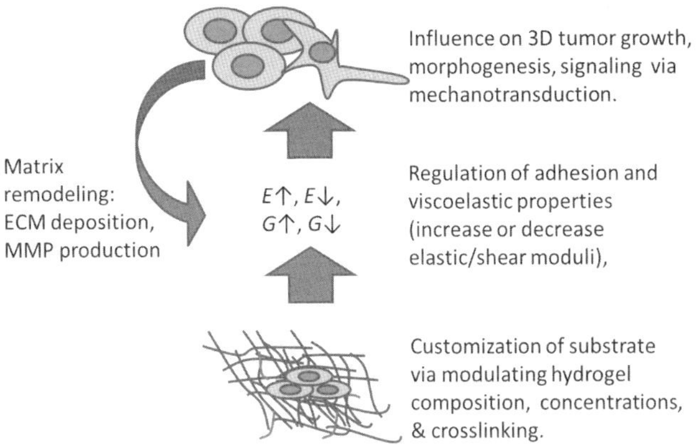

Fig. 2. A schematic representation of the implementation of hydrogel-based 3D tumor models for study of mechanobiology in tumor growth. By customized formulations and admixtures of ECM components, synthetic crosslinked gels functionalized with biocompatible peptide surfaces, use of nanofiber scaffolds, and other means of modeling variable matrix mechanics and adhesion properties, the influence of extracellular rheology on tumor growth, morphogenesis and signaling can be systematically interrogated. Furthermore, depending on the biochemical composition of the model ECM (e.g., if cleavable by MMPs) the subsequent mechanical remodeling of the matrix by the cancer cells themselves can also be evaluated.

and 3D growth behavior, but also regulates expression of key molecular targets with potential impact for matrix rigidity-dependent therapeutic interventions. Further study by the same group has revealed that matrix mechanics also regulate microRNA dependent phosphatase and tensin homolog (PTEN) signaling.[259]

Another variation on the use of reconstituted ECM mixtures was introduced by Ulrich *et al.* using a constant concentration of collagen I mixed with varying amounts of low melting temperature agarose.[260,261] The latter component, introduced as a biochemically inert means of increasing matrix rigidity, is shown to have minimal impact on the collagen gel microstructure over a range of shear modulus values (approximately $1 \text{ Pa} < G < 1\text{kPa}$). This method is explored primarily as a means to examine the role of extracellular rigidity in regulating invasive behavior in embedded cancer cells. A noted experimental challenge here is the necessity to mix and work with two materials with contrasting temperature-dependent gelation behavior. The same approach has also been implemented in examining the size of multicellular 3D pancreatic epithelial cancer cell colonies grown in overlay cultures,[262] with results similar to those of Paszek *et al.* cited earlier in studies of breast cancer cells grown on laminin-rich/collagen I matrix substrata.

In the context of collagen-based 3D culture models, other approaches exist to vary scaffold rigidity by non-enzymatic glycation of collagen, a process in which proteins are crosslinked by reducing sugar.[263] This approach increases collagen gel stiffness four-fold (175 to 730 Pa) without any significant change to the collagen architecture.[263] This method was used to examine the role of matrix stiffness in cell spreading and outgrowth of bovine aortic endothelial cells.

In the discussion of regulating mechanobiology via manipulation of naturally derived matrices such as collagen, the mechanical properties of these materials and how they interact with other materials are highly dependent on details of the purification methods. To this end, Yang *et al.* explored the differences between acid-solubilized (AS) and pepsin-treated (PT) collagen gels.[264] Chondroitin sulfate (CS), a glycosaminoglycan, induces fibril bundling, increases the mesh size, and induces changes in the viscoelastic moduli of AS gels, while it has no significant effect on the PT collagen.[264]

In order to measure rigidity-controlled tumor growth in 3D culture models using naturally-derived matrices it also critical to consider the biochemical and biological, but non-mechanical, effects of the material. It should be clear that different animal derived ECM materials, even if titrated to concentrations with equivalent rigidity, might have contrasting biological characteristics. For example, in the work of Yang *et al.* addressed in the preceding paragraph, collagen gel rigidity was tuned by variation of CS and HA concentrations but with potentially very different impacts on growth behavior depending on which stiffening agent was used. While CS inhibited glioma invasion, the HA had no effect. This example underscores the importance of careful consideration of the roles of biochemical and mechanical effects when drawing conclusions about matrix composition impacts upon tumor behavior. Similarly, a number of studies have observed marked dissemination from the tumor core of 3D tumor spheroids grown inside collagen I that is not observed in laminin rich BM. In this context, Beck *et al.* explored the variation of crosslinking density of PEG to tune rigidity in BM while monitoring the impact upon ECM invasion.[265] The outcome was a limiting of tumor growth with increased stiffness although no dissemination was reported. However, in this study the authors then use peptide-conjugated cyclodextrin (α-CDYRGDS) rings to incorporate adhesive signals into PEG. The latter method permitted control over adhesion and rigidity to reveal dissemination of normal and malignant MECs at intermediate values of adhesion and stiffness.

6.1.2. *Probing cancer mechanobiology in synthetic hydrogels*

The use of reconstituted animal-derived ECM hydrogels for modeling mechanobiology effects in 3D culture poses certain inherent challenges. As noted above earlier, the ECM proteins themselves have biological effects that may be difficult to isolate from mechanical effects. Additionally, the materials themselves exhibit batch-dependent variability that impacts reproducibility of matrix mechanics, microstructure, and other biophysical factors. For these reasons, the development of purely synthetic scaffolds for 3D culture with total control of structural, mechanical, and biochemical function are highly desirable. To this end, a synthetic RADA-16 peptide hydrogel system of self-assembling nanofibers introduced by Zhang *et al.*[189,266] was developed to provide customization of ECM composition by attaching functional groups to peptide cores to mimic collagens, laminins, and other ECM components, and has been shown to support 3D cell growth.[29,194,267,268] The commercially available version of this RADA-16 product produces a very soft hydrogel, and as such has applicability as a substrate with tunable stiffness.

A well-established approach to variation of substrate rigidity, which was developed in the context of traction force analysis of cell motility,[269] entails biocompatible RGD functionalization of crosslinked polyacrylamide (PA) gels. Because the substrate compliance is exquisitely controlled by the crosslinking of the synthetic PA substrate, it is not limited by the concentrations of commercially-available ECM components, and thus able to probe a much wider range of substrate rigidities. However, this approach, which requires processing of the 2D substrate surface to achieve RGD functionalization of PA is inherently limited to models involving overlay on the substrate (in contrast to embedding) and does not allow invasion below the surface. In general there are significant considerations in adopting an appropriate 3D model system geometry that have been discussed elsewhere[270] and this can be an important point depending on the experimental system being probed.

A synthetic matrix material demonstrating many of the same advantages has also been developed using HA-methacrylate polymers based on a previously established method,[271] and functionalized with RGD peptides to promote cell adhesion.[272] This approach has been show to achieve variation of shear modulus in the range of 50 Pa to 35 kPa with corresponding mesh sizes between approximately 50 to 150 nm and independent control of ECM ligand density (RGD density). Building on earlier work by the same group elucidating the role of ECM rigidity in regulating the spreading and motility of glioma cells,[273] these RGD-derivatized HA matrices were used specifically to mimic the biophysical microenvironment of normal and malignant brain tissue. In particular the authors note that the nanometer sized pores exhibited by the HA–RGD hydrogels are similar to the extracellular spaces in brain tissue and that observing the adhesion and rigidity-dependent motility of glioma cells in this environment is significant, in that it contrasts the mode of motility in which cells move along collagen fibers.

The impact of independent variation of mechanical and biochemical properties in 3D tumor model scaffolds have also been examined using PEG gels modified with specific bioactive peptides. Gill *et al.* used synthetic PEG scaffolds activated by bioactive peptides to study murine lung adenocarcinoma.[274] This modified PEG exhibits cell adhesion and cell mediated proteolytic degradation while the stiffness can be adjusted independently. Fibronectin-derived RGDS was incorporated in the scaffold to induce cell adhesion.[274] This approach leads to significant influences on epithelial morphogenesis and proliferation while being able to independently regulate the matrix stiffness and the concentration of the cell adhesive ligand.

A novel approach to examining the role of extracellular mechanical forces in regulating growth in 3D tumor spheroids was recently described using alginate shells that encapsulate tumor spheroids.[275] This approach entails the implementation of a microfluidic system, in which syringe-pumped co-aligned capillaries containing the alginate and cell solutions are combined in such a way as to drive the calcium-mediated gelation of the alginate to form a shell around a multicellular cluster of controlled size. An elegant feature of this approach is that the elastic microcapsules serve to impose mechanical confinement, yet also serve as quantitative mechanical sensors that measure the stress exerted by the expanding spheroids. This approach, though perhaps difficult to implement in laboratories that are not equipped for microfluidics, comprises an elegant way to study the impact of controlled elastic confinement on growth behavior in multicellular spheroids.

6.1.3. *Towards informing tumor stiffness-dependent treatment response and other future directions for mechanobiology in 3D tumor models*

From the growing body of evidence in the scientific literature it is increasingly clear that extracellular tissue mechanics play myriad complex and often counterintuitive roles in tumor growth, signaling, and invasive behavior. While we may have just begun to scratch the surface of understanding cancer mechanobiology, it seems reasonable to project that hydrogel-based 3D cell culture models will continue to play a crucial role in this investigation for all of the reasons cited. Animal model systems may restore more complex heterotypic cellular interactions that play critical mechano-regulatory functions, however, detailed quantitative readouts of growth and biochemical signaling are far less accessible for *in situ* monitoring. Indeed a crucial feature of the *in vitro* 3D systems discussed here is that they provide a unique balance between restoration of physiologically relevant microenvironmental cues with convenience and suitability for detailed imaging-based characterization.[26,27,42,276–278]

A natural extension of the methods for investigating mechanobiology in 3D cancer models described here is the development of a concerted effort to leverage these insights in therapeutic assessment. Building on the considerable progress in using 3D tumor models as robust tools to elucidate growth behavior, these research tools could be extended further to provide a new framework to evaluate therapeutics customized for specific microenvironments. Indeed, the idea of targeting tumor stroma is being explored in the clinic,[279–281] primarily as a means to enhance drug penetration into dense fibrotic and hypovascular lesions. Taking this concept a step further, using *in vitro* 3D tumor models as a platform for therapeutic evaluation with customized matrix mechanics could reveal treatment strategies that explicitly target the rigidity dependent malignant behavior that these systems exhibit. In other words, viewing stromal rigidity as a mechanosensitive signaling partner, in addition to a potential physical obstruction to drug delivery. In conjunction with clinical efforts to explicitly disrupt tumor stroma,[279–281] the identification of rigidity-dependent growth and biomarker status enabled by customized *in vitro* 3D systems could inform the use of adjuvant agents to drive the mechanical microenvironment to favorable conditions for enhanced efficacy, and potentially interrupt tumor promoting stromal remodeling. Furthermore, the development and refinement of approaches that allow monitoring of the changes induced by the cells upon their environment enable these platforms to study both sides of the mechanoregulatory dialogue between cancer cells and the physical microenvironment, which they modify and adapt in response to a variety of extracellular cues.[36] Moving forward, these concepts could be integrated with technologies that are (or are potentially) suitable for quantitative clinical mechanical imaging[282–284] to achieve patient-customized mechanobiology informed therapeutics.

6.2. *The impact of hydrodynamic stress on 3D cancer biology*

Tumor cells encounter many barriers to growth and progression, including shear stress from hydrodynamic flow, as they advance through the various phases that lead to metastatic disease.[285,286] Only a small fraction of tumor cells overcome these barriers and manage to colonize distant sites. Recent studies integrating these physical cues with *in vitro* tumor models have provided valuable insights into the impact of fluidic stress on the biological characteristics of cancer cells.[24,287–290] Elegant studies by Shields and colleagues[290] explored the influence of flow on the mechanisms used by tumor cells to access lymphatics. The authors showed that interstitial flow amplified autologous chemokine gradients and enhanced the polarization and migration of tumor cells through 3D matrices that mimicked lymphatic endothelium. Polacheck and colleagues subsequently demonstrated that cell density and the strength of flow determined the direction of migration of MDA-MB-231 breast carcinoma cells in 3D collagen I scaffolds.[291] The authors confirm that autologous chemotaxis plays a role in migration of tumor cells with flow, but also describe a competing mechanism that is dependent on the activation of focal adhesion kinase (FAK). A more recent follow-up study by the same group demonstrated that the stresses imparted by interstitial flow on cells are balanced by an increase in FAK, F actin, vinculin, and paxillin-dependent protrusions localized to the sites of maximum tension. Interestingly, silencing paxillin, which plays a key role in actin localization, did not impact the speed of cell migration, but did affect rheotaxis.[287] These studies provide important insights into the influence of interstitial flow on the migration of cells through tissue via the activation of mechanotransduction and chemotaxis pathways. A recent study by our group evaluated the impact of flow on the genetic, molecular, and morphological characteristics of 3D tumor nodules.[24] Hydrodynamic stress caused a post-translational increase in EGFR expression and activation, as well as changes in morphology and biomarker expression consistent with EMT. These results suggest that fluidic streams induce genetic and phenotypic changes indicative of aggressive and resistant disease. These and other studies[288,289,292] demonstrate the importance of developing 3D systems that integrate hydrodynamic stress and stromal cues including signaling with heterotypic partners such as endothelial and immune cells. The mechanistic insights gained from these models may inform the design of therapeutic strategies that account for molecular changes due to physical stress to complement conventional therapeutic regimens.

7. Perspective and Conclusion

The value of restoring 3D structure *in vitro* has been under investigation since some of the earliest studies by Costachel and colleagues[55] and Sutherland and colleagues.[43] Sutherland *et al.* reported the generation of non-adherent tumor spheroids using Chinese hamster V79 lung cells with growth kinetics that were similar to corresponding transplants in mice. These spheroids showed zones of proliferation and necrosis that were missing in monolayer cultures, due apparently to gradients of decreasing nutrients and oxygen from the periphery to the center of the nodules.[43] Studies by Teicher and colleagues[293] and Kobayashi and colleagues[294] demonstrated the importance of restoring microenvironmental cues to meaningfully evaluate treatment response. Chemoresistant sublines of EMT-6 mammary tumors that were generated *in vivo* showed no significant resistance when these cells were treated with the same agents in monolayer.[293] Interestingly, the chemoresistance patterns observed *in vivo*, including cross-agent resistance, were restored when these sublines were plated as multicellular spheroids.[294] Subsequent studies by Arteaga[295] and Foster,[296–300] among others, have provided valuable insights into mechanisms of tumor growth and progression, targeting chemoresistance and designing therapeutic regimens using multicellular spheroids.[6,22,116,242–245,301–305]

The Bissell laboratory has pioneered concomitant efforts to develop 3D models for adherent tumor nodules that restore architecture and communication with the ECM.[19,37,63,71,77,83,200,202,248–254,270,306–309] Elegant studies using these models have provided valuable insight into the importance of communication with ECM in regulating cellular organization and polarization and the loss of this communication as a critical factor in creating and maintaining a malignant phenotype.

These and related publications have informed the use of tumor spheroids and adherent 3D models as important tools to elucidate biological mechanisms and to guide the design and evaluation of targeted and combination therapies. Building on these studies, increasingly sophisticated 3D models are being developed to integrate critical cues from the tumor microenvironment. Due in part to findings from these bioengineered systems it is becoming increasingly evident that communication with heterotypic partners and physical influences, such as mechanical and fluidic stress, can play an essential role in the fate of tumors. The ultimate value and feasibility of using 3D models to improve the efficiency of drug discovery will depend on the balance between the biological relevance, throughput and costs of these systems, among other factors. With continued development, bioengineered 3D models could play an increasingly important role in the design and evaluation of customized therapies to improve the management of cancer.

Acknowledgments

This work was supported by the following grants from the National Institutes of Health and the National Cancer Institute: P01 CA084203-06, R01 CA158415, R01 CA160998, and R01 CA156177.

References

1. W. N. Hait, Anticancer drug development: the grand challenges. *Nat. Rev. Drug Discov.*, **9**, 253–254 (2010).
2. S. Breslin and L. O'Driscoll, Three-dimensional cell culture: the missing link in drug discovery. *Drug Discov. Today*, **18**, 240–249 (2013).
3. M. R. Ebrahimkhani, C. L. Young, D. A. Lauffenburger, L. G. Griffith, and J. T. Borenstein, Approaches to *in vitro* tissue regeneration with application for human disease modeling and drug development. *Drug Discov. Today*, **19**,754–762.
4. F. Pammolli, L. Magazzini, and M. Riccaboni, The productivity crisis in pharmaceutical R&D. Nat. *Rev. Drug Discov.*, **10**, 428–438 (2011).
5. J. A. DiMasi and H. G. Grabowski, Economics of new oncology drug development. *J. Clin. Oncol.*, **25**, 209–216 (2007).
6. L. A. Kunz-Schughart, J. P. Freyer, F. Hofstaedter, and R. Ebner, The use of 3-D cultures for high throughput screening: the multicellular spheroid model. *J. Biomol. Screen.*, **9**, 273–285 (2004).
7. A. Ocana, A. Pandiella, L. L. Siu, and I. F. Tannock, Preclinical development of molecular-targeted agents for cancer. *Nat. Rev. Clin. Oncol.*, **8**, 200–209 (2011).

8. T. A. Yap, S. K. Sandhu, P. Workman, and J. S. de Bono, Envisioning the future of early anticancer drug development. *Nat. Rev. Cancer*, **10**, 514–523 (2010).
9. D. Hanahan and R. A. Weinberg, Hallmarks of cancer: the next generation. *Cell*, **144**, 646–674 (2011).
10. D. A. Haber, N. S. Gray, and J. Baselga, The evolving war on cancer. *Cell*, **145**, 19–24 (2011).
11. R. L. Schilsky, Personalized medicine in oncology: the future is now. *Nat. Rev. Drug Discov.*, **9**, 363–366 (2010).
12. A. M. Gonzalez-Angulo, B. T. Hennessy, and G. B. Mills, Future of personalized medicine in oncology: a systems biology approach. *J. Clin. Oncol.*, **28**, 2777–2783 (2010).
13. D. H. Roukos, Systems medicine: a real approach for future personalized oncology? *Pharmacogenomics*, **11**, 283–287 (2010).
14. L.V. Sequist, A. Muzikansky, and J. A. Engelman, A new BATTLE in the evolving war on cancer. *Cancer Discov.*, **1**, 14–16 (2011).
15. E. S. Kim, R. S. Herbst, I. I. Wistuba *et al.*, The BATTLE trial: personalizing therapy for lung cancer. *Cancer Discov.*, **1**, 44–53 (2011).
16. A. M. Tsimberidou, U. Ringborg, and R. L. Schilsky, Strategies to overcome clinical, regulatory, and financial challenges in the implementation of personalized medicine. *American Society of Clinical Oncology educational book/ASCO American Society of Clinical Oncology Meeting*, 118–125 (2013).
17. L. G. Griffith and M. A. Swartz, Capturing complex 3D tissue physiology *in vitro*. *Nat. Rev. Mol. Cell Biol.*, **7**, 211–224 (2006).
18. K. M. Yamada and E. Cukierman, Modeling tissue morphogenesis and cancer in 3D. *Cell*, **130**, 601–610 (2007).
19. B. Weigelt, C. M. Ghajar, and M. J. Bissell, The need for complex 3D culture models to unravel novel pathways and identify accurate biomarkers in breast cancer. *Adv. Drug Deliv. Rev.*, **69–70**, 42–51 (2014).
20. N. T. Elliott and F. Yuan, A review of three-dimensional *in vitro* tissue models for drug discovery and transport studies. *J. Pharm. Sci.*, **100**, 59–74 (2011).
21. L. C. Kimlin, G. Casagrande, and V. M. Virador, *In vitro* three-dimensional (3D) models in cancer research: An update. *Mol. Carcinog.*, **52**, 167–182 (2013).
22. M. Upreti, A. Jamshidi-Parsian, N. A. Koonce *et al.*, Tumor-endothelial cell three-dimensional spheroids: New aspects to enhance radiation and drug therapeutics. *Transl. Oncol.*, **4**, 365–376 (2011).
23. I. Rizvi, J. P. Celli, and C. L. Evans *et al.*, Synergistic enhancement of carboplatin efficacy with photodynamic therapy in a three-dimensional model for micrometastatic ovarian cancer. *Cancer Res.*, **70**, 9319–9328 (2010).
24. I. Rizvi, U. A. Gurkan, S. Tasoglu *et al.*, Flow induces epithelial-mesenchymal transition, cellular heterogeneity and biomarker modulation in 3D ovarian cancer nodules. *Proc. Natl. Acad. Sci. USA*, **110**, E1974–1983 (2013).
25. I. Rizvi, S. Anbil, N. Alagic *et al.*, PDT dose parameters impact tumoricidal durability and cell death pathways in a 3D ovarian cancer model. *Photochem. Photobiol.*, **89**, 942–952 (2013).

26. J. P. Celli, I. Rizvi, C. L. Evans, A. O. Abu-Yousif, and T. Hasan, Quantitative imaging reveals heterogeneous growth dynamics and treatment-dependent residual tumor distributions in a three-dimensional ovarian cancer model. *J. Biomed. Opt.*, **15**, 051603 (2010).
27. J. P. Celli, I. Rizvi, A. R. Blanden *et al.*, An imaging-based platform for high-content, quantitative evaluation of therapeutic response in 3D tumour models. *Sci. Rep.*, **4**, 3751 (2014).
28. R. Rahmanzadeh, P. Rai, J. P. Celli *et al.*, Ki-67 as a molecular target for therapy in an *in vitro* three-dimensional model for ovarian cancer. *Cancer Res.*, **70**, 9234–9242 (2010).
29. A. O. Abu-Yousif, I. Rizvi, C. L. Evans, J. P. Celli, and T. Hasan, PuraMatrix encapsulation of cancer cells. *J. Vis. Exp.* (34), e1692, doi: 100.3791/1692 (2009).
30. H. A. Kenny, T. Krausz, S. D. Yamada, and E. Lengyel, Use of a novel 3D culture model to elucidate the role of mesothelial cells, fibroblasts and extra-cellular matrices on adhesion and invasion of ovarian cancer cells to the omentum. *Int. J. Cancer*, **121**, 1463–1472 (2007).
31. K. M. Nieman, H. A. Kenny, C. V. Penicka *et al.*, Adipocytes promote ovarian cancer metastasis and provide energy for rapid tumor growth. *Nat. Med.*, **17**, 1498–1503 (2011).
32. M. Alemany-Ribes and C. E. Semino, Bioengineering 3D environments for cancer models. *Adv. Drug Deliv. Rev.*, 79, 40–49 (2014).
33. D. Huh, G. A. Hamilton, and D. E. Ingber, From 3D cell culture to organs-on-chips. *Trends Cell Biol.*, **21**, 745–754 (2011).
34. S. Kumar and V. M. Weaver, Mechanics, malignancy, and metastasis: the force journey of a tumor cell. *Cancer Metast. Rev.*, **28**, 113–127 (2009).
35. M. J. Paszek, N. Zahir, K. R. Johnson *et al.*, Tensional homeostasis and the malignant phenotype. *Cancer Cell*, **8**, 241–254 (2005).
36. D. Jones, W. Hanna, H. El-Hamidi, and J. Celli, Longitudinal measurement of extracellular matrix rigidity in 3D tumor models using particle-tracking microrheology. *JoVE*, e53102 (2014).
37. C. M. Ghajar, H. Peinado, H. Mori *et al.*, The perivascular niche regulates breast tumour dormancy. *Nat. Cell Biol.*, **15**, 807–817 (2013).
38. D. Wirtz, K. Konstantopoulos, and P. C. Searson, The physics of cancer: The role of physical interactions and mechanical forces in metastasis. *Nat. Rev. Cancer*, **11**, 512–522 (2011).
39. M. D. Glidden, J. P. Celli, I. Massodi, I. Rizvi, B. W. Pogue, and T. Hasan, Image-based quantification of benzoporphyrin derivative uptake, localization, and photobleaching in 3D tumor models, for optimization of pdt parameters. *Theranostics*, **2**, 827–839 (2012).
40. K. W. Eliceiri, M. R. Berthold, I. G. Goldberg *et al.*, Biological imaging software tools. *Nat. Methods*, **9**, 697–710 (2012).
41. G. Myers, Why bioimage informatics matters. *Nat. Methods*, **9**, 659–660 (2012).
42. C. L. Evans, I. Rizvi, T. Hasan, and J. F. de Boer, *In vitro* ovarian tumor growth and treatment response dynamics visualized with time-lapse OCT imaging. *Opt. Express*, **17**, 8892–8906 (2009).

43. R. M. Sutherland, J. A. McCredie, and W. R. Inch, Growth of multicell spheroids in tissue culture as a model of nodular carcinomas. *J. Natl. Cancer Inst.*, **46**, 113–120 (1971).
44. R. P. W. Schwarz, and D. A. Wolf, inventor The United States of America as Represented By The Administrator of the National Aeronautics and Space Administration, assignee, Rotating bio-reactor cell culture apparatus. United States of America (1991).
45. R. Mitteregger, G. Vogt, E. Rossmanith, and D. Falkenhagen, Rotary cell culture system (RCCS): A new method for cultivating hepatocytes on microcarriers. *Int. J. Artif. Organs*, **22**, 816–822 (1999).
46. B. Grun, E. Benjamin, J. Sinclair *et al.*, Three-dimensional *in vitro* cell biology models of ovarian and endometrial cancer. *Cell Prolif.*, **42**, 219–228 (2009).
47. S. J. Smith, M. Wilson, J. H. Ward *et al.*, Recapitulation of tumor heterogeneity and molecular signatures in a 3D brain cancer model with decreased sensitivity to histone deacetylase inhibition. *PLoS One*, **7**, e52335 (2012).
48. A. L. van Wezel, Growth of cell-strains and primary cells on micro-carriers in homogeneous culture. *Nature*, **216**, 64–65 (1967).
49. R. P. Schwarz, T. J. Goodwin, and D. A. Wolf, Cell culture for three-dimensional modeling in rotating-wall vessels: an application of simulated microgravity. *J. Tissue Cult. Methods*, **14**, 51–57 (1992).
50. N. E. Timmins and L. K. Nielsen, Generation of multicellular tumor spheroids by the hanging-drop method. *Methods Mol. Med.*,**140**, 141–151 (2007).
51. R. G. Harrison, The outgrowth of the nerve fiber as a mode of protoplasmic movement. *J. Exp. Zool.*, **9**, 787–846 (1910).
52. G. M. Keller, *In vitro* differentiation of embryonic stem cells. *Curr. Opin.Cell Biol.*, **7**, 862–869 (1995).
53. R. Foty, A simple hanging drop cell culture protocol for generation of 3D spheroids. *J. Vis. Exp.*, (51), e2720, doi: 10.3791/2720 (2011).
54. M. Zietarska, C. M. Maugard, A. Filali-Mouhim *et al.*, Molecular description of a 3D *in vitro* model for the study of epithelial ovarian cancer (EOC). *Mol. Carcinog.*, **46**, 872–885 (2007).
55. O. Costachel, L. Fadei, and E. Badea, Tumor cell suspension culture on non adhesive substratum. *Z. Krebsforsch*, **72**, 24–31, 1986.
56. O. Wichterle and D. Lim, Hydrophilic gels for biological use. *Nature*, **185**, 117–118 (1960).
57. T. W. Minett, B. J. Tighe, M. J. Lydon, and D. A. Rees, Requirements for cell spreading on polyHEMA coated culture substrates. *Cell Biol. Int. Rep.*, **8**, 151–159 (1984).
58. J. Folkman and A. Moscona, Role of cell shape in growth control. *Nature*, **273**, 345–349 (1978).
59. J. M. Yuhas, A. P. Li, A. O. Martinez, and A. J. Ladman, A simplified method for production and growth of multicellular tumor spheroids. *Cancer Res.*, **37**, 3639–3643 (1977).
60. J. Folkman and M. Hochberg, Self-regulation of growth in three dimensions. *J. Exp. Med.*, **138**, 745–753 (1973).

61. H. K. Kleinman, M. L. McGarvey, J. R. Hassell *et al.*, Basement membrane complexes with biological activity. *Biochemistry*, **25**, 312–318 (1986).
62. Y. Kubota, H. K. Kleinman, G. R. Martin, and T. J. Lawley, Role of laminin and basement membrane in the morphological differentiation of human endothelial cells into capillary-like structures. *J. Cell Biol.*, **107**, 1589–1598 (1988).
63. O. W. Petersen, L. Ronnov-Jessen, A. R. Howlett, and M. J. Bissell, Interaction with basement membrane serves to rapidly distinguish growth and differentiation pattern of normal and malignant human breast epithelial cells. *Proc. Natl. Acad. Sci. USA*, **89**, 9064–9068 (1992).
64. J. Debnath, K. R. Mills, N. L. Collins, M. J. Reginato, S. K. Muthuswamy, and J. S. Brugge, The role of apoptosis in creating and maintaining luminal space within normal and oncogene-expressing mammary acini. *Cell*, **111**, 29–40 (2002).
65. C. S. Szot, C. F. Buchanan, J. W. Freeman, and M. N. Rylander, 3D *in vitro* bioengineered tumors based on collagen I hydrogels. *Biomaterials*, **32**, 7905–7912 (2011).
66. G. H. Altman, F. Diaz, C. Jakuba *et al.*, Silk-based biomaterials. *Biomaterials*, **24**, 401–416 (2003).
67. K. Inouye, M. Kurokawa, S. Nishikawa, and M. Tsukada, Use of *Bombyx mori* silk fibroin as a substratum for cultivation of animal cells. *J. Biochem. Biophys. Methods*, **37**, 159–164 (1998).
68. G. H. Altman, R. L. Horan, H. H. Lu *et al.*, Silk matrix for tissue engineered anterior cruciate ligaments. *Biomaterials*, **23**, 4131–4141 (2002).
69. S. Zhang, T. Holmes, C. Lockshin, and A. Rich, Spontaneous assembly of a self-complementary oligopeptide to form a stable macroscopic membrane. *Proc. Natl. Acad. Sci. USA*, **90**, 3334–3338 (1993).
70. Z. Yang and X. Zhao, A 3D model of ovarian cancer cell lines on peptide nanofiber scaffold to explore the cell-scaffold interaction and chemotherapeutic resistance of anticancer drugs. *Int. J. Nanomed.*, **6**, 303–310 (2011).
71. P. A. Kenny, G. Y. Lee, C. A. Myers *et al.*, The morphologies of breast cancer cell lines in three-dimensional assays correlate with their profiles of gene expression. *Mol. Oncol.*, **1**, 84–96 (2007).
72. L. C. Windus, D. L. Kiss, T. Glover, and V. M. Avery, *In vivo* biomarker expression patterns are preserved in 3D cultures of Prostate Cancer. *Exp. Cell Res.*, **318**, 2507–2519 (2012).
73. S. Ghosh, G. C. Spagnoli, I. Martin *et al.*, Three-dimensional culture of melanoma cells profoundly affects gene expression profile: a high density oligonucleotide array study. *J. Cell Physiol.*, **204**, 522–531 (2005).
74. V. Dangles, V. Lazar, P. Validire *et al.*, Gene expression profiles of bladder cancers: evidence for a striking effect of *in vitro* cell models on gene patterns. *Br. J. Cancer*, **86**, 1283–1289 (2002).
75. K. Lawrenson, E. Benjamin, M. Turmaine, I. Jacobs, S. Gayther, and D. Dafou, *In vitro* three-dimensional modelling of human ovarian surface epithelial cells. *Cell Prolif.*, **42**, 385–393 (2009).

76. Z. Yang, Z. Wang, Y. Fan, and Q. Zheng, Expression of CD133 in SW620 colorectal cancer cells is modulated by the microenvironment. *Oncol. Lett.*, **4**, 75–79 (2012).
77. J. Le Beyec, R. Xu, S. Y. Lee *et al.*, Cell shape regulates global histone acetylation in human mammary epithelial cells. *Exp. Cell Res.*, **313**, 3066–3075 (2007).
78. T. R. Cox and J. T. Erler, Remodeling and homeostasis of the extracellular matrix: implications for fibrotic diseases and cancer. *Dis. Model Mech.*, **4**, 165–178 (2011).
79. F. T. Bosman and I. Stamenkovic, Functional structure and composition of the extracellular matrix. *J. Pathol.*, **200**, 423–428 (2003).
80. P. Lu, V. M. Weaver and Z. Werb, The extracellular matrix: a dynamic niche in cancer progression. *J. Cell Biol.*, **196**, 395–406 (2012).
81. J. T. Erler and V. M. Weaver, Three-dimensional context regulation of metastasis. *Clin. Exp. Metastasis*, **26**, 35–49 (2009).
82. S. Miyamoto, B. Z. Katz, R. M. Lafrenie, and K. M. Yamada, Fibronectin and integrins in cell adhesion, signaling, and morphogenesis. *Ann. NY Acad. Sci.*, **857**, 119–129 (1998).
83. D. Radisky, J. Muschler, and M. J. Bissell, Order and disorder: the role of extracellular matrix in epithelial cancer. *Cancer Invest.*, **20**, 139–153 (2002).
84. I. Stamenkovic, Extracellular matrix remodelling: the role of matrix metalloproteinases. *J Pathol.*, **200**, 448–464 (2003).
85. D. E. Ingber, Can cancer be reversed by engineering the tumor microenvironment? *Semin. Cancer Biol.*, **18**, 356–364 (2008).
86. D. E. Ingber, J. A. Madri, and J. D. Jamieson, Role of basal lamina in neoplastic disorganization of tissue architecture. *Proc. Natl. Acad. Sci. USA*, **78**, 3901–3905 (1981).
87. J. M. Lee, S. Dedhar, R. Kalluri, and E. W. Thompson, The epithelial-mesenchymal transition: new insights in signaling, development, and disease. *J. Cell Biol.*, **172**, 973–981 (2006).
88. S. M. Pupa, S. Menard, S. Forti, and E. Tagliabue, New insights into the role of extracellular matrix during tumor onset and progression. *J. Cell Physiol.*, **192**, 259–267 (2002).
89. H. K. Kleinman, J. Koblinski, S. Lee, and J. Engbring, Role of basement membrane in tumor growth and metastasis. *Surg. Oncol. Clin. N. Am.*, **10**, 329–338, ix (2001).
90. L. Kass, J. T. Erler, M. Dembo, and V. M. Weaver, Mammary epithelial cell: influence of extracellular matrix composition and organization during development and tumorigenesis. *Int. J. Biochem. Cell Biol.*, **39**, 1987–1994 (2007).
91. J. T. Erler, K. L. Bennewith, M. Nicolau *et al.*, Lysyl oxidase is essential for hypoxia-induced metastasis. *Nature*, **440**, 1222–1226 (2006).
92. E. J. Mackie, R. Chiquet-Ehrismann, C. A. Pearson *et al.*, Tenascin is a stromal marker for epithelial malignancy in the mammary gland. *Proc. Natl. Acad. Sci. USA*, **84**, 4621–4625 (1987).
93. G. G. Zhu, L. Risteli, M. Makinen, J. Risteli, A. Kauppila, and F. Stenback, Immunohistochemical study of type I collagen and type I pN-collagen in benign and malignant ovarian neoplasms. *Cancer*, **75**, 1010–1017 (1995).

94. M. Tamamori, H. Ito, M. Hiroe, F. Marumo, and R. I. Hata, Stimulation of collagen synthesis in rat cardiac fibroblasts by exposure to hypoxic culture conditions and suppression of the effect by natriuretic peptides. *Cell Biol. Int.*, **21**, 175–180 (1997).
95. S. Y. Lu, D. S. Wang, M. Z. Zhu, Q. H. Zhang, Y. Z. Hu, and J. M. Pei, Inhibition of hypoxia-induced proliferation and collagen synthesis by vasonatrin peptide in cultured rat pulmonary artery smooth muscle cells. *Life Sci.*, **77**, 28–38 (2005).
96. Y. Horino, S. Takahashi, T. Miura, and Y. Takahashi, Prolonged hypoxia accelerates the posttranscriptional process of collagen synthesis in cultured fibroblasts. *Life Sci.*, **71**, 3031–3045 (2002).
97. J. Kukacka, J. Bibova, H. Ruskoaho, and V. Pelouch, Protein remodeling of extracellular matrix in rat myocardium during four-day hypoxia: The effect of concurrent hypercapnia. *Gen. Phys. Biophys.*, **26**, 133–142 (2007).
98. Y. Takahashi, S. Takahashi, Y. Shiga, T. Yoshimi, and T. Miura, Hypoxic induction of prolyl 4-hydroxylase alpha (I) in cultured cells. *J. Biol. Chem.*, **275**, 14139–14146 (2000).
99. Q. T. Le, N. C. Denko, and A. J. Giaccia, Hypoxic gene expression and metastasis. *Cancer Metast. Rev.*, **23**, 293–310 (2004).
100. N. C. Denko, L. A. Fontana, K. M. Hudson *et al.*, Investigating hypoxic tumor physiology through gene expression patterns. *Oncogene*, **22**, 5907–5914 (2003).
101. K. Csiszar, Lysyl oxidases: A novel multifunctional amine oxidase family. *Prog. Nucleic. Acid Res. Mol. Biol.*, **70**, 1–32 (2001).
102. H. M. Kagan and W. Li, Lysyl oxidase: properties, specificity, and biological roles inside and outside of the cell. *J. Cell Biochem.*, **88**, 660–672 (2003).
103. P. P. Provenzano, D. R. Inman, K. W. Eliceiri, S. M. Trier, and P. J. Keely, Contact guidance mediated three-dimensional cell migration is regulated by Rho/ROCK-dependent matrix reorganization. *Biophys. J.*, **95**, 5374–5384 (2008).
104. P. P. Provenzano, K. W. Eliceiri, J. M. Campbell, D. R. Inman, J. G. White, and P. J. Keely, Collagen reorganization at the tumor-stromal interface facilitates local invasion. *BMC Med.*, **4**, 38 (2006).
105. W. Wang, J. B. Wyckoff, V. C. Frohlich *et al.*, Single cell behavior in metastatic primary mammary tumors correlated with gene expression patterns revealed by molecular profiling. *Cancer Res.*, **62**, 6278–6288 (2002).
106. K. B. Hotary, E. D. Allen, P. C. Brooks, N. S. Datta, M. W. Long, and S. J. Weiss, Membrane type I matrix metalloproteinase usurps tumor growth control imposed by the three-dimensional extracellular matrix. *Cell*, **114**, 33–45 (2003).
107. K. Wolf, Y. I. Wu, Y. Liu *et al.*, Multi-step pericellular proteolysis controls the transition from individual to collective cancer cell invasion. *Nat. Cell Biol.*, **9**, 893–904 (2007).
108. C. Gaggioli, S. Hooper, C. Hidalgo-Carcedo *et al.*, Fibroblast-led collective invasion of carcinoma cells with differing roles for RhoGTPases in leading and following cells. *Nat. Cell Biol.*, **9**, 1392–1400 (2007).
109. F. Sabeh, I. Ota, K. Holmbeck *et al.*, Tumor cell traffic through the extracellular matrix is controlled by the membrane-anchored collagenase MT1-MMP. *J. Cell Biol.*, **167**, 769–781 (2004).

110. P. Friedl and K. Wolf, Tube travel: The role of proteases in individual and collective cancer cell invasion. *Cancer Res.*, **68**, 7247–7249 (2008).
111. E. L. Baker and M. H. Zaman, The biomechanical integrin. *J. Biomech.*, **43**, 38–44 (2010).
112. J. S. Desgrosellier and D. A. Cheresh, Integrins in cancer: biological implications and therapeutic opportunities. *Nat. Rev. Cancer*, **10**, 9–22 (2010).
113. D. I. Bellovin, K. J. Simpson, T. Danilov *et al.*, Reciprocal regulation of RhoA and RhoC characterizes the EMT and identifies RhoC as a prognostic marker of colon carcinoma. *Oncogene*, **25**, 6959–6967 (2006).
114. M. J. Paszek and V. M. Weaver, The tension mounts: Mechanics meets morphogenesis and malignancy. *J. Mammary Gland Biol. Neoplasia.*, **9**, 325–342 (2004).
115. C. Fischbach, R. Chen, T. Matsumoto *et al.*, Engineering tumors with 3D scaffolds. *Nat. Methods*, **4**, 855–860 (2007).
116. R. Chignola, A. Schenetti, G. Andrighetto *et al.*, Forecasting the growth of multicell tumour spheroids: Implications for the dynamic growth of solid tumours. *Cell Prolif.*, **33**, 219–229 (2000).
117. L. A. Gurski, A. K. Jha, C. Zhang, X. Jia, and M. C. Farach-Carson, Hyaluronic acid-based hydrogels as 3D matrices for *in vitro* evaluation of chemotherapeutic drugs using poorly adherent prostate cancer cells. *Biomaterials*, **30**, 6076–6085 (2009).
118. C. Feder-Mengus, S. Ghosh, A. Reschner, I. Martin, and G. C. Spagnoli, New dimensions in tumor immunology: What does 3D culture reveal? *Trends Mol. Med.*, **14**, 333–340 (2008).
119. A. Gorlach, P. Herter, H. Hentschel, P. J. Frosch, and H. Acker, Effects of nIFN beta and rIFN gamma on growth and morphology of two human melanoma cell lines: comparison between two- and three-dimensional culture. *Int. J. Cancer*, **56**, 249–254 (1994).
120. M. T. Santini, G. Rainaldi, R. Romano *et al.*, MG-63 human osteosarcoma cells grown in monolayer and as three-dimensional tumor spheroids present a different metabolic profile: A (1)H NMR study. *FEBS Lett.*, **557**, 148–154 (2004).
121. U. Cheema, R. A. Brown, B. Alp, and A. J. MacRobert, Spatially defined oxygen gradients and vascular endothelial growth factor expression in an engineered 3D cell model. *Cell Mol. Life Sci.*, **65**, 177–186 (2008).
122. M. Valcarcel, B. Arteta, A. Jaureguibeitia *et al.*, Three-dimensional growth as multicellular spheroid activates the proangiogenic phenotype of colorectal carcinoma cells via LFA-1-dependent VEGF: Implications on hepatic micrometastasis. *J. Transl. Med.*, **6**, 57 (2008).
123. C. Fischbach, H. J. Kong, S. X. Hsiong, M. B. Evangelista, W. Yuen, and D. J. Mooney, Cancer cell angiogenic capability is regulated by 3D culture and integrin engagement. *Proc. Natl. Acad. Sci. USA*, **106**, 399–404 (2009).
124. D. Yamazaki, S. Kurisu, and T. Takenawa, Involvement of Rac and Rho signaling in cancer cell motility in 3D substrates. *Oncogene*, **28**, 1570–1583 (2009).

125. M. A. Wozniak, K. Modzelewska, L. Kwong, and P. J. Keely, Focal adhesion regulation of cell behavior. *Biochim. Biophys. Acta*, **1692**, 103–119 (2004).
126. L. David, V. Dulong, D. Le Cerf, L. Cazin, M. Lamacz, and J. P. Vannier, Hyaluronan hydrogel: an appropriate three-dimensional model for evaluation of anticancer drug sensitivity. *Acta biomater.*, **4**, 256–263 (2008).
127. J. L. Horning, S. K. Sahoo, S. Vijayaraghavalu *et al.*, 3-D tumor model for *in vitro* evaluation of anticancer drugs. *Mol. Pharm.*, **5**, 849–862 (2008).
128. M. K. Singh, D. Dadke, E. Nicolas *et al.*, A novel Cas family member, HEPL, regulates FAK and cell spreading. *Mol. Biol. Cell*, **19**, 1627–1636 (2008).
129. M. W. Tibbitt and K. S. Anseth, Hydrogels as extracellular matrix mimics for 3D cell culture. *Biotechnol. Bioeng.*, **103**, 655–663 (2009).
130. D. Liang, B. S. Hsiao, and B. Chu, Functional electrospun nanofibrous scaffolds for biomedical applications. *Adv. Drug Deliv. Rev.*, **59**, 1392–1412 (2007).
131. R. Langer and D. A. Tirrell, Designing materials for biology and medicine. *Nature*, **428**, 487–492 (2004).
132. A. C. Jen, M. C. Wake, and A. G. Mikos, Review: hydrogels for cell immobilization. *Biotechnol. Bioeng.*, **50**, 357–364 (1996).
133. J. Lee, M. J. Cuddihy, and N. A. Kotov, Three-dimensional cell culture matrices: state of the art. *Tissue Eng. Pt. B Rev.*, **14**, 61–86 (2008).
134. H. K. Kleinman and G. R. Martin, Matrigel: basement membrane matrix with biological activity. *Semin. Cancer Biol.*, **15**, 378–386 (2005).
135. J. A. Engbring and H. K. Kleinman, The basement membrane matrix in malignancy. *J. Pathol.*, **200**, 465–470 (2003).
136. L. R. Languino, K. R. Gehlsen, E. Wayner, W. G. Carter, E. Engvall, and E. Ruoslahti, Endothelial cells use alpha 2 beta 1 integrin as a laminin receptor. *J. Cell Biol.*, **109**, 2455–2462 (1989).
137. P. M. Carpenter, A. V. Dao, Z. S. Arain *et al.*, Motility induction in breast carcinoma by mammary epithelial laminin 332 (laminin 5). *Mol. Cancer Res.*, **7**, 462–475 (2009).
138. Z. Zhou, M. Doi, J. Wang *et al.*, Deletion of laminin-8 results in increased tumor neovascularization and metastasis in mice. *Cancer Res.*, **64**, 4059–4063 (2004).
139. H. Miyamoto, T. Murakami, K. Tsuchida, H. Sugino, H. Miyake, and S. Tashiro, Tumor-stroma interaction of human pancreatic cancer: Acquired resistance to anticancer drugs and proliferation regulation is dependent on extracellular matrix proteins. *Pancreas*, **28**, 38–44 (2004).
140. A. Passaniti, R. M. Taylor, R. Pili *et al.*, A simple, quantitative method for assessing angiogenesis and antiangiogenic agents using reconstituted basement membrane, heparin, and fibroblast growth factor. *Lab Invest.*, **67**, 519–528 (1992).
141. A. Albini, Y. Iwamoto, H. K. Kleinman *et al.*, A rapid *in vitro* assay for quantitating the invasive potential of tumor cells. *Cancer Res.*, **47**, 3239–3245 (1987).
142. S. N. Bae, G. Arand, H. Azzam *et al.*, Molecular and cellular analysis of basement membrane invasion by human breast cancer cells in Matrigel-based *in vitro* assays. *Breast Cancer Res. Treat.*, **24**, 241–255 (1993).

143. C. Xu, M. S. Inokuma, J. Denham *et al.*, Feeder-free growth of undifferentiated human embryonic stem cells. *Nat. Biotechnol.*,**19**, 971–974 (2001).
144. C. S. Hughes, L. M. Postovit, and G. A. Lajoie, Matrigel: A complex protein mixture required for optimal growth of cell culture. *Proteomics.*, **10**,1886–1890 (2010).
145. L. A. Gurski, N. J. Petrelli, X. Jia, and M. C. Farach-Carson, 3D matrices for anti-cancer Drug testing and Development. *Oncology*, **25**, 20–25 (2010).
146. F. H. Silver and G. Pins, Cell growth on collagen: A review of tissue engineering using scaffolds containing extracellular matrix. *J. Long Term Eff. Med. Implants*, **2**, 67–80 (1992).
147. C. J. Doillon, E. Gagnon, R. Paradis, and M. Koutsilieris, Three-dimensional culture system as a model for studying cancer cell invasion capacity and anticancer drug sensitivity. *Anticancer Res.*, **24**, 2169–2177 (2004).
148. B. A. Roeder, K. Kokini, J. E. Sturgis, J. P. Robinson, and S. L. Voytik-Harbin, Tensile mechanical properties of three-dimensional type I collagen extracellular matrices with varied microstructure. *J. Biomech. Eng.*, **124**, 214–222 (2002).
149. Y. L. Yang, S. Motte, and L. J. Kaufman, Pore size variable type I collagen gels and their interaction with glioma cells. *Biomaterials*, **31**, 5678–5688 (2010).
150. K. Sisson, C. Zhang, M. C. Farach-Carson, D. B. Chase, and J. F. Rabolt, Evaluation of cross-linking methods for electrospun gelatin on cell growth and viability. *Biomacromolecules*, **10**,1675–1680 (2009).
151. O. Hartman, C. Zhang, E. L. Adams *et al.*, Microfabricated electrospun collagen membranes for 3-D cancer models and drug screening applications. *Biomacromolecules*, **10**, 2019–2032 (2009).
152. J. A. Kiefer and M. C. Farach-Carson, Type I collagen-mediated proliferation of PC3 prostate carcinoma cell line: Implications for enhanced growth in the bone microenvironment. *Matrix Biol.*, **20**, 429–437 (2001).
153. S. M. Ellerbroek, Y. I. Wu, C. M. Overall, and M. S. Stack, Functional interplay between type I collagen and cell surface matrix metalloproteinase activity. *J. Biol. Chem.*, **276**, 24833–24842 (2001).
154. A. Menke, C. Philippi, R. Vogelmann *et al.*, Down-regulation of E-cadherin gene expression by collagen type I and type III in pancreatic cancer cell lines. *Cancer Res.*, **61**, 3508–3517 (2001).
155. C. L. Hall, C. W. Dubyk, T. A. Riesenberger, D. Shein, E. T. Keller, and K. L. van Golen, Type I collagen receptor (alpha2beta1) signaling promotes prostate cancer invasion through RhoC GTPase. *Neoplasia*, **10**, 797–803 (2008).
156. I. Bentov, M. Damodarasamy, S. Plymate, and M. J. Reed, B16/F10 tumors in aged 3D collagen *in vitro* simulate tumor growth and gene expression in aged mice *in vivo*. *In Vitro Cell Dev. Biol. Anim.*, **49**, 395–399 (2013).
157. O. Ilina, G. J. Bakker, A. Vasaturo, R. M. Hofmann, and P. Friedl, Two-photon laser-generated microtracks in 3D collagen lattices: principles of MMP-dependent and -independent collective cancer cell invasion. *Phys. Biol.*, **8**, 015010 (2011).

158. C. M. Kraning-Rush, S. P. Carey, M. C. Lampi, and C. A. Reinhart-King, Microfabricated collagen tracks facilitate single cell metastatic invasion in 3D. *Integr. Biol.*, **5**, 606–616 (2013).
159. C. Chen, M. C. Chang, R. K. Hsieh, Y. F. Chang, J. Lin, and K. W. Tsan, Activation of CD44 facilitates DNA repair in T-cell lymphoma but has differential effects on apoptosis induced by chemotherapeutic agents and ionizing radiation. *Leuk. Lymphoma*, **46**, 1785–1795 (2005).
160. L. Y. Bourguignon, P. A. Singleton, H. Zhu, and F. Diedrich, Hyaluronan-mediated CD44 interaction with RhoGEF and Rho kinase promotes Grb2-associated binder-1 phosphorylation and phosphatidylinositol 3-kinase signaling leading to cytokine (macrophage-colony stimulating factor) production and breast tumor progression. *J. Biol. Chem.*, **278**, 29420–29434 (2003).
161. Y. Fujita, M. Kitagawa, S. Nakamura *et al.*, CD44 signaling through focal adhesion kinase and its anti-apoptotic effect. *FEBS Lett.*, **528**, 101–108 (2002).
162. J. A. Burdick, C. Chung, X. Jia, M. A. Randolph, and R. Langer, Controlled degradation and mechanical behavior of photopolymerized hyaluronic acid networks. *Biomacromolecules*, **6**, 386–391 (2005).
163. X. Xu, C. R. Sabanayagam, D. A. Harrington, M. C. Farach-Carson, and X. Jia, A hydrogel-based tumor model for the evaluation of nanoparticle-based cancer therapeutics. *Biomaterials*, **35**, 3319–3330 (2014).
164. C. Vepari, and D. L. Kaplan, Silk as a biomaterial. *Prog. Polym. Sci.*, **32**, 991–1007 (2007).
165. Y. Wang, H. J. Kim, G. Vunjak-Novakovic, and D. L. Kaplan, Stem cell-based tissue engineering with silk biomaterials. *Biomaterials*, **27**, 6064–6082 (2006).
166. S. Hofmann, S. Knecht, R. Langer *et al.*, Cartilage-like tissue engineering using silk scaffolds and mesenchymal stem cells. *Tissue Eng.*, **12**, 2729–2738 (2006).
167. J. Mauney, T. Nguyen, K. Gillen, C. Kirker-Head, J. Gimble, and D. Kaplan, Engineering adipose-like tissue *in vitro* and *in vivo* utilizing human bone marrow and adipose-derived mesenchymal stem cells with silk fibroin 3D scaffolds. *Biomaterials*, **28**, 5280–5370 (2007).
168. L. Soffer, X. Wang, X. Zhang *et al.*, Silk-based electrospun tubular scaffolds for tissue-engineered vascular grafts. *J. Biomater. Sci. Polym. E.*, **19**, 653–664 (2008).
169. U. J. Kim, J. Park, H. J. Kim, M. Wada, and D. L. Kaplan, Three-dimensional aqueous-derived biomaterial scaffolds from silk fibroin. *Biomaterials*, **26**, 2775–2785 (2005).
170. R. E. Unger, M. Wolf, K. Peters, A. Motta, C. Migliaresi, and C. James Kirkpatrick, Growth of human cells on a non-woven silk fibroin net: a potential for use in tissue engineering. *Biomaterials*, **25**, 1069–1075 (2004).
171. D,. N. Rockwood, R. C. Preda, T. Yucel, X. Wang, M. L. Lovett, and D. L. Kaplan, Materials fabrication from *Bombyx mori* silk fibroin. *Nat. Protoc.*, **6**, 1612–1631 (2011).
172. X. Wang, L. Sun, M. V. Maffini, A. Soto, C. Sonnenschein, and D. L. Kaplan, A complex 3D human tissue culture system based on mammary stromal cells and silk scaffolds for modeling breast morphogenesis and function. *Biomaterials*, **31**, 3920–3929 (2010).

173. S. Talukdar, M. Mandal, D. W. Hutmacher, P. J. Russell, C. Soekmadji, and S. C. Kundu, Engineered silk fibroin protein 3D matrices for *in vitro* tumor model. *Biomaterials*, **32**, 2149–2159 (2011).
174. P. H. Tan, K. Z. Aung, S. L. Toh, J. C. Goh, and S. S. Nathan, Three-dimensional porous silk tumor constructs in the approximation of *in vivo* osteosarcoma physiology. *Biomaterials*, **32**, 6131–6137 (2011).
175. S. S. Rao, J. Dejesus, A. R. Short, J. J. Otero, A. Sarkar, and J. O. Winter, Glioblastoma behaviors in three-dimensional collagen-hyaluronan composite hydrogels. *ACS Appl. Mater. Inter.*, **5**, 9276–9284 (2013).
176. S. J. Florczyk, M. Leung, Z. Li, J. I. Huang, R. A. Hopper, and M. Zhang, Evaluation of three-dimensional porous chitosan-alginate scaffolds in rat calvarial defects for bone regeneration applications. *J. Biomed. Mater. Res. Pt. A*, **101**, 2974–2983 (2013).
177. E. Kaemmerer, F. P. Melchels, B. M. Holzapfel, T. Meckel, D. W. Hutmacher, and D. Loessner, Gelatine methacrylamide-based hydrogels: an alternative three-dimensional cancer cell culture system. *Acta Biomater.*, **10**, 2551–2562 (2014).
178. D. Eyrich, F. Brandl, B. Appel *et al.*, Long-term stable fibrin gels for cartilage engineering. *Biomaterials*, **28**, 55–65 (2007).
179. A. K. Azab, B. Orkin, V. Doviner *et al.*, Crosslinked chitosan implants as potential degradable devices for brachytherapy: *In vitro* and *in vivo* analysis. *J. Control. Release*, **111**, 281–289 (2006).
180. J. E. Barralet, L. Wang, M. Lawson, J. T. Triffitt, P. R. Cooper, and R. M. Shelton, Comparison of bone marrow cell growth on 2D and 3D alginate hydrogels. *J. Mater. Sci. Mater. Med.*, **16**, 515–519 (2005).
181. C. Touboul, R. Lis, H. Al Farsi *et al.*, Mesenchymal stem cells enhance ovarian cancer cell infiltration through IL6 secretion in an amniochorionic membrane based 3D model. *J. Transl. Med.*, **11**, **28** (2013).
182. A. J. Engler, S. Sen, H. L. Sweeney, and D. E. Discher, Matrix elasticity directs stem cell lineage specification. *Cell*, **126**, 677–689 (2006).
183. C. E. Kandow, P. C. Georges, P. A. Janmey, and K. A. Beningo, Polyacrylamide hydrogels for cell mechanics: steps toward optimization and alternative uses. *Method. Cell Biol.*, **83**, 29–46 (2007).
184. B. J. Gill and J. L. West, Modeling the tumor extracellular matrix: tissue engineering tools repurposed towards new frontiers in cancer biology. *J. Biomech.*, **47**, 1969–1978 (2013).
185. M. P. Lutolf and J. A. Hubbell, Synthetic biomaterials as instructive extracellular microenvironments for morphogenesis in tissue engineering. *Nat. Biotechnol.*, **23**, 47–55 (2005).
186. J. K. Tessmar and A. M. Gopferich, Customized PEG-derived copolymers for tissue-engineering applications. *Macromol. Biosci.*, **7**, 23–39 (2007).
187. S. Nemir, H. N. Hayenga, and J. L. West, PEGDA hydrogels with patterned elasticity: novel tools for the study of cell response to substrate rigidity. *Biotechnol. Bioeng.*, **105**, 636–644 (2010).

188. B. J. Gill, D. L. Gibbons, L. C. Roudsari *et al.*, A synthetic matrix with independently tunable biochemistry and mechanical properties to study epithelial morphogenesis and EMT in a lung adenocarcinoma model. *Cancer Res.*, **72**, 6013–6023 (2012).
189. S. Zhang, F. Gelain, and X. Zhao, Designer self-assembling peptide nanofiber scaffolds for 3D tissue cell cultures. *Semin. Cancer Biol.*, **15**, 413–420 (2005).
190. F. Gelain, A. Horii, and S. Zhang, Designer self-assembling peptide scaffolds for 3-d tissue cell cultures and regenerative medicine. *Macromol. Biosci.*, **7**, 544–451 (2007).
191. M. A. Bokhari, G. Akay, S. Zhang, and M. A. Birch, The enhancement of osteoblast growth and differentiation *in vitro* on a peptide hydrogel-polyHIPE polymer hybrid material. *Biomaterials*, **26**, 5198–5208 (2005).
192. J. D. Hartgerink, E. Beniash, and S. I. Stupp, Self-assembly and mineralization of peptide-amphiphile nanofibers. *Science*, **294**, 1684–1688 (2001).
193. J. P. Schneider, D. J. Pochan, B. Ozbas, K. Rajagopal, L. Pakstis, and J. Kretsinger, Responsive hydrogels from the intramolecular folding and self-assembly of a designed peptide. *J. Am. Chem. Soc.*, **124**, 15030–15037 (2002).
194. A. Horii, X. Wang, F. Gelain, and S. Zhang, Biological designer self-assembling Peptide nanofiber scaffolds significantly enhance osteoblast proliferation, differentiation and 3-D migration. *PLoS ONE*, **2**, e190 (2007).
195. M. A. Serban, A. Scott, and G. D. Prestwich, Use of hyaluronan-derived hydrogels for three-dimensional cell culture and tumor xenografts. In *Current Protocols in Cell Biology*, Ed. J. S. Bonifacino *et al.*, 10–14 (2008).
196. G. D. Prestwich, Evaluating drug efficacy and toxicology in three dimensions: using synthetic extracellular matrices in drug discovery. *Acc. Chem. Res.*, **41**, 139–148 (2008).
197. S. E. Grieshaber, A. J. Farran, S. Lin-Gibson, K. L. Kiick, and X. Jia, Synthesis and characterization of elastin-mimetic hybrid polymers with multiblock, alternating molecular architecture and elastomeric properties. *Macromolecules*, **42**, 2532–2541 (2009).
198. H. J. Lee, J. S. Lee, T. Chansakul, C. Yu, J. H. Elisseeff, and S. M. Yu, Collagen mimetic peptide-conjugated photopolymerizable PEG hydrogel. *Biomaterials*, **27**, 5268–5276 (2006).
199. X. Jia and K. L. Kiick, Hybrid multicomponent hydrogels for tissue engineering. *Macromol. Biosci.*, **9**, 140–156 (2009).
200. D. S. Dolberg and M. J. Bissell, Inability of rous sarcoma virus to cause sarcomas in the avian embryo. *Nature*, **309**, 552–556 (1984).
201. N. A. Bhowmick, E. G. Neilson, and H. L. Moses, Stromal fibroblasts in cancer initiation and progression. *Nature*, **432**, 332–337 (2004).
202. C. C. Park, M. J. Bissell, and M. H. Barcellos-Hoff, The influence of the microenvironment on the malignant phenotype. *Trends Mol. Med.*, **6**, 324–329 (2000).
203. R. Kalluri, Basement membranes: structure, assembly and role in tumour angiogenesis. *Nat. Rev. Cancer*, **3**, 422–433 (2003).

204. T. Jacks and R. A. Weinberg, Taking the study of cancer cell survival to a new dimension. *Cell*, **111**, 923–925 (2002).
205. J. MacDougall and L. Matrisian, Contributions of tumor and stromal matrix metalloproteinases to tumor progression, invasion and metastasis. *Cancer Metast. Rev.*, **14**, 351–362 (1995).
206. E. Lengyel, J. Burdette, H. Kenny *et al.*, Epithelial ovarian cancer experimental models. *Oncogene*, **33**, 3619–3633 (2013).
207. V. Weaver, M. Bissell, A. Fischer, and O. Peterson, The importance of the microenvironment in breast cancer progression: Recapitulation of mammary tumorigenesis using a unique human mammary epithelial cell model and a three-dimensional culture assay. *Biochem. Cell Biol.*, **74**, 833–851 (1996).
208. C. L. Evans, A. O. Abu-Yousif, Y. J. Park *et al.*, Killing hypoxic cell populations in a 3D tumor model with EtNBS-PDT. *PLoS One*, **6**, e23434 (2011).
209. A. F. Olumi, G. D. Grossfeld, S. W. Hayward, P. R. Carroll, T. D. Tlsty, and G. R. Cunha, Carcinoma-associated fibroblasts direct tumor progression of initiated human prostatic epithelium. *Cancer Res.*, **59**, 5002–5011 (1999).
210. P. Basset, J. P. Bellocq, C. Wolf *et al.*, A novel metalloproteinase gene specifically expressed in stromal cells of breast carcinomas. *Nature*, **348**, 699–704 (1990).
211. C. Singer, A. Rasmussen, H. S. Smith, M. E. Lippman, H. T. Lynch, and K. J. Cullen, Malignant breast epithelium selects for insulin-like growth factor II expression in breast stroma: evidence for paracrine function. *Cancer Res.*, **55**, 2448–2454 (1995).
212. J. H. Wright, S. McDonnell, G. Portella, G. T. Bowden, A. Balmain, and L. M. Matrisian, A switch from stromal to tumor cell expression of stromelysin-1 mRNA associated with the conversion of squamous to spindle carcinomas during mouse skin tumor progression. *Mol. Carcinog.*, **10**, 207–215 (1994).
213. R. Kalluri and M. Zeisberg, Fibroblasts in cancer. *Nat. Rev. Cancer.*, **6**, 392–401 (2006).
214. S. L. Schor, A. M. Schor, and G. Rushton, Fibroblasts from cancer patients display a mixture of both foetal and adult-like phenotypic characteristics. *J. Cell Sci.*, **90**, 401–407 (1988).
215. S. L. Schor, A. M. Schor, G. Rushton, and L. Smith, Adult, foetal and transformed fibroblasts display different migratory phenotypes on collagen gels: Evidence for an isoformic transition during foetal development. *J. Cell Sci.*, **73**, 221–234 (1985).
216. D. G. Harkin and E. D. Hay, Effects of electroporation on the tubulin cytoskeleton and directed migration of corneal fibroblasts cultured within collagen matrices. *Cell Motil. Cytoskel.*, **35**, 345–357 (1996).
217. R. F. Hwang, T. Moore, T. Arumugam *et al.*, Cancer-associated stromal fibroblasts promote pancreatic tumor progression. *Cancer Res.*, **68**, 918–926 (2008).
218. J. W. Franses, A. B. Baker, V. C. Chitalia, and E. R. Edelman, Stromal endothelial cells directly influence cancer progression. *Sci. Transl. Med.*, **3**, 66ra5 (2011).
219. J. Folkman, P. Cole, and S. Zimmerman, Tumor behavior in isolated perfused organs: *In vitro* growth and metastases of biopsy material in rabbit thyroid and canine intestinal segment. *Ann. Surg.*, **164**, 491–502 (1966).

220. P. Carmeliet and R. K. Jain, Angiogenesis in cancer and other diseases. *Nature*, **407**, 249–257 (2000).
221. N. Ohga, S. Ishikawa, N. Maishi *et al.*, Heterogeneity of tumor endothelial cells: comparison between tumor endothelial cells isolated from high-and low-metastatic tumors. *Am. J. Path.*, **180**, 1294–1307 (2012).
222. K. Hida, N. Ohga, K. Akiyama, N. Maishi, and Y. Hida, Heterogeneity of tumor endothelial cells. *Cancer Sci.*, **104**, 1391–1395 (2013).
223. J. Condeelis and J. Pollard, Macrophages: Obligate partners for tumor cell migration, invasion, and metastasis. *Cell*, **124**, 263–266 (2006).
224. A. Mantovani, Tumor-associated macrophages in neoplastic progression: a paradigm for the *in vivo* function of chemokines. *Lab. Invest.*, **71**, 5 (1994).
225. L. M. Coussens and Z. Werb, Inflammation and cancer. *Nature*, **420**, 860–867 (2002).
226. L. Bingle, N. J. Brown, and C. E. Lewis, The role of tumour-associated macrophages in tumour progression: Implications for new anticancer therapies. *J. Pathol.*, **196**, 254–265 (2002).
227. B. Z. Qian and J. W. Pollard, Macrophage diversity enhances tumor progression and metastasis. *Cell*, **141**, 39–51 (2010).
228. J. W. Pollard, Tumour-educated macrophages promote tumour progression and metastasis. *Nat. Rev. Cancer*, **4**, 71–78, 2004.
229. A. Konur, M. Kreutz, R. Knuchel, S. W. Krause, and R. Andreesen, Three-dimensional co-culture of human monocytes and macrophages with tumor cells: analysis of macrophage differentiation and activation. *Int. J. Cancer*, **66**, 645–652 (1996).
230. K. U. Kim, S. M. Wilson, K. S. Abayasiriwardana *et al.*, A novel *in vitro* model of human mesothelioma for studying tumor biology and apoptotic resistance. *Am. J. Respir. Cell Mol. Biol.*, **33**, 541–548 (2005).
231. L. Bingle, C. Lewis, K. Corke, M. Reed, and N. Brown, Macrophages promote angiogenesis in human breast tumour spheroids *in vivo*. *Brit. J. Cancer*, **94**, 101–107 (2006).
232. Y. Manabe, S. Toda, K. Miyazaki, and H. Sugihara, Mature adipocytes, but not preadipocytes, promote the growth of breast carcinoma cells in collagen gel matrix culture through cancer-stromal cell interactions. *J. Pathol.*, **201**, 221–228 (2003).
233. B. Dirat, L. Bochet, M. Dabek *et al.*, Cancer-associated adipocytes exhibit an activated phenotype and contribute to breast cancer invasion. *Cancer Res.*, **71**, 2455 2465 (2011).
234. E. Lengyel, Ovarian cancer development and metastasis. *Am. J. Pathol.*, **177**, 1053–1064 (2010).
235. K. M. Nieman, H. A. Kenny, C. V. Penicka *et al.*, Adipocytes promote ovarian cancer metastasis and provide energy for rapid tumor growth. *Nat. Med.*, **17**, 1498–1503 (2011).
236. F. R. Huss and G. Kratz, Mammary epithelial cell and adipocyte co-culture in a 3-D matrix: the first step towards tissue-engineered human breast tissue. *Cells Tissues Organs*, **169**, 361–367 (2001).
237. S. E. Mutsaers, The mesothelial cell. *Int. J. Biochem. Cell Biol.*, **36**, 9–16 (2004).

238. A. Ferrandez-Izquierdo, S. Navarro-Fos, M. Gonzalez-Devesa, R. Gil-Benso, and A. Llombart-Bosch, Immunocytochemical typification of mesothelial cells in effusions: *in vivo* and *in vitro* models. *Diagn. Cytopathol.*, **10**, 256–262 (1994).
239. M. Yanez-Mo, E. Lara-Pezzi, R. Selgas *et al.*, Peritoneal dialysis and epithelial-to-mesenchymal transition of mesothelial cells. *N. Engl. J. Med.*, **348**, 403–413 (2003).
240. O. Ohtani, Y. Ohtani, and R. X. Li, Phylogeny and ontogeny of the lymphatic stomata connecting the pleural and peritoneal cavities with the lymphatic system — a review. *Ital. J. Anat. Embryol.*, **106**, 251–259 (2001).
241. H. A. Kenny, K. M. Nieman, A. K. Mitra, and E. Lengyel, The first line of intra-abdominal metastatic attack: breaching the mesothelial cell layer. *Cancer Discov.*, **1**, 100–102 (2001).
242. K. M. Burleson, M. P. Boente, S. E. Pambuccian, and A. P. Skubitz, Disaggregation and invasion of ovarian carcinoma ascites spheroids. *J. Transl. Med.*, **4**, 6 (2006).
243. K. M. Burleson, R. C. Casey, K. M. Skubitz, S. E. Pambuccian, T. R. Oegema Jr., and A. P. Skubitz, Ovarian carcinoma ascites spheroids adhere to extracellular matrix components and mesothelial cell monolayers. *Gynecol. Oncol.*, **93**, 170–181 (2004).
244. K. M. Burleson, L. K. Hansen, and A. P. Skubitz, Ovarian carcinoma spheroids disaggregate on type I collagen and invade live human mesothelial cell monolayers. *Clin. Exp. Metastasis*, **21**, 685–697 (2004).
245. R. C. Casey, K. M. Burleson, K. M. Skubitz *et al.*, Beta 1-integrins regulate the formation and adhesion of ovarian carcinoma multicellular spheroids. *Am. J. Pathol.*, **159**, 2071–2080 (2001).
246. F. Zanella, J. B. Lorens, and W. Link, High content screening: seeing is believing. *Trends Biotechnol.*, **28**, 237–245 (2010).
247. C. Möller and M. Slack, Impact of new technologies for cellular screening along the drug value chain. *Drug Discov. Today*, **15**, 384–390 (2010).
248. M. J. Bissell and D. Radisky, Putting tumours in context. *Nat. Rev. Cancer*, **1**, 46–54 (2001).
249. M. J. Bissell, V. M. Weaver, S. A. Lelievre, F. Wang, O. W. Petersen, and K. L. Schmeichel, Tissue structure, nuclear organization, and gene expression in normal and malignant breast. *Cancer Res.*, **59**, 1757s–1763s; discussion 63s–64s (1999).
250. T. Gudjonsson, L. Ronnov-Jessen, R. Villadsen, M. J. Bissell, and O. W. Petersen, To create the correct microenvironment: three-dimensional heterotypic collagen assays for human breast epithelial morphogenesis and neoplasia. *Methods*, **30**, 247–255 (2003).
251. S. K. Muthuswamy, D. Li, S. Lelievre, M. J. Bissell, and J. S. Brugge, ErbB2, but not ErbB1, reinitiates proliferation and induces luminal repopulation in epithelial acini. *Nat. Cell Biol.*, **3**, 785–792 (2001).
252. C. M. Nelson and M. J. Bissell, Of extracellular matrix, scaffolds, and signaling: tissue architecture regulates development, homeostasis, and cancer. *Annu. Rev. Cell Dev. Biol.*, **22**, 287–309 (2006).
253. C. C. Park, H. Zhang, M. Pallavicini *et al.*, Beta1 integrin inhibitory antibody induces apoptosis of breast cancer cells, inhibits growth, and distinguishes malignant from normal phenotype in three dimensional cultures and *in vivo*. *Cancer Res.*, **66**, 1526–1535 (2006).

254. D. C. Radisky, D. D. Levy, L. E. Littlepage *et al.*, Rac1b and reactive oxygen species mediate MMP-3-induced EMT and genomic instability. *Nature*, **436**, 123–127 (2005).
255. Y. Li, L. Petrovic, J. P. Celli, and C. S. Yelleswarapu, Digital holographic microscopy for imaging growth and treatment response in 3D tumor models. Proc. SPIE 8947, *Imaging, Manipulation, and Analysis of Biomolecules, Cells, and Tissues XII*, 894704 (4 March 2014); doi: 10.1117/12.2040515 894704–894707 (2014).
256. A. D. Bershadsky, N. Q. Balaban, and B. Geiger, Adhesion-dependent cell mechanosensitivity. *Annu. Rev. Cell Dev. Biol.*, **3**;19:677–695.
257. J. S. Desgrosellier and D. A. Cheresh, Integrins in cancer: Biological implications and therapeutic opportunities. *Nat. Rev. Cancer.*, **10**, 9–22 (2010).
258. J. W. Lee and R. Juliano, Mitogenic signal transduction by integrin- and growth factor receptor-mediated pathways. *Mol. Cells.*, **17**, 188–202 (2004).
259. J. K. Mouw, Y. Yui, L. Damiano *et al.*, Tissue mechanics modulate microRNA-dependent PTEN expression to regulate malignant progression. *Nat. Med.*, **20**, 360–367 (2014).
260. T. A. Ulrich, A. Jain, K. Tanner, J. L. MacKay, and S. Kumar, Probing cellular mechanobiology in three-dimensional culture with collagen-agarose matrices. *Biomaterials*, **31**, 1875–1884 (2010).
261. T. A. Ulrich, T. G. Lee, H. K. Shon, D. W. Moon, and S. Kumar, Microscale mechanisms of agarose-induced disruption of collagen remodeling. *Biomaterials.*, **32**, 5633–5642 (2011).
262. J. P. Celli, Stromal interactions as regulators of tumor growth and therapeutic response: A potential target for photodynamic therapy? *Isr. J. Chem.*, **52**, 757–766 (2012).
263. B. N. Mason, A. Starchenko, R. M. Williams, L. J. Bonassar, and C. A. Reinhart-King, Tuning three-dimensional collagen matrix stiffness independently of collagen concentration modulates endothelial cell behavior. *Acta Biomater.*, **9**, 4635–4644 (2013).
264. Y. L. Yang, C. Sun, M. E. Wilhelm, L. J. Fox, J. Zhu, and L. J. Kaufman, Influence of chondroitin sulfate and hyaluronic acid on structure, mechanical properties, and glioma invasion of collagen I gels. *Biomaterials.*, **32**, 7932–7940 (2011).
265. J. N. Beck, A. Singh, A. R. Rothenberg, J. H. Elisseeff, and A. J. Ewald, The independent roles of mechanical, structural and adhesion characteristics of 3D hydrogels on the regulation of cancer invasion and dissemination. *Biomaterials*, **34**, 9486–9495 (2013).
266. S. Zhang, T. C. Holmes, C. M. DiPersio, R. O. Hynes, X. Su, and A. Rich, Self-complementary oligopeptide matrices support mammalian cell attachment. *Biomaterials.*, **16**, 1385–1393 (1995).
267. F. Gelain, D. Bottai, A. Vescovi, and S. Zhang, Designer Self-Assembling Peptide Nanofiber Scaffolds for Adult Mouse Neural Stem Cell 3-Dimensional Cultures. *PLoS ONE*, **e119**, 1–11 (2006).
268. P. Panorchan, J. S. H. Lee, T. P. Kole, Y. Tseng, and D. Wirtz, Microrheology and ROCK Signaling of Human Endothelial Cells Embedded in a 3D Matrix. *Biophys. J.*, **91**, 3499–34507 (2006).

269. C. A. Reinhart-King, M. Dembo, and D. A. Hammer, Endothelial Cell Traction Forces on RGD-Derivatized Polyacrylamide Substrata†. *Langmuir*, **19**, 1573–1579 (2002).
270. G. Y. Lee, P. A. Kenny, E. H. Lee, and M. J. Bissell, Three-dimensional culture models of normal and malignant breast epithelial cells. *Nat. Meth.*, **4**, 359–365 (2007).
271. K. A. Smeds,, A. Pfister-Serres, D. Miki *et al.*, Photocrosslinkable polysaccharides for *in situ* hydrogel formation. *J. Biomed. Mater. Res.*, **54**, 115–121 (2001).
272. B. Ananthanarayanan, Y. Kim, and S. Kumar, Elucidating the mechanobiology of malignant brain tumors using a brain matrix-mimetic hyaluronic acid hydrogel platform. *Biomaterials*, **32**, 7913–7923 (2011).
273. T. A. Ulrich, E. M. de Juan Pardo, and S. Kumar, The mechanical rigidity of the extracellular matrix regulates the structure, motility, and proliferation of glioma cells. *Cancer Res.*, **69**, 4167–4174 (2009).
274. B. J. Gill, D. L. Gibbons, L. C. Roudsari *et al.*, A Synthetic Matrix with Independently Tunable Biochemistry and Mechanical Properties to Study Epithelial Morphogenesis and EMT in a Lung Adenocarcinoma Model. *Cancer Res.*, **72**, 6013–6023 (2012).
275. K. Alessandri, B. R. Sarangi, V. V. Gurchenkov *et al.*, Cellular capsules as a tool for multicellular spheroid production and for investigating the mechanics of tumor progression *in vitro. Proceedings of the National Academy of Sciences of the United States of America*, **110**, 14843–14848 (2013).
276. M. Sameni, A. Anbalagan, M. B. Olive, K. Moin, R. R. Mattingly, and B. F. Sloane, MAME models for 4D live-cell imaging of tumor: microenvironment interactions that impact malignant progression. **e3661** (2012).
277. Y. Jung, A. J. Nichols, O. J. Klein, E. Roussakis, and C. L. Evans, Label-Free, longitudinal visualization of pdt response *in vitro* with optical coherence tomography. *Isr. J. Chem.*, **52**, 728–744 (2012).
278. K. Moin, M. Sameni, B. C. Victor, J. M. Rothberg, R. R. Mattingly, and B. F. Sloane, 3D/4D functional imaging of tumor-associated proteolysis: impact of microenvironment. Methods Enzymol., **506**, 175–194 (2012).
279. K. Garber, Stromal depletion goes on trial in pancreatic cancer. *J. Natl. Cancer Inst.*, **102**, 448–450 (2010).
280. K. P. Olive, M. A. Jacobetz, C. J. Davidson *et al.*, Inhibition of Hedgehog signaling enhances delivery of chemotherapy in a mouse model of pancreatic cancer. *Science*, **324**, 1457–1461 (2009).
281. D. D. Von Hoff, R. K. Ramanathan, M. J. Borad *et al.*, Gemcitabine plus nab-paclitaxel is an active regimen in patients with advanced pancreatic cancer: a phase I/II trial. *J. Clin. Oncol.*, **29**, 4548–4554 (2011).
282. M. Giovannini, L. C. Hookey, E. Bories, C. Pesenti, G. Monges, and J. R. Delpero, Endoscopic ultrasound elastography: The first step towards virtual biopsy? Preliminary results in 49 patients. *Endoscopy*, **38**, 344–248 (2006).
283. S. K. Nadkarni, B. E. Bouma, D. Yelin, A. Gulati, and G. J. Tearney, Laser speckle imaging of atherosclerotic plaques through optical fiber bundles. *J. Biomed. Opt.*, 13, 054016 (2008).

284. G. Scarcelli and S. H. Yun, Confocal Brillouin microscopy for three-dimensional mechanical imaging. *Nat. Photon.*, **2**, 39–43 (2008).
285. G. P. Gupta and J. Massague, Cancer metastasis: building a framework. *Cell*, **127**, 679–695 (2006).
286. D. X. Nguyen, P. D. Bos, and J. Massague, Metastasis: from dissemination to organ-specific colonization. *Nat. Rev. Cancer*, **9**, 274–284 (2009).
287. W. J. Polacheck, A. E. German, A. Mammoto, D. E. Ingber, and R. D. Kamm, Mechanotransduction of fluid stresses governs 3D cell migration. *Proc. Natl. Acad. Sci. USA*, **111**, 2447–2452 (2014).
288. W. J. Polacheck, R. Li, S. G. Uzel, and R. D. Kamm, Microfluidic platforms for mechanobiology. *Lab Chip*, **13**, 2252–2267 (2013).
289. W. J. Polacheck, I. K. Zervantonakis, and R. D. Kamm, Tumor cell migration in complex microenvironments. *Cell Mol. Life Sci.*, **70**, 1335–1356 (2013).
290. J. D. Shields, M. E. Fleury, C. Yong, A. A. Tomei, G. J. Randolph, and M. A. Swartz, Autologous chemotaxis as a mechanism of tumor cell homing to lymphatics via interstitial flow and autocrine CCR7 signaling. *Cancer Cell*, **11**, 526–538 (2007).
291. W. J. Polacheck, J. L. Charest, and R. D. Kamm, Interstitial flow influences direction of tumor cell migration through competing mechanisms. *Proceedings of the National Academy of Sciences of the United States of America*, **108**, 11115–111120 (2011).
292. M. A. Swartz and A. W. Lund, Lymphatic and interstitial flow in the tumour microenvironment: linking mechanobiology with immunity. *Nat. Rev. Cancer*, **12**, 210–219 (2012).
293. B. A. Teicher, T. S. Herman, S. A. Holden *et al.*, Tumor resistance to alkylating agents conferred by mechanisms operative only *in vivo*. *Science*, **247**, 1457–1461 (1990).
294. H. Kobayashi, S. Man, C. H. Graham, S. J. Kapitain, B. A. Teicher, and R. S. Kerbel, Acquired multicellular-mediated resistance to alkylating agents in cancer. *Proc. Natl. Acad. Sci. USA*, **90**, 3294–3298 (1993).
295. T. Ohmori, J. L. Yang, J. O. Price, and C. L. Arteaga, Blockade of tumor cell transforming growth factor-betas enhances cell cycle progression and sensitizes human breast carcinoma cells to cytotoxic chemotherapy. *Exp. Cell Res.*, **245**, 350–359 (1998).
296. T. H. Foster, D. F. Hartley, M. G. Nichols, and R. Hilf, Fluence rate effects in photodynamic therapy of multicell tumor spheroids. *Cancer Res.*, **53**, 1249–1254 (1993).
297. M. G. Nichols and T. H. Foster, Oxygen diffusion and reaction kinetics in the photodynamic therapy of multicell tumour spheroids. *Phys. Med. Biol.*, **39**, 2161–2181 (1994).
298. C. E. Bigelow, S. Mitra, R. Knuechel, and T. H. Foster, ALA- and ALA-hexylester-induced protoporphyrin IX fluorescence and distribution in multicell tumour spheroids. *Br. J. Cancer*, **85**, 727–734 (2001).
299. S. Coutier, S. Mitra, L. Bezdetnaya *et al.*, Effects of fluence rate on cell survival and photobleaching in meta-tetra-(hydroxyphenyl)chlorin-photosensitized Colo 26 multicell tumor spheroids. *J. Photochem Photobiol.*, **73**, 297–303 (2001).
300. J. C. Finlay, S. Mitra, M. S. Patterson, and T. H. Foster, Photobleaching kinetics of Photofrin *in vivo* and in multicell tumour spheroids indicate two simultaneous bleaching mechanisms. *Phys. Med. Biol.*, **49**, 4837–4860 (2004).

301. L. Bingle, C. E. Lewis, K. P. Corke, M. W. Reed, and N. J. Brown, Macrophages promote angiogenesis in human breast tumour spheroids *in vivo. Br. J. Cancer*, **94**, 101–107 (2006).
302. I. Dufau, C. Frongia, F. Sicard *et al.*, Multicellular tumor spheroid model to evaluate spatio-temporal dynamics effect of chemotherapeutics: Application to the gemcitabine/CHK1 inhibitor combination in pancreatic cancer. *BMC Cancer*, **12**, 15 (2012).
303. J. Friedrich, R. Ebner, and L. A. Kunz-Schughart, Experimental anti-tumor therapy in 3-D: spheroids — old hat or new challenge? *Int. J. Radiat. Biol.*, **83**, 849–871 (2007).
304. J. Friedrich, C. Seidel, R. Ebner, and L. A. Kunz-Schughart, Spheroid-based drug screen: considerations and practical approach. *Nat. Protoc.*, **4**, 309–324 (2009).
305. Y. C. Tung, A. Y. Hsiao, S. G. Allen, Y. S. Torisawa, M. Ho, and S. Takayama, High-throughput 3D spheroid culture and drug testing using a 384 hanging drop array. *Analyst.*, **136**, 473–478 (2011).
306. M. J. Bissell, H. G. Hall, and G. Parry, How does the extracellular matrix direct gene expression? *J. Theor. Biol.*, **99**, 31–68 (1982).
307. D. Radisky, C. Hagios, and M. J. Bissell, Tumors are unique organs defined by abnormal signaling and context. *Semin. Cancer Biol.*, **11**, 87–95 (2001).
308. K. L. Schmeichel and M. J. Bissell, Modeling tissue-specific signaling and organ function in three dimensions. *J. Cell Sci.*, **116**, 2377–2388 (2003).
309. A. Correia and M. Bissell, The tumor microenvironment is a dominant force in multidrug resistance. *Drug Resist. Update*, **15**, 39–49 (2012).

Chapter 10

Hydrogel-Mediated Patterning of Cellular and Biomolecular Microarrays for Screening Assays and Biosensing

You Jung Kang, Soo Hyun Park, and Sheereen Majd
Cellular Biophysics and Biotechnology Lab,
Department of Biomedical Engineering, Pennsylvania State University, PA 16802, USA

Microarrays of cells and biomolecules such as deoxyribonucleic acid (DNA) and proteins have become essential components of today's academic and industrial research. These microarrays are widely applied for gene sequencing, screening of molecular interactions, drug discovery, and sensing of biomolecules. Increasing demand for high-density arrays of bioactive molecules has led to the development of variety of fabrication techniques. Wet hydrogel materials that provide hydrated and biocompatible environments for storage or delivery of fragile biological components have found different applications in fabrication of cellular and biomolecular arrays. In this chapter, we provide an overview of different types of bioarrays and their applications in biosensing and screening assays and discuss the applications of hydrogels in fabrication of these microarrays.

1. Introduction

Micro- and nano-scale arrays of bioactive molecules and cells, also known as bioarrays or biochips, have become increasingly common in biological, medical, and pharmaceutical research. Deoxyribonucleic acid (DNA) chips were the first type of microarrays developed and their significant impact on the field of genomics led to the development of a number of other bioarrays such as protein and lipid microarrays. These miniaturized libraries of molecules on small chips, have enabled high-throughput screening of molecular interactions and thus, gained popularity in medical diagnostics, drug discovery, and sensing.

In traditional form, these arrays are composed of molecules that are immobilized on flat substrates. Immobilization of cells or fragile biomolecules on surfaces, however, typically leads to the loss of their functions and limits the use of these biochips. To overcome this limitation, more recent approaches take advantage of biocompatible hydrogels for better protection of biomolecules and cells during and after immobilization. Hydrogels have unique properties that make them ideal for interfacing biological materials as they closely mimic the soft and hydrated nature of these materials. They have hence, become increasingly popular in biomedical applications. In the area of bioarrays for biosensing and screening assays, hydrogels have been employed as carriers for biomolecules and cells during their transfer to the host substrates, protecting these components and improving the quality of the resultant arrays. Hydrogels have also been utilized as substrates for these arrays, providing a more natural environment and prolonging the lifetime of these biological materials.

In this chapter, we provide an overview of the application of hydrogels for fabrication of biological microarrays. In the following sections, first, a number of techniques that utilize hydrogels either as carriers for cells/biomolecules (in the form of ink solutions and stamps) or as substrates for these arrays are described. Next, different types of bioarrays, including DNA, protein, membrane, and cell arrays and their applications in screening assays or biosensing are discussed.

2. Hydrogel-Based Fabrication of Bioarrays

Numerous methods for fabrication of microarrays of biomolecules (including DNA, proteins, and cellular membranes), cells, and tissues have been developed to date. All these microarray fabrication methods have a common objective of reducing the material consumption and cost while facilitating data collection and analysis. To this end, such arrays should ideally have uniform and high density of biomolecules that retain their bioactivity. In this chapter, we focus on those fabrication techniques that apply hydrogels to pattern arrays of biomolecules or cells, and satisfy the above requirements. We categorize these fabrication methods into two main groups in terms of the usage of hydrogels: (i) the methods that employ hydrogels as carriers for biomolecules (i.e., in a form of inks, beads, or stamps) and (ii) the methods that employ hydrogels as substrates for patterning the bimolecular and cellular arrays. Table 1 summarizes these fabrication techniques with their advantages and limitations.

Table 1. Hydrogel-based techniques for fabrication of bioarrays.

Technique		Hydrogel usage	Advantages	Disadvantages
Direct-writing	Pin printing	Ink solution	Accurate patterning with various configurations and multi-probes, 3D patterning	Expensive, slow, contamination, low density probes
	Ink-jet printing			Clogging problems, causing shear stress
Capillary-based hydrogel stamping		Stamp	Cheap, multiple-stamping	Slow
Microcontact printing (μCP)		Stamp	Cheap, fast, multiple-stamping, parallel-stamping, stamping on curved surface	Patterning with few different inks
Laser direct-writing (LDW)		Bead	Mild condition. High-resolution	Time-consuming
Micro-molding		Substrate	Cheap, fast, 3D patterning, stamping on curved surface	Varying with methods

2.1. *Hydrogels as carriers*

2.1.1. *Direct-writing method*

In this method, hydrogel solutions are applied as ink solutions that contain desired biomolecules or cells. The ink solutions are printed or spotted onto a substrate in an array format followed by curing the hydrogel to solidify the gel spots with biomolecules. The curing process varies, depending on the type of hydrogel applied. For instance, crosslinking in polyethylene glycol (PEG)-based hydrogels is typically initiated with an exposure to ultraviolet (UV) light while crosslinking in agarose-based hydrogels is a temperature dependent, and crosslinking in alginate-based hydrogels is dependent on the presence of calcium chloride ($CaCl_2$).

2.1.1.1. Pin printing

Pin printing system was first developed by Pat Brown and his colleagues, and has been widely applied for the production of DNA microarrays.[1] Recently, this method was applied to fabricate arrays with hydrogel inks, to better protect biomolecules and cells during array fabrication.[2] In this technique, a sharp and narrow needle pin (with or without a capillary), manufactured from stainless steel, tungsten, or titanium, is first dipped into a hydrogel solution containing desired biomolecules or cells. In the case of solid-pin (without capillary) printing, which is the simplest microarray printing (Fig. 1), a single inking step can transfer only a few droplets onto the substrate due to limited amounts of loaded solution on the tips. Therefore, multiple inking steps are required for sequential solid-pin printings in order to achieve similar size of deposits at the beginning and the end of process. In addition, multiple rinsing and re-filling steps required for producing large arrays of various biomolecules may increase the chance of cross-contamination within the array. The concept of solid-pin printing has also been explored for patterning at nano-scale, by replacing the larger needle pins with the atomic force microscopy (AFM) tips, a technique referred to as "dip-pen nanolithography" (DPN).[3]

In contrast, the split-pin (with a capillary) printing and the quill-pin (with a capillary and a reservoir) printing are capable of printing many spots repetitively and serially with less intermediate re-inking steps. In this case, larger amount of the sample is loaded into the gab of the pin by capillary forces and is transferred onto the substrate sequentially.

To achieve higher throughput production of biomolecular arrays using these printing methods, multiple-pin printing systems has been developed (Fig. 2). In these systems, a large number of pins are floating vertically on the pin head-translation system, transferring the biomolecule-loaded hydrogel droplets onto the substrate in a parallel fashion. Currently, 96- and 394-pins are the commercial configurations that are commonly used in the microarray fabrication. Multiple-pin system is beneficial for reducing the printing time, which is crucial for maintaining bimolecular activities or cellular viabilities during patterning the arrays. For instance, the Biomek 2000 system with multiple-pins is capable of fabricating

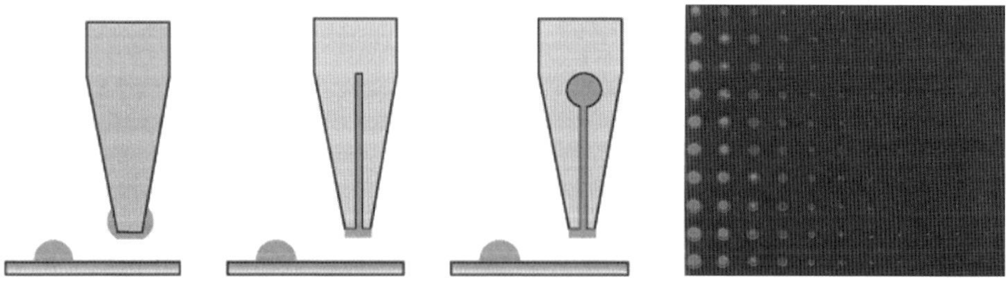

(A) (B)

Fig. 1. Patterning biomolecular arrays using pin printing with hydrogel-based inks. (A) Schematic of inking and subsequent transferring of inks to the substrate using a solid-pin, split-pin, or quill-pin. (B) A representative microarray produced by the SSP015 solid-pin printing method with spots of polyethylene glycol diacrylate (PEG-DA) hydrogel containing neonatal human dermal fibroblasts labeled with fluorescent dye (Cy3). Figure adapted from the Ref. 2 with permission.

two slides with 100 spots within 5 min.[2] However, the arrays produced by these multi-pin systems tend to have lower qualities compared to those produced by the single-pin systems; the size distributions of spots were reported to be larger.[4] Moreover, changes in environmental humidity and contaminations have detrimental effects on the quality of the resultant arrays since the exposed-tips tend to dry out quickly. Most importantly, printing imprecision has been reported as an issue in the multi-pin systems due to wobbly pins, which are not aligned together during the contact with the surface.[5]

2.1.1.2. Ink-jet printing

Ink-jet printing is another simple and reliable form of direct-writing method. This method does not require highly sensitive and expensive robotic systems; commercially available printers are modified and used to transfer biomolecules or cells onto the substrate in an array format.[8] There are two types of inkjet printing systems: thermal and piezoelectric printing systems. For patterning bioactive ink solutions, which are temperature sensitive, patterning is typically performed by piezoelectric inkjet printing with hydrogel-based inks.[9] In piezoelectric inkjet printing, the ejector is first filled with an ink solution (biomolecules or cells/aqueous hydrogel suspension) via a connection to an ink reservoir. Upon application of a voltage, a mechanical valve ejector generates pressure pulse in the piezoelectric nozzle. This pressure pulse overcomes the surface tension of viscous hydrogel precursors and ejects the droplets of cell (or biomolecule)-laden hydrogels onto the substrate. A programmed stage controller allows the biomolecule-containing hydrogel droplets to be precisely patterned in an array format on the substrate. Once the hydrogel drops are patterned, gelation process is initiated immediately to minimize the spreading of the

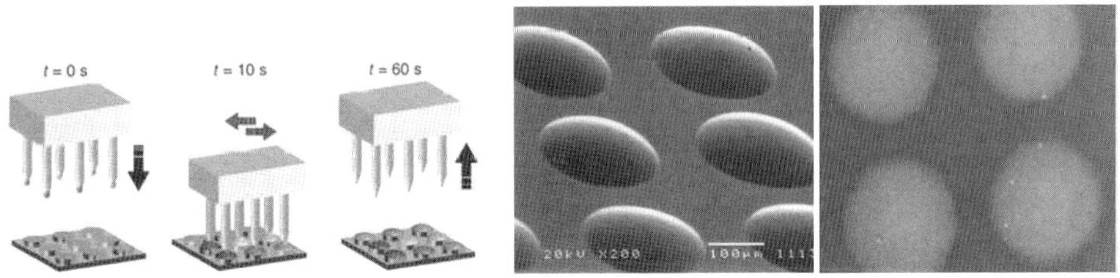

(A) (B)

Fig. 2. Patterning biomolecular arrays using multi-pin printing system with hydrogel-based inks. (A) Schematic of a multi-pin printing system. Figure adapted from reference[6] with permission. (B) Scanning Electron Microscopy (SEM) image (left) and fluorescence micrograph (right) of a representative array of printed PEG-DA hydrogel containing fluorescently-labeled acetylcolinesterase. Figures adapted from the reference[7] with permission.

droplets onto the substrate. Pataky *et al.* reported an interesting approach for the fast gelation of alginate drops by using a Ca^{2+} containing gelatin-coated substrate that provided calcium ions for immediate gelation upon transfer of alginate hydrogel drops onto the substrate.[10] Although ink-jet printing is more cost- and time-effective than the pin-printing, it also has a few drawbacks including clogging of the nozzle part and generating mechanical shear stress on the biomolecules or cells during fabrication that could affect their activities.

2.1.2. *Capillary-based hydrogel stamping*

Capillary-based hydrogel microstamper (Fig. 3) was first developed by Martin and colleagues in 1998.[11] These authors applied a polyacrylamide-based hydrogel as a stamp that absorbed the solution of biomolecules and transferred them onto a substrate via conformal contact. The hydrogel stamper used in this approach, was constructed and cured in the narrow end of a capillary tube. The crosslinked hydrogel in the capillary tube was initially dried, and an extra polymer part was removed to create a flat tip. The capillary tip with the hydrogel was incubated in the protein solution to reach equilibrium. The swelled hydrogel was then slightly pushed out of the capillary end, allowing a conformal contact between the gel and the substrate surface. A solution of Immunoglobulin G (IgG) protein was transferred to the substrate in a sequential manner using this capillary-filled hydrogel. This technique enabled multiple transfers from the hydrogel stamper without intermediate re-inking steps and

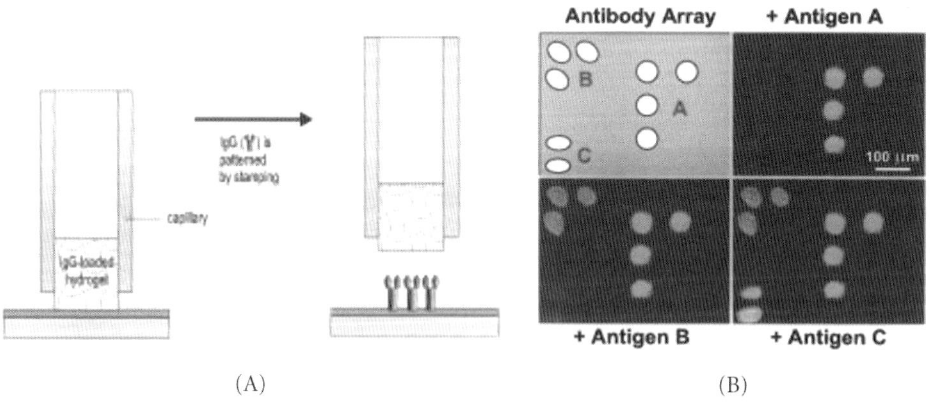

Fig. 3. Immobilization of antibodies in an array format using a capillary-based hydrogel stamper. (A) Schematic representation of a capillary-based hydrogel stamper. (B) Representative arrays of antibodies printed by this technique after exposure to three different antigens that specifically bind to the three different antibodies on the surface: antigen A for rabbit anti-goat, antigen B for rabbit anti-mouse, and antigen C for goat anti-human. Figures adapted from Ref. 11 with permission.

therefore, provided an efficient route for relatively rapid and inexpensive patterning of proteins in an array format compared to the ink-jet printing systems. Moreover, this work initiated the development of hydrogel-based microcontact printing methods that will be discussed next.

2.1.3. *Microcontact printing method*

The concept of microcontact printing (Fig. 4) was first developed by Whitesides group in 1993 for patterning of self-assembled monolayers (SAMs) of alkanethiols onto the gold substrates.[12] In this technique, poly(dimethylsiloxane) (PDMS) stamps are prepared by replica molding (REM) from a micro-patterned master (Fig. 4(A)), which is fabricated by photolithography methods.[13] In brief, PDMS is casted on, cured, and released from the patterned master, producing a topographically patterned PDMS stamp. The resultant elastomer PDMS stamps with hundreds or thousands of spots are then inked with desired solutions and transfer the ink solution to the substrate with two-dimensional patterns over large areas (30 cm × 40 cm) upon making conformal contact with the substrate (Fig. 4(B)). This simple technique has proven extremely powerful as it enables reproducible and inexpensive manufacturing of microarrays without any sophisticated equipment. However, the use of PDMS has limitations for patterning of fragile biomolecules and cells due to inherent hydrophobicity of this material.[14] The surface of PDMS stamps, thus, should be modified to become hydrophilic by oxygen plasma pretreatment or additional functionalizing processes for the application of water-soluble ink solutions of biomolecules.[15,16]

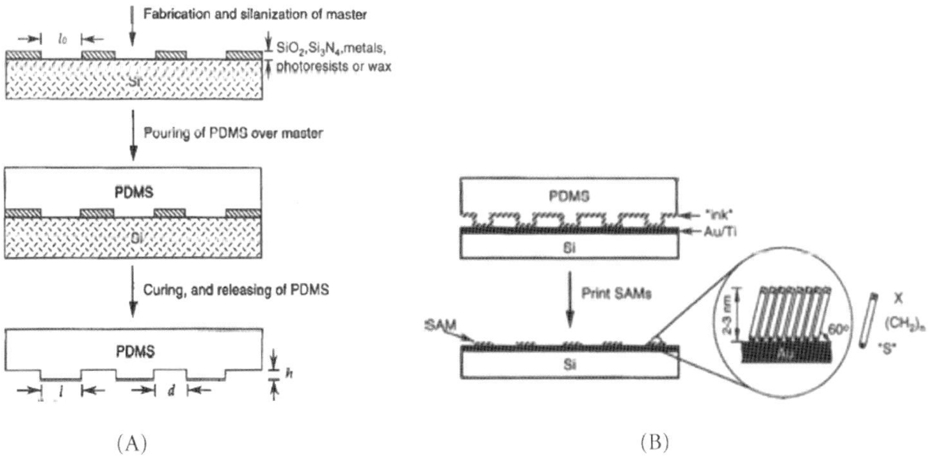

Fig. 4. The schematic illustration of microcontact printing method. (A) Fabrication of a PDMS stamp via replica molding, and (B) subsequent microcontact printing of SAMs onto a gold surface. Figures adapted from Ref. 12 with permission.

To address the abovementioned limitations, others and we have explored and applied wet hydrogel stamps for patterning of various bioactive molecules including lipids, soluble proteins, membrane proteins, as well as cells.[11,17–21] Figure 5, for example, depicts the concept of agarose hydrogel-based microcontact printing method that was developed for printing of proteins, lipids, and cells onto different substrates. In this approach, the hydrogel stamp is prepared by casting hot and liquid agarose solution on a mold containing a PDMS master with desired features. The agarose solution is then crosslinked by lowering the temperature, producing a stamp with desired positive features. The patterned hydrogel stamp absorbs the ink solution of biomolecules by directly pipetting solution onto the posts or by immersing the posts in the solution, and is then placed in contact with the substrate to pattern the biomolecules. During this process, the biomolecules are stored in the hydrated and biocompatible agarose gel matrix and thus, retain their activity. This method is able to generate spatially addressable arrays of few different biomolecules.[17,19] Similar to the capillary-based hydrogel stamping, hydrogel-based microcontact printing permits multiple transfer without the need for intermediate re-inking. This characteristic, in combination with the large number of posts on a stamp that transfer materials in each stamping event, make this technique extremely efficient for production of multiple copies of a large array of

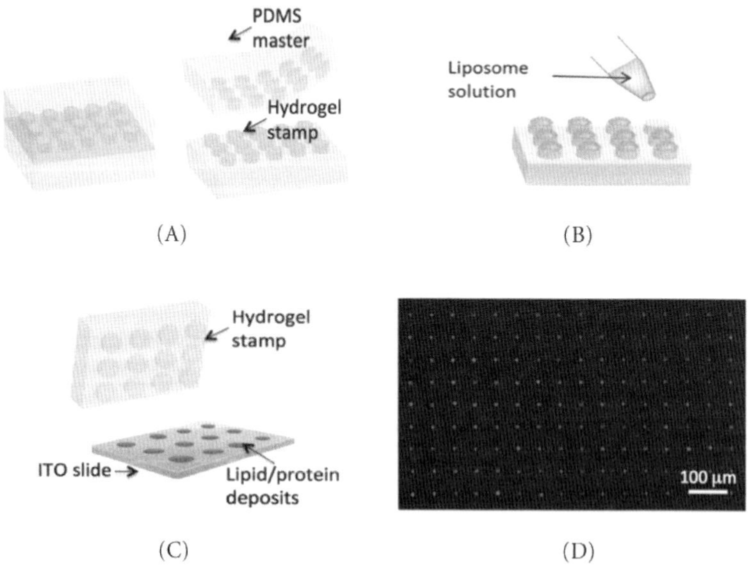

Fig. 5. Schematic representation of hydrogel-based microcontact printing. Steps of (A) stamp fabrication via replica molding, (B) stamp inking, and (C) printing of biomolecules onto a substrate. (D) A fluorescent micrograph of an array of hydrogel-stamped lipid deposits containing fluorescent-labeled lipids. Figures adapted from Ref. 21 with permission.

biomolecules within few minutes.[17,18,21] Most importantly, the use of hydrogel (here, agarose) as a stamp provides a low-cost and facile method that enables fabrication of biomolecular arrays with high efficiency.

2.1.4. *Laser direct-writing (LDW)*

Hydrogel microbeads that encapsulate cells or biomaterials have been applied in bottom-up tissue engineering as building blocks due to their well-defined shape and physiologically relevant three-dimensional (3D) microenvironments.[22–24] The matrix-assisted pulsed laser evaporation direct-writing (MAPLE DW) is a non-contacting, nozzle-free, and computer-aided method that enables precise placement of the microbeads on the planar substrate surface (Fig. 6).[25–27] In this method, the transparent quartz support is first coated with biomaterial-containing hydrogel film (called a donor substrate), and then paired with an uncoated substrate (an acceptor substrate). Next, a laser beam is focused onto the donor hydrogel film, melting selected parts of the hydrogel. Finally, the melted parts are ejected from a donor hydrogel and deposited onto the acceptor substrate. Recently, Phamduy *et al.* showed an improved LDW, which controlled deposition of cell-laden hydrogel microbeads on a substrate with high spatial resolution ($\pm 5\ \mu$m) and improved cellular viability.[28] In this technique, two different hydrogels were used; gelatin hydrogel was used for the donor substrate while alginate was used for cell-laden microbeads in the donor substrate. The mouse embryonic stem (ES) cells were first mixed with the alginate-based hydrogel solution, and then the alginate was cured in a PDMS mold with a medium containing Ca^{2+} ions resulting in the cell-encapsulating hydrogel microbeads. The donor substrate was then spin-coated with a mixture of liquid gelatin and the cell-loaded hydrogel microbeads. Once the gelatin was crosslinked and formed

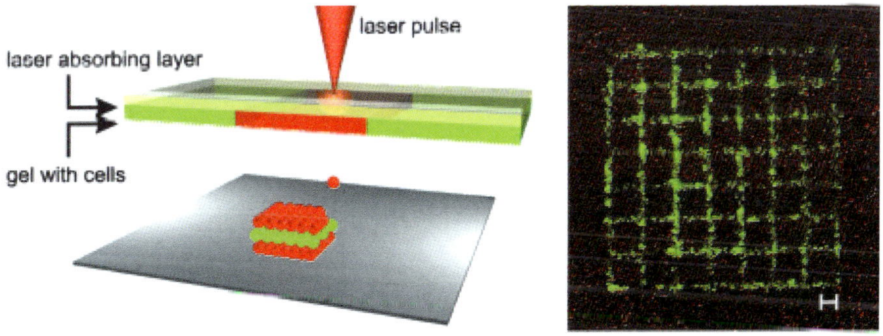

Fig. 6. Laser direct-writing method with hydrogels. Schematic representations of laser direct-writing system (left) and a representative grid pattern (right). The red represents keratinocytes expressing mCherry while the green represents fibroblasts expressing GFP on the substrate. Figure adapted from Ref. 27 with permission.

gelatin-alginate hydrogel film (a donor substrate) on the support, a single argon fluoride laser (ArF) laser pulse focused on the substrate melted specific parts of gelatin and ejected alginate microbeads onto the acceptor substrate. This two-hydrogel system was beneficial to sustain the activity of the biomolecules since alginate remained in gel form during the deposition. It has been reported that this method can produce droplets as small as picoliter-scale, allowing high-resolution patterning.[25] Although the use of a single laser-beam for transferring individual microbeads leads to longer printing time, this technique does not require the use of nozzles. Thus, this technique does not suffer from nozzle clogging as well as contamination problems and offers better protection of biomolecules from shear or thermal stresses.

2.2. *Hydrogels as substrates*

2.2.1. *Micro-molding method*

Fabrication of patterns with complex topologies can be achieved by micro-molding methods, such as REM (Fig. 5(A)).[29] This technique has been adapted for the fabrication of hydrogel nano-/micro-wells (substrates with negative features) in order to spatially organize biomolecules or cells in an array format.[30–32] Kobel *et al.* introduced a method that combined REM with microcontact printing, referred as soft embossing, for fabricating PEG hydrogel microwells (Fig. 7(A)).[31] In this method, liquid hydrogel is casted on a flat substrate, and PDMS mold is quickly pressed and embossed into the hydrogel-coated substrate. After gelation, the PDMS mold is carefully removed to release the patterned microwells. In addtion, Kawai group introduced a nanoimprint lithography (NIL) method, which is one of advanced embossing techniques for fabricating poly(vinyl alcohol) (PVA) hydrogel nanowells for membrane arrays (Fig. 7(B)).[32] In this method, PVA-coated substrate

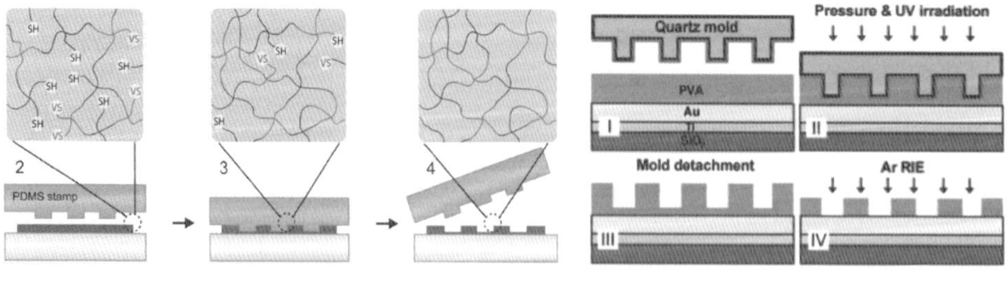

(A) (B)

Fig. 7. Advanced micro-molding technique combined with microcontact printing method. Schematic illustration of embossing method (A) and imprinting method (B). Figures adapted from Refs. 31 and 32 with permission.

(Au) is patterned against a quartz mold via embossing method, and cured by UV irradiation. The mold is peeled off from the patterned PVA surface, and hydrogel nanowells with exposed Au surface are obtained after reactive ion etching (RIE) using Argon. Such hydrogel nano-/micro-wells provide compartments for precise and highly selective immobilization of biomolecules as well as cells with hydrated environments.

Recently, Moeller *et al.* applied PEG-DA hydrogel micro-wells for 3D culture of stem cells (Fig. 8).[30] In this study, the hydrogel micro-wells were molded on an elastomeric PDMS template with positive features and crosslinked under UV light. Next, the PDMS template was peeled off, and pluripotent murine ES cells were seeded into each micro-well. The hydrogel micro-wells allowed controlling of size, shape, and homogeneity of cell population and hence, enabled a precise control over the differentiation of pluripotent stem cells.

With the abovementioned overview on the application of hydrogels in fabrication of cellular and biomolecular arrays, next, we discuss these fabrication methods in four main groups in terms of the type of probes immobilized on the substrate: (i) DNA, (ii) protein, (iii) lipid, and (iv) cell microarrays, with an emphasis on their applications in sensing and screening assays

3. Bioarrays and their Applications

3.1. *DNA microarrays*

Initiated from DNA hybridization technology developed by Edwin Southern in 1975 and DNA sequencing chemistry developed by Frederick Sanger and Walter Gilbert,

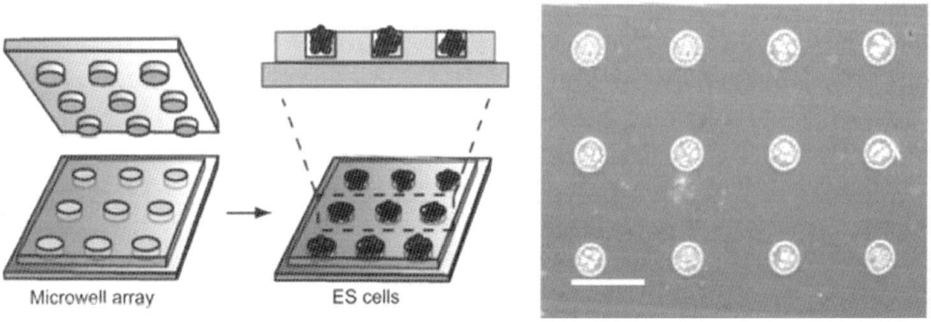

(A) (B)

Fig. 8. The use of micro-molding method in fabrication of hydrogel micro-wells. (A) Schematic illustration of hydrogel micro-wells applied for 3D cell culture. (B) Representative array of PEG-DA micro-wells containing pluripotent murine ES cells. Figures adapted from Ref. 30 with permission.

who won the Nobel Prize in 1980, DNA microarrays have rapidly evolved to sensitively detect gene expression levels, single nucleotide polymorphisms (SNPs), and short tandem repeats (STRs).[33–35] DNA biochips contain an array of short oligonucleotide sequences or DNA probes, which allow high-throughput screening of a large number of genes in parallel using hybridization of the complementary nucleotide sequences to DNA probes, immobilized on the chip.[36] Thus, DNA microarrays have powerful applications in industries where massive gene screening is required for drug development and in academia where identification of unknown signaling pathways via gene expression profiling and sequencing target oligonucleotides are desired.

3.1.1. *Immobilizing DNA probes on hydrogels*

DNA probes in microarrays are immobilized onto solid substrates by either covalent bonding or non-covalent bonding. A common covalent conjugation of oligonucleotides on glass substrates is achieved by carbodiimide coupling reagent. For instance, a glass slide treated with N-hydroxysuccinimide (NHS) will make a covalent bonding with the amine group at the 5′ end prime of oligonucleotides.[36] In order to fabricate DNA microarrays, small amount of oligonucleotides is deposited onto these substrates using robot-assisted spotting techniques. These two-dimensional (2D) arrays can contain 10^6 test spots with as small as a few picomoles to micromoles of DNA droplets.[35]

While the conventional planar DNA microarrays have been broadly applied to screen multi-genes in parallel, they have some limitations in detection of SNPs or STRs. For instance, the planar-nature of these 2D microarrays restricts the oligonucleotide loading capacity per spot surface area, as the narrow free-space between spots increases the risk of cross-contamination between neighboring spots and modifies binding kinetics of DNA hybridization. In order to overcome some of these limitations of the 2D DNA arrays, hydrogel-based DNA microarrays, in which oligonucleotides are conjugated to 3D hydrogel networks, have been explored for the past two decades and will be discussed in this section.

Mirzabekov group pioneered the application of polyacrylamide gel spots to immobilize oligonucleotides, creating micro array of gel-immobilized compounds on chip (MAGIC chips) in 1998.[37,38] The poor reproducibility and irregular deposition were, however, the main drawbacks for massive production of these MAGIC chips. To improve the reproducibility of MAGIC chips, the same group later proposed an alternative approach and developed the immobilized micro array of gel elements on chip (IMAGE chips) (Fig. 9(A)).[39] The main difference between the

MAGIC chip and new IMAGE chip was that in the latter technique, the acrylamide gel was polymerized after it was spotted onto the glass, while in the former technique, a pad of gel precursor was photopolymerized to produce patterns, followed by washing of the uncured gel, which resulted in inconsistent gel patterns.[38,39] Briefly, a glass substrate was treated with Bind Silane to attach polyacrylamide gels. Then, a mixture of acrylamide and bis-acrylamide were spotted onto the substrate by a single-pin robot and co-polymerized by UV photopolymerization. Polyacrylamide hydrogel was then activated to conjugate oligonucleotides, and finally DNA molecules were immobilized in the hydrogel matrix spots with a semi-spherical shape. Using this method, microarrays of 20 × 20 spots were fabricated with 2 μL of the mixture per spot. The IMAGE chip showed an improved consistency in the spot size and thus, enhanced reproducibility compared to the previous MAGIC chip.[39]

Several other groups have employed similar techniques to those used for the production of IMAGE chips to further extend the quality and capability of 3D DNA arrays. For example, Liu *et al.* used PEG cushion on the glass substrate in order to strengthen the adhesion of polyacrylamide gel on the glass substrate (Fig. 9(C)).[40] These authors suggested that biocompatible PEG-coated substrates would provide a biomolecule-resistant background where DNA immobilized polyacrylamide gels were not spotted and therefore, would provide enhanced sensitivity for the detection of SNPs. The resultant spot size of the polyacrylamide hydrogel was approximately 500 μm, providing a greater surface area for DNA binding than the planar array of the same spot size. In these arrays, detection of complementary sequence was achieved by the fluorescence intensity difference between the perfect match compared to the mismatch.[40] Compared to similar fluorescence detection techniques where the authors were able to differentiate the perfect match by 50% fluorescence intensity difference, the study by Liu *et al.* demonstrated 20-fold higher fluorescence intensity from the perfect match.[40]

A different approach taken by Bromberg *et al.* incorporated the DNA immobilization technique using UV photopolymerization into a microfluidic system (Fig. 9(B)).[41] In this method, fluorescence tagged oligonucleotides with polyacrylamide gel precursors were injected into a microfluidic channel and photopolymerized at the cross-sectional regions of the fluidic channels as depicted in Fig. 9(B). Electrophoresis drove the flow of DNA molecules within the fluidic channels where the complementary sequences hybridized with oligonucleotides captured in polyacrylamide gels at the channel intersections and emitted fluorescence. This technique allowed a rapid sorting of complex DNA sequences to identify SNPs. Additionally, the hybridized matching oligonucleotides were easily harvested from the hydrogel for sequencing.[41]

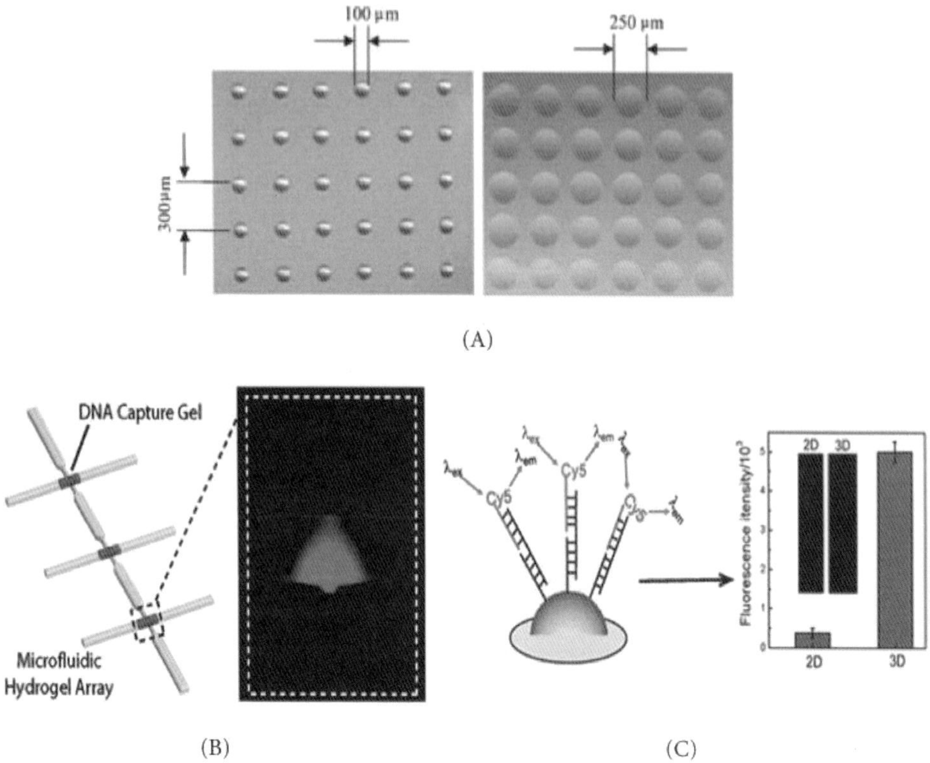

Fig. 9. Polyacrylamide hydrogel-based 3D DNA arrays. (A) Bright field images of DNA immobilized on polyacrylamide hydrogel droplets first developed by the Mirzabekov group (IMAGE chips). Figure adapted from the reference with permission.[39] (B) A schematic drawing of DNA capture polyacrylamide gel, injected into a microfluidic system (left) and the fluorescence signal detected at the cross-section of the microfluidic channels by DNA hybridization (right). Figure adapted from the Ref. 41 with permission. (C) A diagram of DNA immobilized onto polyacrylamide hydrogel droplets and fluorescence detection system designed by Liu *et al.* This 3D DNA arrays showed significantly higher fluorescence signal for SNP sensing. Figure adapted from Ref. 40 with permission.

3.1.2. *Encapsulation of DNA probes in hydrogel network*

While many of the hydrogel-conjugated DNA microarrays were based on polyacrylamide and 5′ prime modified DNA chemistry, a simpler approach reported by the Mirkin group utilized agarose gel as an ink for DNA delivery via DPN (Fig. 10).[3] While DPN is one of the most advanced direct-writing techniques and has been applied to fabricate bimolecular arrays in nano-scale spots, the consistency in the delivery of biomolecules to the solid substrate with this technique has been troublesome for the production of such arrays.[42] The use of agarose hydrogel as a carrier for the biomolecules improved the uniformity in deposition of oligonucleotides and

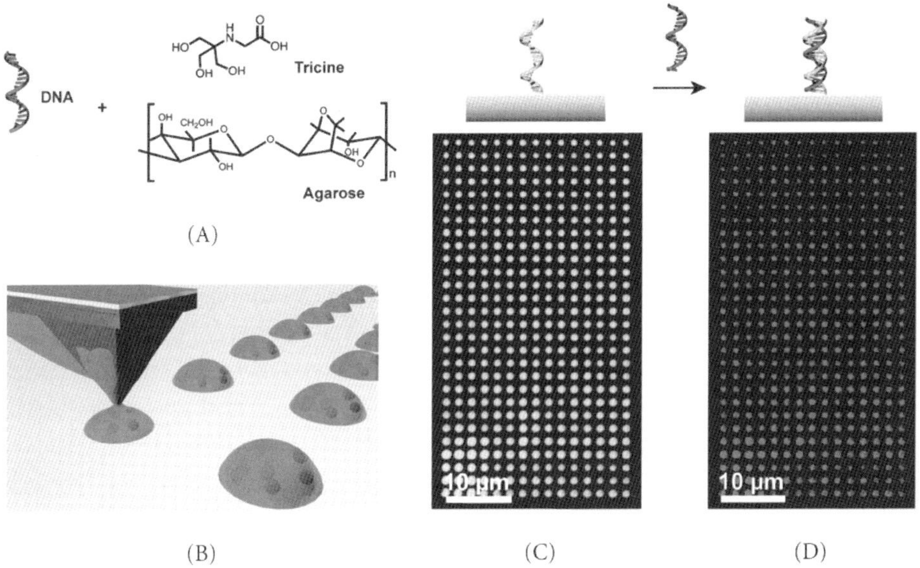

Fig. 10. Agarose-assisted DPN of oligonucleotides to fabricate DNA nanoarrays. (A) An illustration of DNA and matrix materials. (B) A demonstration of AFM tip and the matrix complex deposited on the glass surface. (C) A fluorescence image of DNA nanoarrays before hybridization with complementary sequences. (D) A fluorescence image of DNA nanoarrays after hybridization. The figures adapted from Ref. 3 with permission.

enabled fabrication of nano-scale DNA arrays, which are considered more cost-effective and high-throughput compared with the conventional high-density DNA microarrays. In the study by Senesi *et al.*, 0.15% agarose in 10 mM tricine was used as the ink to assist the deposition of fluorescently labeled (Cy3) oligonucleotide and detection of the complementary sequences was achieved by DNA hybridization. The size of spots in these DNA chips ranged from 50 nm to 1 μm for 0.5 to 5 dwell time of AFM tips at the substrate surface.[3] With the shortest dwell time, the authors were able to fabricate 450 spots in 1 min, enabling the fast production of stable DNA nanoarrays.

Therefore, the hydrogel-based DNA microarrays offer several advantages over the conventional 2D DNA microarrays. First, 3D structure of the hydrogel droplet increases the DNA loading capacity per spot and enhances the sensitivity of detection for SNPs or STRs. Second, the hydrated network of the hydrogel further stabilizes DNA molecules during the screening process and allows lengthier detection steps, such as polymerase chain reactions. Lastly, the limitation of the spot size and space between the spots to avoid contamination is overcome by larger loading capacities of 3D DNA arrays or hydrogel-mediated DPN.

3.1.3. *Hydrogel-mediated DNA aptamer arrays*

Besides DNA microarrays, hydrogels have been utilized in creating microarrays of DNA and RNA aptamers.[43] Aptamers are short DNA or RNA sequences folded in 3D structure to present binding motifs for other biological molecules, such as proteins and even cells. Aptamers are commonly employed for sensing target molecules using different methods such as aptamer microarrays,[44] aptamer-coated microbead flow cytometry,[45] and single channel recording through aptamer-decorated nanopores.[46] An interesting approach by Revzin and Tuleuova applied a substrate with patterned PEG hydrogel to create an array of DNA aptamers (Fig. 11) in order to detect low concentrations of interferon-γ (IFN-γ), a key cytokine secreted during immune response.[43] As Fig. 11(A) demonstrates, PEG hydrogel was patterned by UV exposure through a photomask to create microwells with a ~35 μm diameter, corresponding to the size of a single immune cell. Biotinylated DNA aptamers were then immobilized within the PEG hydrogel micro-wells and applied for detection of

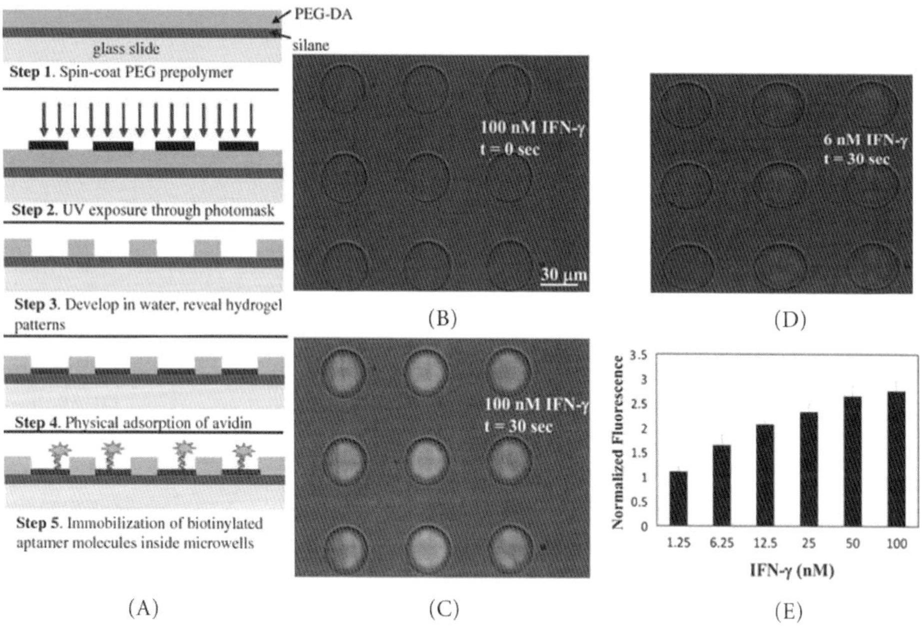

Fig. 11. Detection of cytokine molecules via micropatterned DNA aptamers between PEG hydrogel barriers. (A) A schematic diagram of PEG-hydrogel separated micro-wells and immobilization of DNA aptamers inside microwells. (B–D) Bright field/fluorescence overlay images of micro-wells incubated with IFN-γ. (B) No FRET signal was detected at t = 0 sec. (C) Bright FRET signal was detected after 30 secs. incubation with 100 nM IFN-γ. (D) FRET signal was apparent with lower concentration of IFN-γ, 6 nM. (E) Normalized fluorescence intensity was plotted against IFN-γ concentration and showed a linear correlation. Figures adapted Ref. 43 with permission.

IFN-γ at concentrations as low as 1.25 nM using Förster resonance energy transfer (FRET) technique. Recently, the Revzin group employed a similar PEG hydrogel microwell platform along with fluorescently labeled aptamers to detect protease released from cells.[47]

3.1.4. *Sensing applications of DNA microarrays*

A number of techniques for detection of DNA hybridization on DNA biochips have been developed to facilitate DNA-based sensing. While details of the detection mechanisms are thoroughly discussed in an excellent review by Sassolas *et al.*,[36] the detection mechanisms can be either reagent-labeled or label-free. The reagent-labeled detections include optical sensing, such as fluorescence-based detection where target molecules are labeled with fluorescence probes and where molecular beacons for FRET are applied.[48,49] Chemiluminescence is another reagent-labeled detection technique where, for example, electrochemiluminescent (ECL) reactions are induced by catalysis of HRP-conjugated probe, which binds to DNA.[50] Conductive nanoparticles such as gold,[51] silver,[52] carbon nanotubes,[53] conducting polymers,[54] and quantum dots,[55–57] can also be employed to label the target DNA for electrochemical detection.[58]

The label-free detection methods are attractive as direct and powerful analytical methods. For instance, surface plasmon resonance (SPR) imaging measures polariton propagation at the substrate interface and, as a result, senses fine changes in electrochemical properties of the target molecule.[59] Thus, SPR imaging has been applied for label-free sensing of DNA hybridization on electrically conductive substrates.[60–64] Also, cantilever based sensing, where DNA hybridization alters the mass on the cantilever array and displays the difference in distance, is another example of label-free sensing.[65] Both reagent-labeled sensing and label-free sensing have evolved to detect smaller amounts of the target DNA or minute changes in DNA sequence, and various methods now offer nano- to femto-molar range detections.[36] However, the density of the immobilized DNA on the array substrate remains to be the limiting factor for the conventional planar DNA microarrays, which can be overcome by 3D hydrogel DNA arrays discussed previously.

The application of the DNA microarrays is not limited to the detection of DNA hybridization.[66] Functionalized oligonucleotide strands have also been applied for detection various biochemical reactions, such as enzyme reactions,[18] protein-DNA binding events,[63,64,67] and pH changes.[68] Recently, the development of the aptamers technology has facilitated the application of DNA arrays in wide-range of biosensing (e.g., protein, hormone, or chemicals).[69]

3.2. *Protein microarrays*

Many aspects of cellular functions cannot be explained at the nucleic acid level since these functions are modulated through protein translation and post-translational modifications.[70–72] In order to understand cellular functions at the protein level, the concept of protein arrays was first introduced in a form of antibody matrix in 1983, and was later miniaturized into microarrays.[73] When combined with the high-throughput screening technology, the protein microarrays provide powerful tools for monitoring expression levels of target proteins and interactions of proteins with other biomolecules. These arrays have, thus, become popular for biomarker discoveries, biosensing, medical diagnostics, and monitoring disease states.[74] In general, there are two types of protein microarrays: one for predicting protein abundance and another for identifying protein interaction/function.[75] In the abundance-based arrays, desired antibodies or aptamers are selected and patterned in an array format due to their high specificity and sensitivity to the specific antigens of interest. In the function-based arrays, functional proteins (e.g., enzymes, adaptor proteins, cytokines, etc.) are extracted from cells or patient tissue and immobilized on the arrays for functional studies.[76]

3.2.1. *Fabrication of protein microarrays using hydrogels*

Protein microarrays provide high-throughput platforms with low consumption of samples that are ideal for determination of protein abundance and function. Therefore, development of new fabrication and detection methods for high quality protein microarrays have attracted an increasing attention.[77] Methods for immobilization of proteins or capture molecules on the substrates are the main determinants of the quality of protein microarrays. Non-covalent adsorption, covalent coupling with functional groups (e.g., aldehyde, epoxy, etc.), and affinity-based binding (e.g., streptavidin-biotin interaction) methods are most common for immobilization of proteins and have been modified for production of high quality protein microarrays.[78–80] For instance, in ProteoChip, ProLinker™, a bifunctional coupling linker between protein and gold substrate allows patterning of concentrated probes (4×10^{12} probes/cm^2) in an array format with well-controlled density (25–5000 spots/cm^2) and orientation.[77] However, sustaining the bioactivity of proteins during immobilization has remained a challenge in this process.[81] The concentration of the probes should ideally be high enough to allow detection of analytes at femto-molar concentrations but proteins tend to be aggregated under such conditions. Moreover, harsh conditions such as coming into contact with low-energy interfaces, dry environments, high ionic strength, and extreme pH conditions cause denaturation of proteins during immobilization. The use of wet hydrogels in protein microarrays can address many of these issues and can provide proper microenvironments in which proteins retain their native 3D folded states. Also, 3D hydrogel matrix, when used as a platform for immobilization of probes, provides higher accessibility as well as excellent sensitivity of the immobilized probes toward targeting molecules from all sides.

Hydrogels can be applied for the transfer of proteins to substrates in an array format. For instance, Mayer *et al.* reported a simple microcontact printing method using an agarose hydrogel stamp for the fabrication of protein microarrays.[19] In this method, the hydrogel stamps were prepared by REM method against PDMS master described in previous section (see details in 2.2.1. Micro-molding method). The patterned agarose stamps absorbed the protein solutions and later printed them onto large areas of the substrate. This method allowed production of high-density protein arrays (~600 spots/cm^2) with eight different proteins. Moreover, these authors demonstrated that the hydrogel stamp has the ability to store the ink solution and allows for creation of up to 100 copies of these protein arrays without the need for re-inking (Fig. 12(A)). However, the agarose stamps applied in this study required a solid support or very thick stamp structure for robust stamping and to avoid sagging and deformation of printed features during printing the proteins. To overcome this issue, Coq *et al.* introduced a rigid self-supporting hydrogel stamp, which was composed of a mixture of 2-hydroxyethyl acrylate (HEA) and PEG-DA.[82] In these stamps, the concentration of PEG-DA as the crosslinker in the hydrogel changed the physical properties of the stamps: the hydrogel with higher concentration of PEG-DA exhibited improved mechanical robustness. For instance, 18 wt.% PEG-DA showed a Young's modulus (Fig. 12(B)) comparable to that of

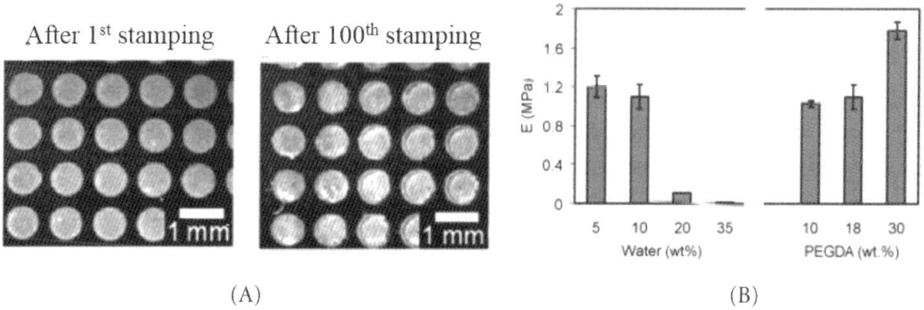

Fig. 12. Main advantages of hydrogel stamps for patterning proteins. (A) Patterning 100 copies of protein arrays using an agarose stamp without re-inking. A stamp loaded with a solution of Tetramethylrhodamine-Bovine serum albumin (TRITC-BSA) immobilized the labeled protein onto glass slides with no significant difference between the 1st (left) and 100th (right) stamped arrays. Figure adapted from the reference [19] with permission. (B) An improved mechanical property in the self-supporting hydrogel. Bar graph showing the Young's modulus (E) of hydrogel stamps with various content of water (left) or PEG-DA (right). Figure adapted from Ref. 82 with permission.

conventional PDMS (1.5-2 MPa). These hydrogel stamps showed a minimal deformation of hydrogel under normal printing pressures. Thus, such improved hydrogel stamps not only provided ideal environments for the transfer of fragile proteins but also offered excellent printing resolutions.

Hydrogel stamps have also been employed to pattern films of conducting polymers with encapsulated proteins that can be applied for biosensing.[83,84] In a recent study, we used patterned agarose stamps as a carrier for the polymer precursor solution that contained proteins such as biotin.[85] Once the stamp was placed on a gold electrode, the conducting polymer was electropolymerized by a three-electrode method as shown in Fig. 13(A). Upon application of electrical current through electrodes (gold: working electrode, hydrogel: counter and reference electrode), a thin film of conducting polymer with encapsulated biotin was deposited only in the contact areas between the patterned stamp and gold substrate, generating positive patterns of conducting polymer films with biotin. The presence of biotin was confirmed via immunofluorescence imaging or Fourier transform Infrared Spectroscopy (FTIR). This versatile technique enabled fabrication of a substrate with multiple distinct chemistries, indicating a potential that multiple distinct proteins can be incorporated onto a single substrate via a simple one-step process.[85] Such electrically conductive films provide excellent platforms for electrochemical sensing using protein microarrays.

Hydrogels have also been utilized as the ink solution for patterning proteins on substrates by either patterning of protein-containing hydrogel solutions or patterning functionalized hydrogel solutions for later addition of proteins. For instance, Rubina *et al.* applied the first approach for fabrication of protein microarrays by an inkjet-printing method (Fig. 14(A)).[81, 86] They prepared an ink solution comprising Gel Elements and antibodies against various proteins, and spotted the inks onto a glass substrate using a QArray pin robot. Residual spaces between spots were filled with hydrogels so that the immobilized proteins in the gel would retain its biological activity for at least six months. Polymerization of gel arrays was

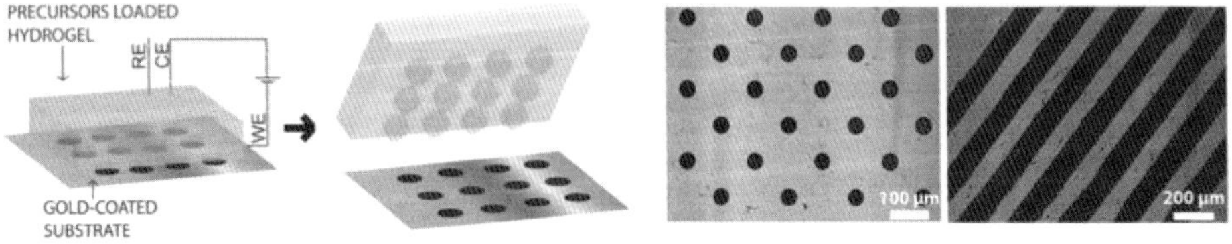

(A) (B)

Fig. 13. (A) A schematic illustration of the hydrogel-mediated electrodeposition of conducting polymer films with proteins encapsulated. (B) Optical images of the conducting polymers patterned via this method in various sizes and geometries. Figures adapted from Ref. 85 with permission.

carried out under the UV light. Immobilized antibodies in this microarray enabled detection of very low-abundance biotoxins (below nanomolar concentrations). In a different study, Stiles *et al.* applied the second approach and reported direct deposition of functionalized hydrogels using DPN (Fig. 14(B)).[87] The mixture of thiol group-conjugated PEG (PEG-thiol) and PEG-dimethacrylate (PEG-DMA) was used as a hydrogel precursor. DPN tips were loaded with the ink solution and applied to pattern the solution onto the substrate (Fig. 14(C), left). After curing the hydrogel arrays, maleimide-conjugated proteins were allowed to react with the four-arms of PEG-thiol and decorated the surface of hydrogel spots (Fig. 14(C), right). By separating the decoration step from the deposition of hydrogels onto the substrate, this technique reduced the exposure of proteins to air and prevented protein denaturation.

Coating substrates with hydrogels has been an alternative strategy for immobilization of proteins in production of high-quality protein microarrays. Piehler and colleagues introduced glass surfaces coated with various hydrogels, such as chitosan-, dextran-, polyoxyethylene-, polyethyleneimine-, and polyacrylamide-based hydrogels, in order to improve the functionality of proteins immobilized on the substrate.[88] Other groups also coated the surface of microarray substrates with hydrophilic monolayers (few nm thin) of PEG or grafted copolymer (e.g., PLL-g-PEG) immediately

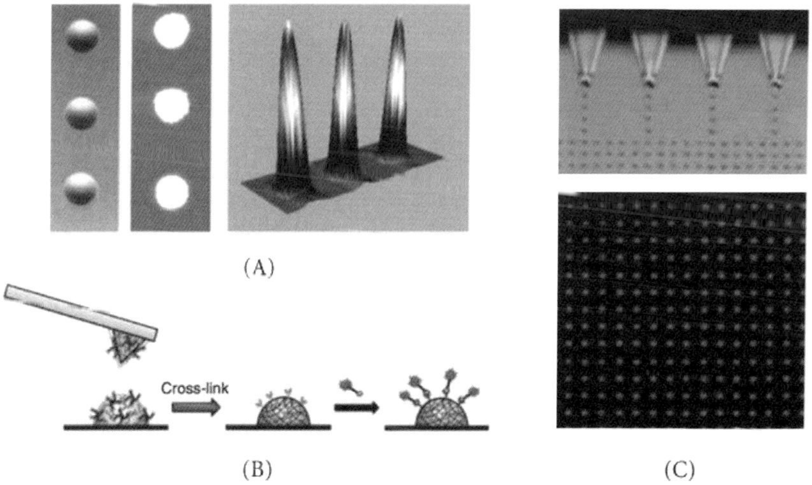

Fig. 14. (A) Phase contrast (left), 2D fluorescence (middle), and 3D fluorescence image of the printed hydrogel arrays containing anti-risin antibodies. The presence of antibodies was confirmed by fluorescently-labeled risin. Figure adapted from Ref. 86 with permission. (B) A schematic illustration of deposition, crosslinking, and tagging process in DPN with a hydrogel ink. Figure adapted from the reference [87] with permission. (C) Photograph of deposition process via DPN and the resultant arrays of the functionalized hydrogels after the reaction with rhodamine-conjugated maleimide. Figure adapted from Ref. 87 with permission.

after oxygen-plasma treatment of the substrates.[89, 90] These hydrogel-coated surfaces were then functionalized and crosslinked with proteins (Fig. 15A). These 2D hydrogel-coated surfaces are beneficial for protecting the native folded state of proteins and increasing their analyte capture capacity. However, the density of immobilized proteins on these substrates could not exceed the conventional methods.[91] To address this issue, 3D hydrogel coatings have been explored.[92, 93] For instance, polyacrylamide hydrogel-coated substrates (Hydrogel™) are commercially available from Perkin-Elmer Life Science and applied in fabrication of protein microarrays (Fig. 15B). More recently, thinner hydrogel supports (~60 nm thick) such as Slide H, OptArray, and Accelr8 have been developed to reduce the diffusion time of analytes through the hydrogel matrix and to minimize the required incubation times, especially for low-abundance proteins. The 3D hydrogel supports are not only able to reduce protein denaturation and non-specific bindings but they are also able to increase the protein loading capacity as well as detection sensitivity.[93]

3.2.2. *Sensing applications of hydrogel-based protein microarrays*

Despite the advantages of applying hydrogels in protein microarrays, such as promoting bioactivity and probe-analyte interactions in the gel matrix, the reduced sensitivity of these microarrays due to the required diffusion of analytes through the hydrogel matrix has been an issue. Thus, many approaches have been developed for increasing the sensitivity of probes toward the low-abundance proteins. The main detection strategies for protein microarray include label-based and label-free techniques. In the label-based sensing techniques, it is necessary to label the molecule of interest with fluorescent dyes, radioisotopes, epitope, and etc. Among these methods, immunostaining, which employs fluorescently labeled antibodies, is the most

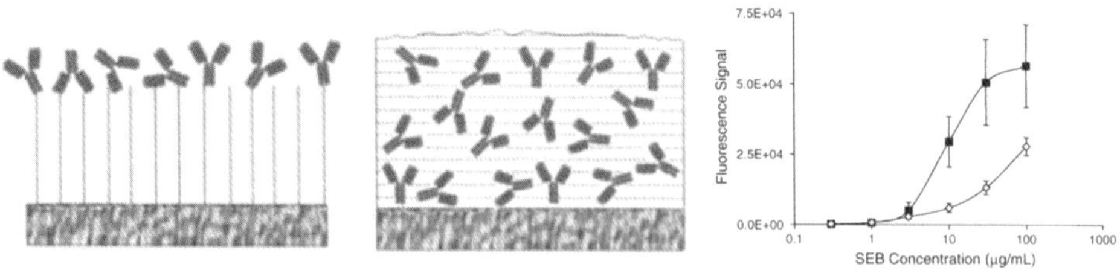

(A) (B) (C)

Fig. 15. Schematic representations of (A) 2D and (B) 3D hydrogel coating on the surface of protein microarrays. Figures adapted from Ref. 91 with permission. (C) Comparison of sensing efficiency by measuring the fluorescent signals from a binding of Cy5 anti-Staphylococcal enterotoxin B (SEB) to the immobilized SEB in a 3D or 2D acrylamide-based hydrogel. Figure adapted from the Ref. 93 with permission.

popular method for the detection of target binding on these microarrays. Although label-based methods have some benefits such as simple instrument requirements, these methods require labeling of the probes and thus, affect the function of immobilized molecules. In addition, molecules used for labeling the proteins may lower sensitivity and specificity in these arrays as they may interfere with the interactions between the probes and their target molecules.[76]

Recently, a number of label-free sensing techniques have emerged for applications in protein microarrays as alternatives for the label-based sensing. Especially, SPR-based biosensors have been combined with hydrogel-based protein microarrays. In an interesting study, Tanaka *et al.* applied SPR to monitor the binding of analytes to proteins in the 3D nanostructured protein hydrogel (3D NPH)-coated gold substrates (Fig. 16(A)).[92] These authors showed that by combining 3D structured hydrogel microarrays and SPR sensing technique, the amount of non-specific binding of streptavidin to the substrates was significantly reduced. Another appealing technology that has been explored in this context is SPR imaging (SPRi), which combines SPR and high-throughput screening. In this method, a sensitive camera monitors the changes in the reflected light on an array and converts these changes into an image. Lee *et al.* demonstrated the applicability of SPRi for direct and multiplexed detection of proteins with low molecular weights.[94] In another study, Zhou *et al.* used SPRi in order to detect the adsorption of various analytes to the patterned probes in the carboxymethylated dextran hydrogel-coated gold substrates (Fig. 4(B)).[95,96] These combined techniques, thus, enable real-time quantitative measurements with high sensitivity (nanomolar range) on protein microarrays.

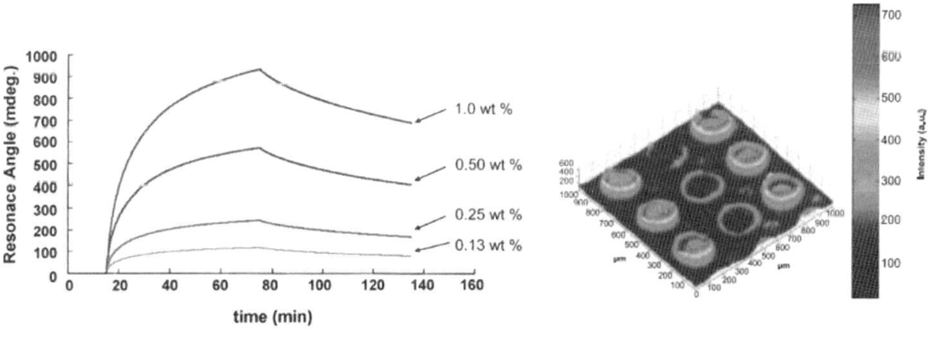

(A) (B)

Fig. 16. (A) SPR signals from analyte-probe binding events. With an increasing concentration of mouse IgG in 3D NPH, the sensitivity of detection toward anti-mouse IgG was increased. Figure adapted from Ref. 92 with permission. (B) 3D SPR imaging of hydrogel microarrays containing various concentration of anti-myoglobin on a gold chip at incident angle of 66.4°, at which a maximum differential reflectivity was observed. Figure adapted from Ref. 95 with permission.

Various scanning probe microscopes (SPMs) have also been used to monitor the binding of analytes to the hydrogel-based protein microarrays. For example, in AFM, the changes in vertical and horizontal deflection of the cantilever upon interactions of the analyte (attached to the cantilever) with the surface are detected and converted to an image of scanned surface. Recently, Tanaka and colleagues showed improved AFM images of binding analyte to the surface when hydrogel-coated substrates were used.[92] Instead of the AFM tip, microcantilevers can also be used for sensing of protein microarrays. While the AFM tip provides information on the surface topography of microarrays, the microcantilever measures the changes in reflection angles on the surface. It has been combined with protein microarrays for screening analyte-probe interactions, detection of biomarkers, and conformational changes of proteins.[97–99] Moreover, other label-free methods such as elipsometry,[100] interference-based techniques (SRIB, Dual-channel BioCD, AIR),[101–103] scanning Kelvin nanoprobe method,[104] and electrochemical impedance spectroscopy[105] have been used in combination with protein microarrays. These label-free sensing techniques exhibit excellent sensitivity and enable multiplexing assays.[76]

3.3. *Lipid microarrays*

Lipid and membrane microarrays have been spotlighted as alternative and versatile biomimetic platforms to study lipid-protein interactions, receptor-ligand interactions, and drug-membrane interactions in high-throughput fashion.[106,107] Membrane arrays are composed of self-assembled layers of synthetic or natural phospholipid molecules, with or without integral membrane proteins, mimicking membranes of cells or organelles.[108] These arrays are particularly appealing for drug development and discovery as membranes not only provide an effective physical barrier around cells, but also host an array of membrane proteins that are linked to various diseases. Therefore, screening platforms, which enable studying the interactions between membrane-incorporated proteins and drugs, are on high demands. For targeting membrane proteins, it is important to reconstitute these proteins in lipid environments similar to that of real cell membranes. Lipid or membrane microarrays provide such platforms with capabilities for efficient screening. In this section, we will review three different types of membrane arrays, which are liposomes, supported lipid bilayers (SLBs), and black lipid membranes (BLMs) microarrays, and how hydrogels have been incorporated in the fabrication of these arrays.

3.3.1. *Hydrogel-mediated fabrication of lipid microarrays*

Hydrogels have been applied in membrane arrays for a number of purposes. For example, hydrogel stamps were used to pattern arrays of SLBs with micron-sized features,[17] hydrogel cushions were employed to stabilize fragile BLMs,[109] and hydrogel barriers were utilized to pattern lipid rafts in nanoscale and to prevent non-specific binding of target proteins.[32]

For patterning arrays of SLBs, Majd and colleagues employed a topographically patterned agarose gel to load small lipid vesicles and to transfer these materials to glass substrates with desired patterns.[17] The transferred lipid molecules on the glass substrate self-assembled into lipid bilayers and formed SLBs. Fluorescence recovery after photobleaching experiment on fluorescently labeled SLBs confirmed the lateral fluidity of the lipids within the bilayer with a diffusion coefficient of 0.2 $\mu m^2/s$. These authors further advanced this hydrogel-based microcontact printing technique by individually inking the hydrogel with solutions of proteoliposomes or cellular membrane fragments to minimize the material consumption and to maximize the concentration of inking materials as Fig. 17A illustrates.[18] This technique enabled simple and rapid fabrication of many copies of membrane arrays with or without functional protein and with a number of distinct compositions for parallel drug or protein-interaction screening.[17,18,107,110]

We recently presented an interesting modification of the abovementioned hydrogel-based microcontact printing in order to create microarrays of giant unilamellar vesicles (GUVs).[17,18,21] Compared to SLBs or BLMs, which are two-dimensional, GUVs are particularly attractive for studying membrane interactions, as these liposomes closely mimic the curvature and size of natural cell membranes. Therefore, incorporations of membrane proteins or channels into GUVs and fabrication of GUV arrays will allow high-throughput studies of membrane interactions as well as transport across membranes. However, the most common technique applied to form GUVs, electroformation, is not favorable for the incorporation of proteins as it requires organic solvents such as chloroform, and has no control over the size of GUVs. Therefore, multiple strategies have been developed to form GUVs in solvent-free environment with fine-control of their size,[111] but most of these strategies involve complicated techniques or fancy instruments that an ordinary laboratory could not afford.

Yet, the versatile technique presented by our group showed very simple steps toward the electroformation of GUVs with size control. In this method, as Fig. 17B describes, lipid molecules were patterned onto conductive indium tin oxide (ITO) substrates using the agarose hydrogel as a stamp. The patterned lipid films were then exposed to an AC electric field that induced formation of large vesicles and their subsequent fusion to grow into GUVs. The size of these GUVs was well controlled by the feature size of the agarose hydrogel stamps. The authors achieved single GUV formation per spot using an agarose stamp with 40 μm features. Furthermore, this method enabled the incorporation of membrane proteins such as aquaporin Z water channel into GUVs. Together, microcontact printing of membranes or lipid vesicles via agarose hydrogel stamps provide a facile method to fabricate membrane arrays such as SLB arrays and GUV arrays, which are excellent platforms to screen lipid-protein and membrane-drug interactions.

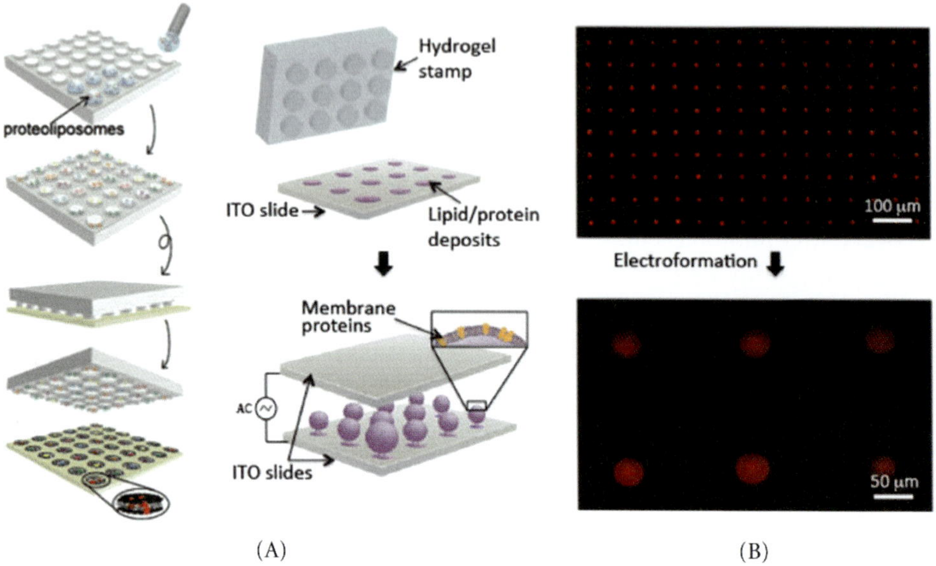

Fig. 17. Microcontact Printing of lipid membrane arrays via agarose hydrogel stamps. (A) Schematic drawings of the fabrication of giant proteoliposome arrays by loading small proteoliposomes onto an agarose stamp and subsequent transfer of the ink materials onto glass by stamping. Figure adapted from the reference [18] with permission. (B) An illustration of electroformation of GUVs from stamped lipid films on the left and the fluorescent micrograph of an array of single GUVs on the right. Figure adapted from Ref. 21 with permission.

Hydrogels has not only been employed in fabrication of membrane microarrays as carriers of lipids, but have also been applied as a barrier between array spots. In an interesting study, the Kawai group patterned PVA hydrogel on a solid substrate by NIL creating nanowells in the size range of 300–500 nm as Fig. 18A illustrates.[32] The patterned PVA hydrogels served as barriers for tethered BLMs containing high concentrations of cholesterol and sphingomyelin, mimicking lipid rafts in actual size, and the authors confirmed the formation of lipid rafts by fluorescence microscopy (Fig. 18(B)). Lipid rafts are microdomains within the cellular membrane, which are rich in cholesterol and sphingolipids and are known to play important roles in membrane protein trafficking.[112] Thus, many receptor-ligand interactions are suspected to occur at lipid raft domains. Kawai and colleagues suggested that the membrane arrays with lipid rafts would provide an excellent platform to study signal transduction *in vitro*.

While SLBs have good mechanical stability as they are supported by solid substrates, freestanding BLMs are well-known for their short lifetime. Yet, BLMs have advantages for studying membrane channels and pore forming proteins, which can be applied to sense the interactions of oligonucleotides, proteins, and drugs at the

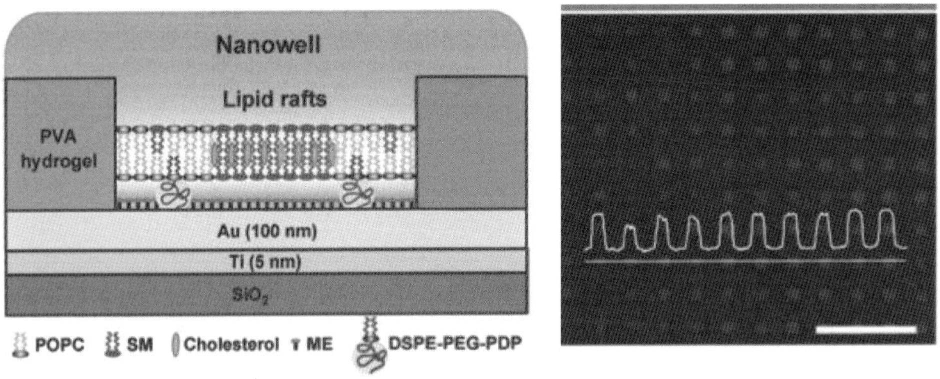

(A) (B)

Fig. 18. Fabrication of nanowell arrays of PVA hydrogel via NIL for single-cell analysis. (A) A schematic drawing of a single nanowell with cholesterol-rich BLM that is formed in the middle of the nanowell. These cholesterol-rich BLMs were able to mimic the lipid rafts in biological scales. (B) fluorescence images of the 1 μm PVA hydrogel nanowells. 1% of the lipid raft contents was TexasRed-DHPE for fluorescence imaging. The scale bar represents 10 μm. Figures adapted from the reference with permission from Ref. 32.

membrane surface.[110] However, the short lifetime of BLMs has delayed their wide application in biotechnology, and the lack of longevity and mechanical stability of BLMs are critical issues that need to be addressed in the model membrane studies.

In order to improve the lifetime of BLMs, the Schmidt group developed hydrogel encapsulated membranes (HEMs) and later conjugated hydrogel encapsulated membranes (cgHEMs).[109,113,114] HEM was prepared by first forming freestanding BLMs in teflon aperture with a 200 μm diameter hole while both sides of the lipid bilayer were filled with polyethylene glycol-dimethacrylate (PEG-DMA) monomer and photoinitiator. Upon polymerization of PEG-DMA, the hydrogel formed and supported the fragile BLM while allowing insertion of pore-forming proteins such as α-hemolysin. Modified HEM, cgHEM had PEG-DA hydrogel only on one side of the lipid bilayer.[115] In this case, the conjugation of lipid molecules to the hydrogel reduced the diffusivity of the lipid molecules, but significantly improved the lifetime and mechanical stability of the BLMs.

In another study, Helix-Nielsen and co-workers employed ethylene tetrafluoroethylene (ETFE)/hydrogel sandwich to encapsulate membrane arrays.[116] These 8 by 8 arrays had a support from ETFE on the top and PEG-DMA hydrogel on the bottom of BLMs, which were prepared in between two solid supports and at the center of each array with a teflon aperture (Fig. 19). These arrays had relatively low capacitance. More importantly, these BLMs were almost "solvent free" as the support from ETFE/hydrogel sandwich induced membrane thinning to the point that

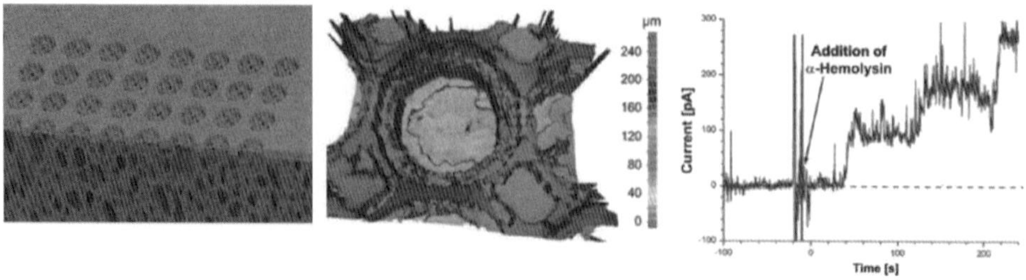

(A) (B) (C)

Fig. 19. (A) A schematic presentation of BLM encapsulated ETFE/hydrogel sandwich in 8 by 8 arrays. (B) The optical scanning image of a single well where the color gradient represents the depth of the well. (C) A plot demonstrating current changes upon insertion of the pore-forming membrane protein, α-HL. Figures adapted from Ref. 116 with permission.

the surrounding solvents were absent at least at the center of the BLMs. This setting made the incorporation of pore-forming proteins possible. Moreover, the membrane lifetime was increased from ~15 min. for freestanding BLMs to ~60 min for ETFE/hydrogel supported BLMs.

Overall, encapsulation of BLM arrays in hydrogels significantly improved the mechanical stability of BLMs and elongated their lifetime, resulting in more durable BLMs for lengthy sensing applications. Particular application of BLM arrays for single-channel recording or single molecule sensing will bring about powerful multi-dimensional sensors.

3.3.2. *Applications of membrane microarrays*

SLBs are readily utilized in screening of lipid-protein and receptor-ligand interactions as the bilayer is supported or cushioned by a solid substrate and hence, is mechanically stable.[117] The advantage of SLB microarrays for sensing originates from the accessibility of one leaflet of bilayer, which can be easily tethered to other biological ligands, such as transmembrane receptors, membrane-proteins, or aptamers. For example, the Groves group functionalized patterned SLBs to reconstitute immunological synapses and applied these SLBs for screening the interactions between the membrane of immune cells and other cells (Fig. 20(A)).[118] In another study, the Love group created PDMS nanowells to screen the interactions of single cells with immunoreceptor-decorated SLB arrays as illustrated in Fig. 20(B).[119] An interesting approach by Maldarelli *et al.* employed SLB-coated glass microbeads called Lipobeads, and applied these functionalized Lipobeads for the detection of the target ligands and quantitative measures of the receptor-ligand binding kinetics within a microfluidic chamber, which was equipped with barricades to capture single Lipobeads (Fig. 20(C)).[120]

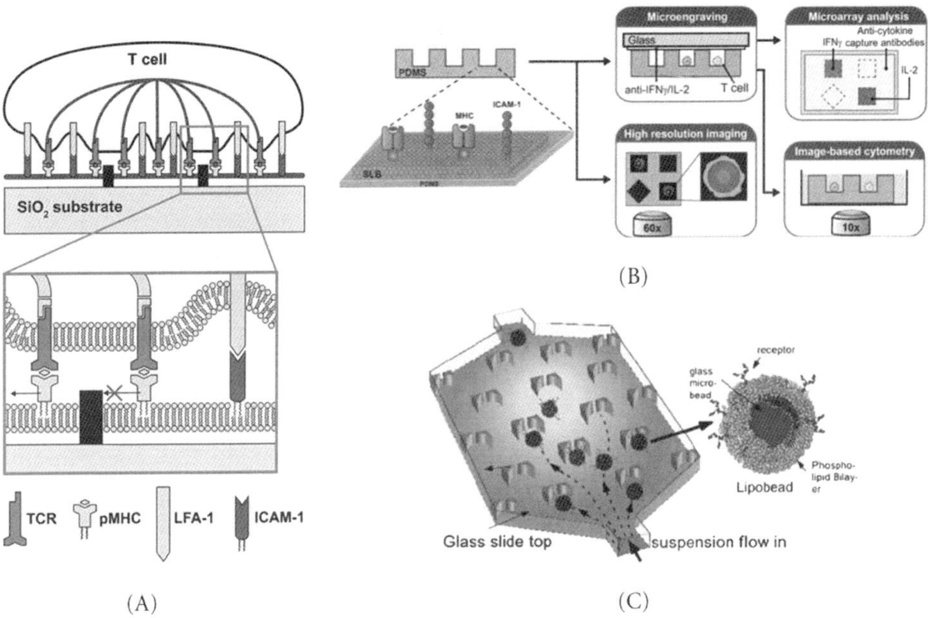

Fig. 20. The application of patterned SLBs, decorated with biomolecules, for studies of protein–cell interactions and ligand-receptor interactions. (A) Partitioning of SLBs by chromium lines (black square in the diagram) limited the lateral diffusion of immunoreceptors decorated on the SLBs. T-cells incubated on top of the patterned SLB platform allowed studies of immunological synapse formation at the molecular level. Figure adapted from the Ref. 118 with permission. (B) The PDMS nanowells coated with immunoreceptor functionalized SLBs were able to capture the formation of single cell immunological synapse and allowed studying cytokine release by fluorescence imaging. Figure adapted from the Ref. 119 with permission. (C) Receptor functionalized lipid bilayer was coated on the glass microbeads (Lipobead) and introduced into the microfluidic chamber with traps to capture a single Lipobead. These Lipobeads provided a sensing platform for quantitative analysis of the ligand-receptor binding kinetics. Figure adapted from the Ref. 120 with permission.

The application of patterned SLBs is not limited to the studies of receptor-ligand and membrane protein-cell interactions. These platforms can also be applied to study the effect of changes in the extra- or intracellular environment on plasma membranes. In this context, the Cremer group has studied the effect of metal ions and pH on the lipid bilayers.[121] For that, the negatively charged and non-charged lipid bilayers were patterned using microfluidic channels and were exposed to copper ions in acidic and basic environments. The authors hypothesized that the binding of copper ions and the negatively charged lipids (phosphatidylserine (PS)) occurs in 1:2 ratio, and the binding event causes the quenching of the fluorescence signal around the charged lipid. Figures 21(A) and 21(B) demonstrates the quenching effect on the SLBs with negatively charged lipids (PS) in pH 8.0.[121] This study indicated that the metal ion activity, such as copper, might be monitored SLB arrays where the lipid composition can be easily manipulated.

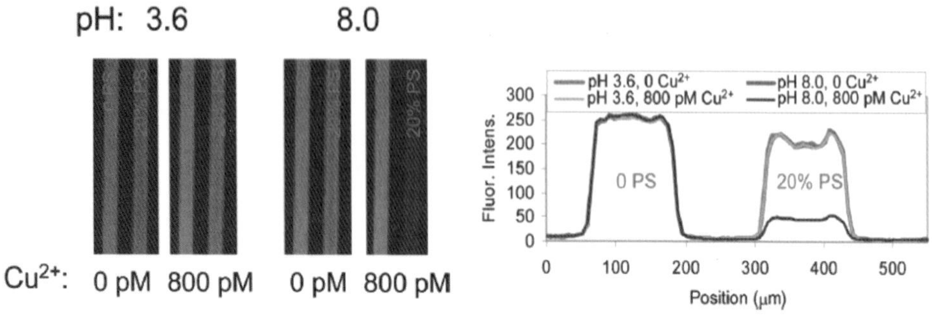

(A) (B)

Fig. 21. (A) Fluorescence images of the SLBs with 0% and 20% PS contents at pH 3.6 and 8.0 in the presence of copper ions. The fluorescence quenching was observed in 20% PS bilayer in the basic environment in the presence of 800 pM copper. (B) The plot represents the fluorescence intensity from the patterned SLBs in microfluidic channels in various conditions. Figures adapted from Ref. 121 with permission.

Despite the advantages of SLBs for some applications, the planar nature of these membranes and the presence of a solid substrate beneath them can be a limiting factor in other applications, where the membrane curvature and accessibility of both membrane leaflets are important. For these applications, liposome arrays are ideal alternatives for SLB arrays. Liposomes are spherical vesicles made of lipid bilayers, and thus overcome the limitation of planar SLBs and enable, for example, studies of drug transport across membranes. Liposomes are often categorized into small (<500 nm), large (500 nm–5 μm), and giant (5–100 μm) vesicles depending on their sizes,[122] with tunable lipid compositions that can also be decorated with biomolecules.[123] These liposomes are commonly applied in studies of protein-lipid interactions, receptor-ligand interactions, transmembrane transport, and drug delivery. For example, a study by the Sharom group employed small unilamellar vesicles (SUVs) with reconstituted P-glycoprotein, also referred as multi-drug resistant pump as this pump may be responsible for the efflux of many cancer-effective drugs, to study the transmembrane transport mechanism in the presence of various drugs and substrates.[124,125] Similarly, the patterned liposomes on a substrate provide excellent platforms to study abovementioned phenomena in a confined volume and with the ease of detection via sophisticated fluorescence microscopy, such as confocal[126] and total internal reflection fluorescence[127] microscopies and SPR.[128,129]

In contrast to SLBs, BLMs are freestanding bilayers that span across a small hole between two aqueous compartments.[130] Thus, BLMs are often applied in electrophysiology. These freestanding lipid bilayers are almost completely impermeable when there is no defect, and thus no current or electrolyte may pass through the membranes. However, an incorporation of pore-forming or transmembrane

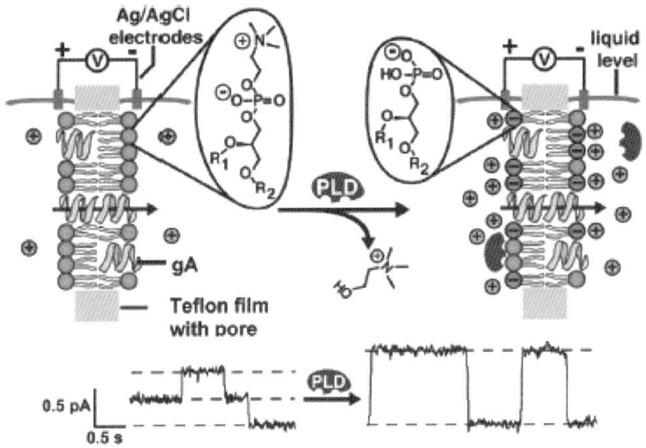

Fig. 22. A schematic illustration of the BLM with gramicidin pores incorporated. The assembly of two gramicidin (gA) pores on each lipid leaflet allows the translocation of ions across the freestanding bilayers. The enzymatic activity of PLD hydrolyzed neutral lipids and resulted in increasing current through the pores. The bottom plot represents a current measured through the nanopores. Figure adapted from Ref. 137 with permission.

channel proteins to BLMs enables small measurable current flow between the two compartments separated by BLMs.[110] Biological nanopores such as, α-hemolysin, ompA, gramicidins are frequently employed in BLM platforms for single-channel recordings.[131, 132] Beside simple self-insertion into bilayers, these protein pores may be further functionalized with aptamers or other binding-motifs for capture of target molecules in single-molecules sensing applications[132,133] or DNA sequencing.[134, 135] For instance, the Baily group advanced the nanopore-based DNA sequencing technology by controlling the translocation speed of the DNA through α-hemolysin and extending the sensing period per base to allow resolving the correspondent current fluctuations.[136] In another study, Majd *et al.* utilized gramicidin pores to monitor the activity of membrane-active enzymes, phospholipase D (PLD) and phospholipase C (PLC).[137] In this work, the enzymatic activity of phospholipase D resulted in hydrolysis of neutral lipids in the membrane and production of negatively charged lipids, which led to the accumulation of ions near the membrane surface and thus altered conductance of the gramicidin pore (Fig. 22). However, the longevity and mechanical stabilities of BLMs have remained obstacles for the universal application of these platforms for sensing purposes.

3.4. *Cellular microarrays*

Arranging cells on surfaces in a controlled array format has become popular in a number of research areas including tissue engineering, 3D cell culture systems, cell-based sensors, drug screening systems, and other cell and molecular biology studies. In order to fabricate the cell arrays with relevant and consistent functions in response to the extracellular stimuli, it is crucial to recapitulate the native microenvironments around the cells, in terms of local extracellular matrix (ECM) cues such as density of cell adhesion ligands, soluble or immobilized growth factors, and mechanical properties. It is also important to maintain mild and physiological conditions including pH, humidity, salt concentration, and temperature in these arrays. In this context, the use of biocompatible hydrogels either as hydrated substrates that host the cell arrays or as materials that transfer cells onto substrates has become a popular strategy. Most importantly, hydrogels can be conjugated with ECM molecules and hence, can control and modulate cell behaviors, such as cell adhesion, proliferation, and differentiation.

3.4.1. *Indirect patterning of cells in an array*

Cell arrays can be prepared by introducing cell populations to substrates with patterned adhesion ligands. This indirect patterning technique relies on the ligand-mediated cell adhesion on the substrate. ECM molecules or ECM-derived cell adhesive peptides (CAPs) have been used as ligands for enhancing cell specific adhesion and viability during patterning. Particularly, short peptide sequences (e.g., RGD peptide) have been widely applied in patterning cell arrays since these short peptides are stable and flexible for modification, tunable for cell binding, and easy to be synthesized in a large scale. This method provides highly efficient and robust transfer of cells onto the substrate.

Hydrogel-based microcontact printing of ECM molecules is the most popular method to pattern cells indirectly. For instance, Hynd *et al.* patterned and functionalized PAA/PEG-DA hydrogel surface with the ECM molecules to control cell attachment and to promote growth on the hydrogel surface.[138] Another interesting technique employed for patterning ECM is electrospinning of nanofibers made of hydrogels that contain ECM molecules. The electrospun scaffolds made of ECM molecules, called ECM nanofiber mats, are widely used as substrates in cell arrays since they mimic the surface chemistry and topology of the natural cell microenvironment and promote cellular adhesion, migration, proliferation, and differentiation.[139,140] Recently, Huang *et al.* presented a simple method to selectively deposit the electrospun alginate-based nanofibers by masking the PEG-DA hydrogel surface, and these authors used the resultant substrate in further patterning of the cells on the hydrogel surface.[141] In all these studies, patterning of ECM molecules allows precise patterning of cells in array format.[141, 142]

Topographical features on hydrogels such as microwells can also be applied for patterning of cells. In this context, PEG hydrogels were widely adopted for the

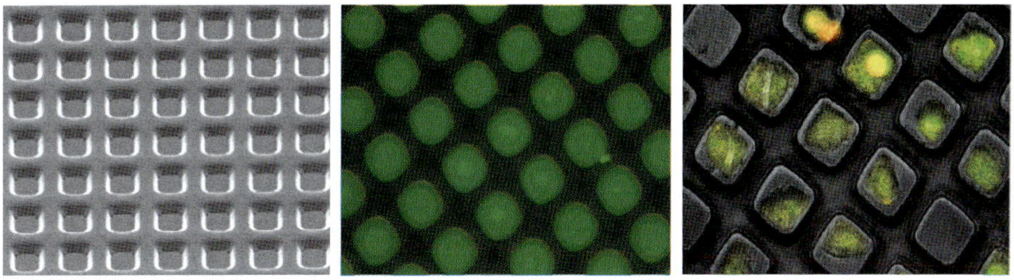

Fig. 23. (A) SEM image and (B) fluorescence micrograph of collagen (green)-coated PEG-DA hydrogel microwells. (C) Viability of 3T3 fibroblast cells in PEG microwells. The determination of cell viability was carried out by Live/Dead assay: Green color represents live cells while dead cells are red. Figures adapted from Ref. 143 with permission.

fabrication of 3D cell arrays based on their cell-repellent properties; they are less adhesive for cells compared to other hydrogel polymers. It has been demonstrated that cells settled down in the bottom of hydrogel microwells due to the cell-repellent properties of PEG, forming aggregates in each well once they were seeded on the surface of microwells. Relying on this property, Revzin *et al.* achieved homogeneous and high-density arrays of cells in the hydrogel microwells.[143] These authors used the photolithography to create PEG-DA hydrogel-coated glass substrate with microstructures. In this method, the masked parts of the hydrogel prepolymer were removed after washing steps while the unmasked parts were polymerized into hydrogel microwells upon exposure to the UV light (Fig. 23A). The exposed glass parts on the bottom of microwells were coated with type I collagen to promote the attachment of hepatocytes (Fig. 23(B)). Both cell-adhesive part (collagen-coated surface) and cell-repellent part (hydrogel walls) enabled precise patterning of the cells (Fig. 23(C)). In order to pattern non-adhesive stem cells in an array format, Moeller *et al.* used heavier PEG polymer (PEG 1000) since the PEG polymer with higher molecular weight has shown less adhesion to cells.[30] Other combined methods, such as soft embossing and imprinting techniques, have also been applied in the formation of hydrogel microwells.[31,144]

3.4.2. *Direct patterning of cells in an array*

Despite the effectiveness of indirect cell patterning approaches, it has been reported that ECM molecules used for these patterning methods may cause unexpected cellular signal pathways by bi-directional cell-ligand interactions, and consequently influence many aspects of cell behaviors, such as cellular morphology, adhesion, migration, proliferation and differentiation.[145,146] One approach to avoid this problems is to pattern the cells directly onto non-functionalized substrates. The microcontact printing has been employed to pattern cells directly onto the substrate without the need for ligand coating. For instance, Weibel *et al.* created arrays of bacterial cells using microcontact printing with agarose gel (Fig. 24(A)).[147] In another study, Stevens *et al.* reported patterning of normal human osteoblasts directly onto the porous hydroxyapatite (HA) surfaces using agarose stamp-mediated microcontact printing (Fig. 24(B)).[20] The use of wet hydrogel stamps in these studies enabled keeping cells hydrated and hence, viable during the printing process. Additionally, the hydrogel stamp was able to deliver different inks (solutions containing different types of cells) simultaneously and to pattern hundred times without intermediate re-inking. This simple technique that offers control over the spacing and geometry of the patterned cells is also attractive for tissue engineering application where, for example, fabrication of hierarchically organized scaffolds is desired.

Another applied approach for direct patterning of cells is encapsulation of the cells in the hydrogel-based scaffolds, which has attracted an increasing attention in tissue engineering and 3D cell culture areas. In this approach, cells are encapsulated in the hydrogels first, and are subsequently patterned onto the desired platform.[22] Micro-molding,[22] ink-jet printing,[9,148] hanging drop,[149,150] and microfluidic[151] techniques are the current popular methods for production of cell-laden hydrogels (Figure 25).

Hydrogel scaffolds allow the transport of oxygen, nutrients, soluble factors and waste, and also provide mild conditions for the encapsulated cells. Therefore, hydrogels have been widely incorporated in fabrication of high quality cell arrays

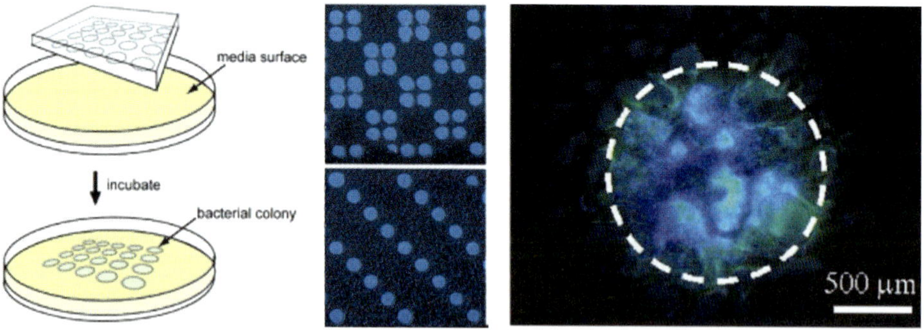

(A) (B)

Fig. 24. (A) Schematic illustrations (left) and the optical micrographs of arrays (right) of bacterial cells patterned using an agarose hydrogel. Figure adapted from the reference with permission.[147] (B) Fluorescence image of one part of patterned human osteoblasts in an array format on the HA by agarose hydrogel stamping. Cellular actin (green) and DNA (blue) were stained for better visualization of cells on the HA. Figure adapted from Ref. 20 with permission.

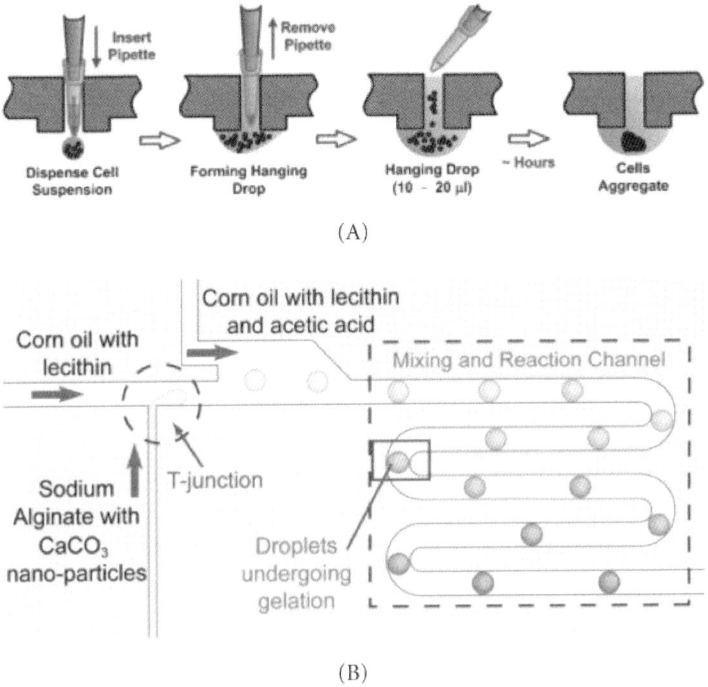

Fig. 25. Schematic illustrations of two methods of (A) hanging drop and (B) microfluidics, used for preparation of spherical cell-laden hydrogel building blocks. (A) In the hanging drop method, the pipette tip dispensed cell containing hydrogel solution on the bottom surface of a well-plate. The hydrophilic surface of the plate attracted the cell suspension by removing the tip quickly, and then a hanging drop was formed simultaneously. After curing the hydrogel, the droplets can be harvested and applied for patterning. Figure adapted from the reference with permission.[152] (B) In the microfluidic-based method, a mixture of cells and alginate were microinjected into the channel and formed into droplets at a T-junction and the spherical hydrogels formed upon exposure to Ca^{2+} ions within the fluidic channel. Figure adapted from Ref. 148 with permission.

particularly for cell-based sensing platforms. However, several conditions used in the gelation process, such as elevated temperatures, UV light, and chemical cross-linkers, have been reported to cause cell damage and consequently cell death. As a result, several physical crosslinking interactions such as ionic interactions (e.g., alginate hydrogels),[153] coiled-coil interactions (e.g., protein hydrogels),[154] and hydrophobic interactions (e.g., chitosan hydrogel)[155] have been explored to form hydrogels under mild conditions. DNA hydrogels crosslinked by enzymatic reactions have also been developed recently.[156] Another challenge in using hydrogels in this context is optimizing their mechanical properties such as stiffness that can potentially affect behaviors of encapsulated cells in response to external stimuli.[157] The stiffness of PEG hydrogel has been controlled and optimized for the fabrication of hydrogel-based cell arrays by changing the length or density of crosslinker.[82,158] Biodegradability of hydrogels is another issue considered for the application of hydrogels in cell encapsulation. Recently, enzyme-sensitive peptides (ESPs) have been incorporated into hydrogel polymers to enable enzyme-mediated biodegradation. For instance, Chau and colleagues synthesized hydrogels containing peptides that can be cleaved by matrix metalloproteinase-2 (MMP-2) in order to gradually provide more space for the growing cells and to allow motility of the cells during the growth.[159]

3.4.3. *Sensing applications of hydrogel-based cell arrays*

Traditional screening procedures for the 2D cell arrays depend on the immunostaining of query molecules and detecting them by flow cytometry, western blotting, enzyme-linked immunosorbent assay (ELISA), and fluorescence or confocal microscopy. However, the cell arrays have been evolved into 3D configurations in order to closely mimic the natural 3D cellular microenvironments, so such label-based methods cannot afford detection and interpretation of the numerous cellular functions in these arrays. In this regard, various label-free techniques have been used as alternative or complimentary methods for the label-based methods. For instance, the label-free impedance or SPR measurements have been applied not only for supporting the quantitative data obtained from immunostaining methods such as expression levels of query molecules, but also for analyzing qualitative cellular states such as degree of cell-cell interactions, cell-matrix adhesion, and formation of tight junctions.[151,160] In an intriguing study, Tran and colleagues showed that integration of an electric cell-substrate impedance sensing (ECIS) system into the hydrogel-based cell chip enabled the high-throughput and real-time cell viability screening.[161] In another recent study, Choi *et al.* introduced a combined technique of label-based and label-free methods, referred to as opto-electric cellular biosensor that enabled both cell imaging and impedance measurements simutaneously.[162] Moreover, Otsuka and colleagues applied a field effect transistor (FET) in the hydrogel-based cell arrays for detection of extracellular potential changes as a result of interactions between cell and other molecules in the culture medium.[163] In summary, label-free sensing techniques could serve as good alternatives to the traditional label-based methods, and when applied on 3D and hydrogel-encapsulated cell arrays, these methods can provide much improved capabilities, sensitivity, and reproducibility.

4. Conclusions

Hydrogels, both natural and synthetic, have been widely utilized for the transfer and storage of cells and biological molecules in bioarrays. Encapsulation of fragile biological materials in hydrogels protects them during immobilization steps. Hydrogels have, therefore, significantly improved the quality of these microarrays and have made them more versatile and powerful for applications in drug screening, medical diagnostics, and biosensing.

References

1. M. Schena, D. Shalon, R. W. Davis, and P. O. Brown, Quantitative monitoring of gene-expression patterns with a complementary-DNA microarray. *Science*, **270**(5235), 467–470 (1995).
2. I. S. Baird, A. Y. Yau, and B. K. Mann, Mammalian cell-seeded hydrogel microarrays printed via dip-pin technology. *Biotechniques*, **44**(2), 249–256 (2008).
3. A. J. Senesi, D. I. Rozkiewicz, D. N. Reinhoudt, and C. A. Mirkin, Agarose-assisted dip-pen nanolithography of oligonucleotides and proteins. *Acs Nano*, **3**(8), 2394–2402 (2009).
4. G. Hardiman, Microarray innovations: technology and experimentation. *Drug Discov.*, **11**, 1–314 (2009).
5. V. Romanov, S. N. Davidoff, A. R. Miles, D. W. Grainger, B. K. Gale, and B. D. Brooks, A critical comparison of protein microarray fabrication technologies. *Analyst*, **139**(6), 1303–1326 (2014).
6. S. Gobaa, S. Hoehnel, M .Roccio, A. Negro, S. Kobel, and M. P. Lutolf, Artificial niche microarrays for probing single stem cell fate in high throughput. *Nat. Methods*, **8**(11), 949–955 (2011).
7. V. K. Yadavalli, W. G. Koh, G. J. Lazur, and M. V. Pishko, Microfabricated protein-containing poly(ethylene glycol) hydrogel arrays for biosensing. *Sensor. Actuat. B Chem.*, **97**(2–3), 290–297 (2004).
8. L. R. Allain, M. Askari, D. L. Stokes, and T. Vo-Dinh, Microarray sampling-platform fabrication using bubble-jet technology for a biochip system. *Fresenius' J. Anal. Chem.*, **371**(2), 146–150 (2001).
9. S. Moon, S. K. Hasan, Y. S. Song, F. Xu, H. O. Keles, F. Manzur, S. Mikkilineni, J. W. Hong, J. Nagatomi, E. Haeggstrom, A. Khademhosseini, and U. Demirci, Layer by layer three-dimensional tissue epitaxy by cell-laden hydrogel droplets. *Tissue Eng. Pt. C Meth.*, **16**(1), 157–166 (2010).
10. K. Pataky, T. Braschler, A. Negro, P. Renaud, M. P. Lutolf, and J. Brugger, Microdrop printing of hydrogel bioinks into 3d tissue-like geometries. *Adv. Mater.*, **24**(3), 391–396 (2012).
11. B. D. Martin, B. P. Gaber, C. H. Patterson, and D. C. Turner, Direct protein microarray fabrication using a hydrogel "stamper". *Langmuir*, **14**(15), 3971–3975 (1998).
12. A. Kumar and G. M. Whitesides, Features of gold having micrometer to centimeter dimensions can be formed through a combination of stamping with an elastomeric stamp and an alkanethiol ink followed by chemical etching. *Appl. Phys. Lett.*, **63**(14), 2002–2004 (1993).

13. W. M. Moreau, *Semiconductor Lithography: Principles, Practices, and Materials. Plenum Press*, NY (1988).
14. A. Perl, D. N. Reinhoudt, and J. Huskens, Microcontact printing: limitations and achievements. *Adv. Mater.*, **21**(22), 2257–2268 (2009).
15. M. Morra, E. Occhiello, R. Marola, F. Garbassi, P. Humphrey, and D. Johnson, On the aging of oxygen plasma-treated polydimethylsiloxane surfaces. *J. Colloid Interf. Sci.*, **137**(1), 11–24 (1990).
16. E. Delamarche, C. Donzel, F. S. Kamounah, H. Wolf, M. Geissler, R. Stutz, P. Schmidt-Winkel, B. Michel, H. J. Mathieu, and K. Schaumburg K, Microcontact printing using poly(dimethylsiloxane) stamps hydrophilized by poly(ethylene oxide) silanes. *Langmuir*, **19**(21), 8749–8758 (2003).
17. S. Majd and M. Mayer, Hydrogel stamping of arrays of supported lipid bilayers with various lipid compositions for the screening of drug-membrane and protein-membrane interactions. *Angew. Chem. Int. Ed.*, **44**(41), 6697–6700 (2005).
18. S. Majd and M. Mayer, Generating arrays with high content and minimal consumption of functional membrane proteins. *J. Am. Chem. Soc.*, **130**(47), 16060–16064 (2008).
19. M. Mayer, J. Yang, I. Gitlin, D. H. Gracias, and G. M. Whitesides, Micropatterned agarose gels for stamping arrays of proteins and gradients of proteins. *Proteomics*, **4**(8), 2366–2376 (2004).
20. M. M. Stevens, M. Mayer, D. G. Anderson, D. B. Weibel, G. M. Whitesides, and R. Langer, Direct patterning of mammalian cells onto porous tissue engineering substrates using agarose stamps. *Biomaterials*, **26**(36), 7636–7641 (2005).
21. Y. J. Kang, H. S. Wostein, and S. Majd, A simple and versatile method for the formation of arrays of giant vesicles with controlled size and composition. *Adv. Mater.*, **25**(47), 6834–6838 (2013).
22. Y. A. Du, E. Lo, S. Ali, and A. Khademhosseini, Directed assembly of cell-laden microgels for fabrication of 3d tissue constructs. *Proceedings of the National Academy of Sciences of the United States of America*, **105**(28), 9522–9527 (2008).
23. N. N. Kachouie, Y. A. Du, H. Bae, M. Khabiry, A. F. Ahari, B. Zamanian, J. Fukuda, and A. Khademhosseini, Directed assembly of cell-laden hydrogels for engineering functional tissues. *Organogenesis*, **6**(4), 234–244 (2010).
24. A. Ovsianikov, M. Malinauskas, S. Schlie, B. Chichkov, S. Gittard, R. Narayan, M. Lobler, K. Sternberg, K. P. Schmitz, and A. Haverich, Three-dimensional laser micro- and nano-structuring of acrylated poly(ethylene glycol) materials and evaluation of their cytoxicity for tissue engineering applications. *Acta Biomater.*, **7**(3), 967–974 (2011).
25. B. R. Ringeisen, P. K. Wu, H. Kim, A. Pique, R. Y. C. Auyeung, H. D. Young, D. B. Chrisey, and D. B. Krizman, Picoliter-scale protein microarrays by laser direct write. *Biotechnol. Progr.*, **18**(5), 1126–1129 (2002).
26. N. R. Schiele, D. B. Chrisey, and D. T. Corr, Gelatin-based laser direct-write technique for the precise spatial patterning of cells. *Tissue Eng. Pt. C-Meth.*, **17**(3), 289–298 (2011).

27. L. Koch, A. Deiwick, S. Schlie, S. Michael, M. Gruene, V. Coger, D. Zychlinski, A. Schambach, K. Reimers, P. M. Vogt, ad B. Chichkov, Skin tissue generation by laser cell printing. *Biotechnol. Bioeng.*, **109**(7), 1855–1863 (2012).
28. T. B. Phamduy, N. A. Raof, N. R. Schiele, Z. J. Yan, D. T. Corr, Y. Huang, Y. B. Xie, and D. B. Chrisey, Laser direct-write of single microbeads into spatially-ordered patterns. *Biofabrication*, **4**(2), 025006 (2012).
29. Y. N. Xia, E. Kim, X. M. Zhao, J. A. Rogers, M. Prentiss, and G. M. Whitesides, Complex optical surfaces formed by replica molding against elastomeric masters. *Science*, **273**(5273), 347–349 (1996).
30. H. C. Moeller, M. K. Mian, S. Shrivastava, B. G . Chung, and A. Khademhosseini, A microwell array system for stem cell culture. *Biomaterials*, **29**(6), 752–763 (2008).
31. S. Kobel, M. Limacher, S. Gobaa, T. Laroche, and M. P. Lutolf, Micropatterning of hydrogels by soft embossing. *Langmuir*, **25**(15), 8774–8779 (2009).
32. B. K. Lee, H. Y. Lee, P. Kim, K. Y. Suh, and T. Kawai, Nanoarrays of tethered lipid bilayer rafts on poly(vinyl alcohol) hydrogels. *Lab Chip*, **9**(1), 132–139 (2009).
33. E. M. Southern, Detection of specific sequences among DNA fragments separated by gel electrophoresis. *J. Mol. Biol.*, **98**(3), 503–517 (1975).
34. E. M. Southern, DNA microarrays. History and overview. *Method. Mol. Biol.*, **170**, 1–15 (2001).
35. M. J. Heller, DNA microarray technology: devices, systems, and applications. *Annu. Rev. Biomed. Eng.*, **4**, 129–153 (2002).
36. A. Sassolas, B. D. Leca-Bouvier, and L. J. Blum, DNA biosensors and microarrays. *Chem. Rev.*, **108**(1), 109–139 (2008).
37. K. R. Khrapko, Y. P. Lysov, A. A. Khorlyn, V. V. Shick, V. L. Florentiev, and A. D. Mirzabekov, An oligonucleotide hybridization approach to DNA sequencing. *FEBS Lett.*, **256**(1–2), 118–122 (1989).
38. D. Proudnikov, E. Timofeev, and A. Mirzabekov, Immobilization of DNA in polyacrylamide gel for the manufacture of DNA and DNA-oligonucleotide microchips. *Anal. Biochem.*, **259**(1), 34–41 (1998).
39. A. Y. Rubina, S. V. Pan'kov, E. I. Dementieva, D. N. Pen'kov, A. V. Butygin, V. A. Vasiliskov, A. V. Chudinov, A. L. Mikheikin, V. M. Mikhailovich, and A. D. Mirzabekov, Hydrogel drop microchips with immobilized DNA: Properties and methods for large-scale production. *Anal. Biochem.*, **325**(1), 92–106 (2004).
40. X. Liu, T. Li, D. J. Liu, and Z. X. Wang, Fabricating three-dimensional hydrogel oligonucleotide microarrays to detect single nucleotide polymorphisms. *Anal. Method.*, **5**(1), 285–290 (2013).
41. A. Bromberg, E. C. Jensen, J. Kim, Y. K. Jung, and R. A. Mathies, Microfabricated linear hydrogel microarray for single-nucleotide polymorphism detection. *Anal. Chem.*, **84**(2), 963–970 (2012).
42. H. Tran, K. L. Killops, and L. M. Campos, Advancements and challenges of patterning biomolecules with sub-50 nm features. *Soft Matter*, **9**(29), 6578–6586 (2013).
43. N. Tuleuova and A. Revzin, Micropatterning of aptamer beacons to create cytokine-sensing surfaces. *Cell Mol. Bioeng.*, **3**(4), 337–344 (2010).

44. L. Gold, D. Ayers, J. Bertino, C. Bock, A. Bock, E. N. Brody, J. Carter, A. B. Dalby, B. E. Eaton, T. Fitzwater, D. Flather, A. Forbes, T. Foreman, C. Fowler, B. Gawande, M. Goss, M. Gunn, S. Gupta, D. Halladay, J. Heil, J. Heilig, B. Hicke, G. Husar, J. Janjic, T. Jarvis, S. Jennings, E. Katilius, T. R. Keeney, N. Kim, T. H. Koch, S. Kraemer, L. Kroiss, N. Le, D. Levine, W. Lindsey, B. Lollo, W. Mayfield, M. Mehan, R. Mehler, S. K. Nelson, M. Nelson, D. Nieuwlandt, M. Nikrad, U. Ochsner, R. M. Ostroff, M. Otis, T. Parker, S. Pietrasiewicz, D. I. Resnicow, J. Rohloff, G. Sanders, S. Sattin, D. Schneider, B. Singer, M. Stanton, A. Sterkel, A. Stewart, S. Stratford, J. D. Vaught, M. Vrkljan, J. J. Walker, M. Watrobka, S. Waugh, A. Weiss, S. K. Wilcox, A. Wolfson, S. K. Wolk, C. Zhang, and D. Zichi, Aptamer-based multiplexed proteomic technology for biomarker discovery. *Plos One*, **5**(12), e15004 (2010).
45. P. J. J. Huang and J. W. Liu, Flow cytometry-assisted detection of adenosine in serum with an immobilized aptamer sensor. *Anal. Chem.*, **82**(10), 4020–4026 (2010).
46. D. Rotem, L. Jayasinghe, M. Salichou, and H. Bayley, Protein detection by nanopores equipped with aptamers. *J. Am. Chem. Soc.*, **134**(5), 2781–2787 (2012).
47. K. J. Son, D. –S. Shin, T. Kwa, Y. Gao, and A. Revzin, Micropatterned sensing hydrogels integrated with reconfigurable microfluidics for detecting protease release from cells. *Anal. Chem.*, **85**(24), 11893–11901 (2013).
48. N. Dai and E. T. Kool, Fluorescent DNA-based enzyme sensors. *Chem. Soc. Rev.*, **40**(12), 5756–5770 (2011).
49. X. Mao, J. H. Jiang, X. M. Xu, X. Chu, Y. Luo, G. L. Shen, and R. Q. Yu, Enzymatic amplification detection of DNA based on "molecular beacon" biosensors. *Biosens. Bioelectron.*, **23**(10), 1555–1561 (2008).
50. M. L. Yang, C. Z. Liu, K. J. Qian, P. G. He, and Y. Z. Fang, Study on the electrochemiluminescence behavior of abei and its application in DNA hybridization analysis. *Analyst*, **127**(9), 1267–1271 (2002).
51. K. Sato, K. Hosokawa, and M. Maeda, Colorimetric biosensors based on DNA-nanoparticle conjugates. *Anal. Sci.*, **23**(1), 17–20 (2007).
52. H. C. Yeh, J. Sharma, J. J. Han, J. S. Martinez, and J. H. Werner, A DNA-silver nanocluster probe that fluoresces upon hybridization. *Nano Lett.*, **10**(8), 3106–3110 (2010).
53. J. Li, H. T. Ng, A. Cassell, W. Fan, H. Chen, Q. Ye, J. Koehne, J. Han,and M. Meyyappan, Carbon nanotube nanoelectrode array for ultrasensitive DNA detection. *Nano Lett.*, **3**(5), 597–602 (2003).
54. J. Wang and M. Jiang, Toward genolelectronics: nucleic acid doped conducting polymers. *Langmuir*, **16**(5), 2269–2274 (2000).
55. C. Y. Zhang, H. C. Yeh, M. T. Kuroki, and T. H. Wang, Single-quantum-dot-based DNA nanosensor. *Nat. Mater.*, **4**(11), 826–831 (2005).
56. S. Su, J. W. Fan, B. Xue, L. H. Yuwen, X. F. Liu, D. Pan, C. H. Fan, and L. H. Wang, DNA-conjugated quantum dot nanoprobe for high-sensitivity fluorescent detection of DNA and micro-rna. *Acs Appl. Mater. Interf.*, **6**(2), 1152–1157 (2014).
57. X. W. He and N. Ma, A general strategy for label-free sensitive DNA detection based on quantum dot doping. *Anal.Chem.*, **86**(7), 3676–3681 (2014).

58. S. A. Martins, M. D. Prazeres, L. P. Fonseca, and G. Monteiro, DNA biosensors: towards a microparticle based-platform for genomic DNA detection. *New Biotechnol.*, **25**, S19–S20 (2009).
59. V. Kodoyianni, Label-free analysis of biomolecular interactions using spr imaging. *Biotechniques*, **50**(1), 32–40 (2011).
60. S. Scarano, M. Mascini, A. P. F. Turner, and M. Minunni, Surface plasmon resonance imaging for affinity-based biosensors. *Biosens. Bioelectron.*, **25**(5), 957–966 (2010).
61. Y. Li, H. J. Lee, and R. M. Corn, Detection of protein biomarkers using rna aptamer microarrays and enzymatically amplified surface plasmon resonance imaging. *Anal. Chem.*, **79**(3), 1082–1088 (2007).
62. J. Spadavecchia, M. G. Manera, F. Quaranta, P. Siciliano, and R. Rella, Surface plamon resonance imaging of DNA based biosensors for potential applications in food analysis. *Biosens. Bioelectron.*, **21**(6), 894–900 (2005).
63. N. Yang, X. D. Su, V. Tjong, and W. Knoll, Evaluation of two- and three-dimensional streptavidin binding platforms for surface plasmon resonance spectroscopy studies of DNA hybridization and protein-DNA binding. *Biosens. Bioelectron.*, **22**(11), 2700–2706 (2007).
64. J. S. Shumaker-Parry, M. H. Zareie, R. Aebersold, and C. T. Campbell, Microspotting streptavidin and double-stranded DNA arrays on gold for high-throughput studies of protein-DNA interactions by surface plasmon resonance microscopy. *Anal. Chem.*, **76**(4), 918–929 (2004).
65. F. Lucarelli, S. Tombelli, M. Minunni, G. Marrazza, and M. Mascini, Electrochemical and piezoelectric DNA biosensors for hybridisation detection. *Anal. Chim. Acta*, **609**(2),139–159 (2008).
66. J. H. Zhai, H. Cui, and R. F. Yang, DNA based biosensors. *Biotechnol. Adv.*, **15**(1), 43–58 (1997).
67. M. L. Bulyk, DNA microarray technologies for measuring protein-DNA interactions. *Curr. Opin. Biotechnol.*, **17**(4), 422–430 (2006).
68. T. Ohmichi, Y. Kawamoto, P. Wu, D. Miyoshi, H. Karimata, and N. Sugimoto, DNA-based biosensor for monitoring ph in vitro and in living cells. *Biochemistry*, **44**(19), 7125–7130 (2005).
69. A. Sassolas, L. J. Blum, and B. D. Leca-Bouvier, Optical detection systems using immobilized aptamers. *Biosens. Bioelectron.*, **26**(9), 3725–3736 (2011).
70. R. B. Parekh and C. Rohlff, Post-translational modification of proteins and the discovery of new medicine. *Curr. Opin. Biotechnol.* **8**(6), 718–723 (1997).
71. E. M. Marcotte, Measuring the dynamics of the proteome. *Genome Res.*, **11**(2), 191–193 (2001).
72. J. E. G. McCarthy, Posttranscriptional control of gene expression in yeast. *Microbiology and Mol. Biol. Rev.*, **62**(4), 1492–1553 (1998).
73. T. W. Chang, Binding of cells to matrices of distinct antibodies coated on solid-surface. *J. Immunol. Methods.*, **65**(1–2), 217–223 (1983).
74. S. F. Kingsmore, Multiplexed protein measurement: Technologies and applications of protein and antibody arrays. *Nat. Rev. Drug Discov.*, **5**(4), 310–320 (2006).

75. J. LaBaer and N. Ramachandran, Protein microarrays as tools for functional proteomics. *Curr. Opin. Chem. Biol.*, **9**(1), 14–19 (2005).
76. S. Ray, G. Mehta, and S. Srivastava, Label-free detection techniques for protein microarrays: prospects, merits and challenges. *Proteomics*, **10**(4), 731–748 (2010).
77. Y. Lee, E. K. Lee, Y. W. Cho, T. Matsui, I. C. Kang, T. S. Kim, and M. H. Han, Proteochip: a highly sensitive protein microarray prepared by a novel method of protein, immobilization for application of protein-protein interaction studies. *Proteomics*, **3**(12), 2289–2304 (2003).
78. K. Nakanishi, H. Muguruma, and I. Karube, A novel method of immobilizing antibodies on a quartz crystal microbalance using plasma-polymerized films for immunosensors. *Anal. Chem.*, **68**(10), 1695–1700 (1996).
79. A. S. Blawas and W. M. Reichert, Protein patterning. *Biomaterials*, **19**(7–9), 595–609 (1998).
80. W. Mueller, H. Ringsdorf, E. Rump, G. Wildburg, X. Zhang, L. Angermaier, W. Knoll, M. Liley, and J. Spinke, Attempts to mimic docking processes of the immune system: Recognition-induced formation of protein multilayers. *Science*, **262**(5140), 1706–1708 (1993).
81. A. Y. Rubina, E. I. Dementieva, A. A. Stomakhin, E. L. Darii, S. V. Pa'kov, V. E. Barsky, S. M. Ivanov, E. V. Konovalova, and A. D. Mirzabekov, Hydrogel-based protein microchips: Manufacturing, properties, and applications. *Biotechniques*, **34**(5), 1008–1014 (2003).
82. N. Coq, T. van Bommel, R. A. Hikmet, H. R. Stapert, and W. U. Dittmer, Self-supporting hydrogel stamps for the microcontact printing of proteins. *Langmuir*, **23**(9), 5154–5160 (2007).
83. M. Ates, A review study of (bio)sensor systems based on conducting polymers. *Mater. Sci. Eng.*, **C33**(4), 1853–1859 (2013).
84. M. Gerard, A. Chaubey, and B. D. Malhotra, Application of conducting polymers to biosensors. *Biosens. Bioelectron.*, **17**(5), 345–359 (2002).
85. S. Park, G. Yang, N. Madduri, M. R. Abidian, and S. Majd, Hydrogel-mediated direct patterning of conducting polymer films with multiple surface chemistries. *Adv. Mater.*, **26**(18), 2782–2787 (2014).
86. A. Y. Rubina, V. I. Dyukova, E. I. Dementieva, A. A. Stomakhin, V. A. Nesmeyanov, E. V. Grishin, and A. S. Zasedatelev, Quantitative immunoassay of biotoxins on hydrogel-based protein microchips. *Anal. Biochem.*, **340**(2), 317–329 (2005).
87. P. L. Stiles, Direct deposition of micro- and nanoscale hydrogels using dip pen nanolithography (dpn). *Nat. Methods.*, **7**(8), I–II (2010).
88. J. Piehler, A. Brecht, K. E. Geckeler, and G. Gauglitz, Surface modification for direct immunoprobes. *Biosens. Bioelectron.*, **11**(6–7), 579–590 (1996).
89. S. Jo and K. Park, Surface modification using silanated poly(ethylene glycol)s. *Biomaterials*, **21**(6), 605–616 (2000).
90. N. P. Huang, R. Michel, J. Voros, M. Textor, R. Hofer, A. Rossi, D. L. Elbert, J. A. Hubbell, and N. D. Spencer, Poly(l-lysine)-g-poly(ethylene glycol) layers on metal

oxide surfaces: Surface-analytical characterization and resistance to serum and fibrinogen adsorption. *Langmuir*, **17**(2), 489–498 (2001).

91. W. Kusnezow and J. D. Hoheisel, Solid supports for microarray immunoassays. *J. Mol. Recognit.*, **16**(4), 165–176 (2003).
92. H. Tanaka, M. Hanasaki, T. Isojima, H. Takeuchi, T. Shiroya, and H. Kawaguchi, Enhancement of sensitivity of spr protein microarray using a novel 3d protein immobilization. *Colloids Surf. B.*, **70**(2), 259–265 (2009).
93. P. T. Charles, E. R. Goldman, J. G. Rangasammy, C. L. Schauer, M. S. Chen, and C. R. Taitt, Fabrication and characterization of 3d hydrogel microarrays to measure antigenicity and antibody functionality for biosensor applications. *Biosens. Bioelectron.*, **20**(4), 753–764 (2004).
94. H. J. Lee, D. Nedelkov, and R. M. Corn, Surface plasmon resonance imaging measurements of antibody arrays for the multiplexed detection of low molecular weight protein biomarkers. *Anal. Chem.*, **78**(18), 6504–6510 (2006).
95. Y. Zhou, O. Andersson, P. Lindberg, and B. Liedberg, Protein microarrays on carboxymethylated dextran hydrogels: immobilization, characterization and application. *Microchim. Acta*, **147**(1–2), 21–30 (2004).
96. Y. Zhou, O. Andersson, P. Lindberg, and B. Liedberg, Reversible hydrophobic barriers introduced by microcontact printing: Application to protein microarrays. *Microchim. Acta*, **146**(3–4), 193–205 (2004).
97. N. Backmann, C. Zahnd, F. Huber, A. Bietsch, A. Pluckthun, H. P. Lang, H. J. Guntherodt, M. Hegner, and C. Gerber, A label-free immunosensor array using single-chain antibody fragments. *Proceedings of the National Academy of Sciences of the United States of America*, **102**(41), 14587–14592 (2005).
98. R. McKendry, J. Y. Zhang, Y. Arntz, T. Strunz, M. Hegner, H. P. Lang, M. K. Baller, U. Certa, E. Meyer, H. J. Guntherodt, and C. Gerber, Multiple label-free biodetection and quantitative DNA-binding assays on a nanomechanical cantilever array. *Proceedings of the National Academy of Sciences of the United States of America*, **99**(15), 9783–9788 (2002).
99. T. Braun, N. Backmann, M. Vogtli, A. Bietsch, A. Engel, H. P. Lang, C. Gerber, and M. Hegner, Conformational change of bacteriorhodopsin quantitatively monitored by microcantilever sensors. *Biophys. J.*, **90**(8), 2970–2977 (2006).
100. C. Qi, Y. Y. Chen, Z. H. Wang, and G. Jin, Investigation of interactions between two monoclonal antibodies and sars virus with a label-free protein array. *IEEE Eng. Med.*, **2**, 1305–1308 (2005).
101. E. Oezkumur, J. W. Needham, D. A. Bergstein, R. Gonzalez, M. Cabodi, J. M. Gershoni, B. B. Goldberg, and M. S. Uenlue, Label-free and dynamic detection of biomolecular interactions for high-throughput microarray applications. *Proceedings of the National Academy of Sciences of the United States of America*, **105**(23), 7988–7992 (2008).
102. X. Wang, M. Zhao, and D. D. Nolte, Area-scaling of interferometric and fluorescent detection of protein on antibody microarrays. *Biosens. Bioelectron..*, **24**(4), 981–987 (2008).

103. C. R. Mace, C. C. Striemer, and B. L. Miller, Detection of human proteins using arrayed imaging reflectometry. *Biosens. Bioelectron.*, **24**(2), 334–337 (2008).
104. M. Thompson, L. F. Cheran, M. Q. Zhang, M. Chacko, H. Huo, and S. Sadeghi, Label-free detection of nucleic acid and protein microarrays by scanning kelvin nanoprobe. *Biosens. Bioelectron.*, **20**(8), 1471–1481 (2005).
105. H. Xu, X. Mao, Q. Zeng, S. Wang, A.-N. Kawde, and G. Liu, Aptamer-functionalized gold nanoparticles as probes in a dry-reagent strip biosensor for protein analysis. *Anal. Chem.*, **81**(2), 669–675 (2009).
106. M. Bally, K. Bailey, K. Sugihara, D. Grieshaber, J. Voros, and B. Stadler, Liposome and lipid bilayer arrays towards biosensing applications. *Small*, **6**(22), 2481–2497 (2010).
107. S. Majd, D. J. Estes, and M. Mayer, Assays for studying annexin binding to artificial bilayers. *Calcium Binding Protein.*, **1**(1), 26–29 (2006).
108. Y. X. Shen, P. O. Saboe, I. T. Sines, M. Erbakan, and M. Kumar, Biomimetic membranes: A review. *J. Membr. Sci.*, **454**, 359–381 (2014).
109. T. J. Jeon, N. Malmstadt, and J. J. Schmidt, Hydrogel-encapsulated lipid membranes. *J. Am. Chem. Soc.*, **128**(1), 42–43 (2006).
110. S. Majd, E. C. Yusko, Y. N. Billeh, M. X. Macrae, J. Yang, and M. Mayer, Applications of biological pores in nanomedicine, sensing, and nanoelectronics. *Curr. Opin. Biotechnol.*, **21**(4), 439–476 (2010).
111. S. F. Fenz and K. Sengupta, Giant vesicles as cell models. *Integr. Biol.*, **4**(9), 982–995 (2012).
112. D. Lingwood, and K. Simons, Lipid rafts as a membrane-organizing principle. *Science*, **327**(5961), 46–50 (2010).
113. T. J. Jeon, J. L. Poulos, and J.J. Schmidt, Long-term storable and shippable lipid bilayer membrane platform. *Lab Chip*, **8**(10), 1742–1744 (2008).
114. N. Malmstadt, L. J. Jeon, and J. J. Schmidt, Long-lived planar lipid bilayer membranes anchored to an *in situ* polymerized hydrogel. *Adv. Mater.*, **20**(1), 84–89 (2008).
115. S. Ibragimova, K. Stibius, P. Szewczykowski, M. Perry, H. Bohr, and C. Helix-Nielsen, Hydrogels for *in situ* encapsulation of biomimetic membrane arrays. *Polymer. Adv. Tech.*, **23**(2), 182–189 (2012).
116. M. R. Lander, S. Ibragimova, C. Rein, J. Vogel, K. Stibius, O. Geschke, M. Perry, and C. Helix-Nielsen, Biomimetic membrane arrays on cast hydrogel supports. *Langmuir*, **27**(11), 7002–7007 (2011).
117. V. Yamazaki, O. Sirenko, R. J. Schafer, L. Nguyen, T. Gutsmann, L. Brade, and J. T. Groves, Cell membrane array fabrication and assay technology. *Bmc Biotechnol.*, **518** (2005).
118. K. D. Mossman, G. Campi, J. T. Groves, and M. L. Dustin, Altered tcr signaling from geometrically repatterned immunological synapses. *Science*, **310**(5751), 1191–1193 (2005).
119. A. J. Torres, R. L. Contento, S. Gordo, K. W. Wucherpfennig, and J. C. Love, Functional single-cell analysis of t-cell activation by supported lipid bilayer-tethered ligands on arrays of nanowells. *Lab Chip*, **13**(1), 90–99 (2013).

120. X. X. Chen, S. Shojaei-Zadeh, M. L. Gilchristc, and C. Maldarelli, A lipobead microarray assembled by particle entrapment in a microfluidic obstacle course and used for the display of cell membrane receptors. *Lab Chip*, 13(15), 3041–3060 (2013).
121. C. F. Monson, X. Cong, A. D. Robison, H. P. Pace, C. M. Liu, M. F. Poyton, and P. S. Cremer, Phosphatidylserine reversibly binds Cu^{2+} with extremely high affinity. *J. Am. Chem. Soc.*, **134**(18), 7773–7779 (2012).
122. D. van Swaay, and A. deMello, Microfluidic methods for forming liposomes. *Lab Chip*, **13**(5), 752–767 (2013).
123. S. M. Christensen, and D. Stamou, Surface-based lipid vesicle reactor systems: Fabrication and applications. *Soft Matter*, **3**(7), 828–836 (2007).
124. P. D. Eckford and F. J. Sharom, Interaction of the p-glycoprotein multidrug efflux pump with cholesterol: Effects on atpase activity, drug binding and transport. *Biochemistry*, **47**(51), 13686–13698 (2008).
125. P. H. Lu, R. H. Liu, and F. J. Sharom, Drug transport by reconstituted p-glycoprotein in proteoliposomes — effect of substrates and modulators, and dependence on bilayer phase state. *Eur. J. Biochem.*, **268**(6), 1687–1697 (2001).
126. M. Morita, T. Hamada, Y. Tendo, T. Hata, M. C. Vestergaard, and M. Takagi, Selective localization of alzheimer's amyloid beta in membrane lateral compartments. *Soft Matter*, **8**(10), 2816–2819 (2012).
127. P. Kuhn, K. Eyer, S. Allner, D. Lombardi, and P. S. Dittrich, A microfluidic vesicle screening platform: Monitoring the lipid membrane permeability of tetracyclines. *Anal. Chem.*, **83**(23), 8877–8885 (2011).
128. M. Onishi and H. Kamimori, High-throughput and sensitive assay for amphotericin b interaction with lipid membrane on the model membrane systems by surface plasmon resonance. *Biol. Pharm. Bull.*, **36**(4), 658–663 (2013).
129. M. Gobbi, F. Re, M. Canovi, M. Beeg, M. Gregori, S. Sesana, S. Sonnino, D. Brogioli, C. Musicanti, P. Gasco, M. Salmona, and M. E. Masserini, Lipid-based nanoparticles with high binding affinity for amyloid-beta(1-42) peptide. *Biomaterials*, **31**(25), 6519–6529 (2010).
130. E. T. Castellana and P. S. Cremer, Solid supported lipid bilayers: From biophysical studies to sensor design. *Surface Science Reports*, **61**(10), 429–444 (2006).
131. H. Bayley, O. Braha, and L. Q. Gu, Stochastic sensing with protein pores. *Adv. Mater.*, **12**(2), 139–142 (2000).
132. S. Howorka and Z. Siwy, Nanopore analytics: sensing of single molecules. *Chem. Soc. Rev.*, **38**(8), 2360–2384 (2009).
133. A. Kocer, L. Tauk, and P. Dejardin, Nanopore sensors: from hybrid to abiotic systems. *Biosens. Bioelectron.*, **38**(1), 1–10 (2012).
134. B. M. Venkatesan and R. Bashir, Nanopore sensors for nucleic acid analysis. *Nat. Nanotechnol.*, **6**(10), 615–624 (2011).
135. D. Deamer, Nanopore analysis of nucleic acids bound to exonucleases and polymerases. *Annu. Rev. Biophys.*, **39**, 79–90 (2010).

136. M. Rincon-Restrepo, E. Milthallova, H. Bayley, and G. Maglia, Controlled translocation of individual DNA molecules through protein nanopores with engineered molecular brakes. *Nano Lett.*, **11**(2), 746–750 (2011).
137. S. Majd, E. C. Yusko, A. D. MacBriar, J. Yang, and M. Mayer, Gramicidin pores report the activity of membrane-active enzymes. *J. Am. Chem. Soc.*, **131**(44), 16119–16126 (2009).
138. M. R. Hynd, J. P. Frampton, N. Dowell-Mesfin, J. N. Turner, and W. Shain, Directed cell growth on protein-functionalized hydrogel surfaces. *J. Neurosci. Methods*, **1 62**(1–2), 255–263 (2007).
139. U. Boudriot, R. Dersch, A. Greiner, and J. H. Wendorff, Electrospinning approaches toward scaffold engineering — a brief overview. *Artif. Organs*, **30**(10), 785–792 (2006).
140. R. A. Neal, S. G. McClugage III, M. C. Link, L. S. Sefcik, R. C. Ogle, and E. A. Botchwey, Laminin nanofiber meshes that mimic morphological properties and bioactivity of basement membranes. *Tissue Eng. Pt. C-Meth.*, **15**(1), 11–21 (2009).
141. S.-H. Huang, T.-C. Chien, and K.-Y. Hung, Selective deposition of electrospun alginate-based nanofibers onto cell-repelling hydrogel surfaces for cell-based microarrays. *Curr. Nanosci.*, **7**(2), 267–274 (2011).
142. M. Y. Li, M. J. Mondrinos, M. R. Gandhi, F. K. Ko, A. S. Weiss, and P. I. Lelkes, Electrospun protein fibers as matrices for tissue engineering. *Biomaterials*, **26**(30), 5999–6008 (2005).
143. A. Revzin, R. G. Tompkins, and M. Toner, Surface engineering with poly(ethylene glycol) photolithography to create high-density cell arrays on glass. *Langmuir*, **19**(23), 9855–9862 (2003).
144. C. J. Campbell, S. K. Smoukov, K. J. M. Bishop, and B. A. Grzybowski, Reactive surface micropatterning by wet stamping. *Langmuir*, **21**(7), 2637–2640 (2005).
145. R. O. Hynes, Integrins: Bidirectional, allosteric signaling machines. *Cell*, **110**(6), 673–687 (2002).
146. M. G. Coppolino, and S. Dedhar, Bi-directional signal transduction by integrin receptors. *Int. J. Biochem. Cell Biol.*, **32**(2), 171–188 (2000).
147. D. B. Weibel, A. Lee, M. Mayer, S. F. Brady, D. Bruzewicz, J. Yang, W. R. DiLuzio, J. Clardy, and G. M. Whitesides, Bacterial printing press that regenerates its ink: Contact-printing bacteria using hydrogel stamps. *Langmuir*, **21**(14), 6436–6442 (2005).
148. W. -H. Tan and S. Takeuchi, Monodisperse alginate hydrogel microbeads for cell encapsulation. *Adv. Mater.*, **19**(18), 2696–2701(2007).
149. D. Yip and H. Cho, A multicellular 3d heterospheroid model of liver tumor and stromal cells in collagen gel for anti-cancer drug testing. *Biochem. Biophys. Res. Comm.*, **433**(3), 327–332 (2013).
150. M.-J. Yang, C.-H. Chen, P.-J. Lin, C.-H. Huang, W. Chen, and H.-W. Sung, Novel method of forming human embryoid bodies in a polystyrene dish surface-coated with a temperature-responsive methylcellulose hydrogel. *Biomacromolecules*, **8**(9), 2746–2752 (2007).

151. B. Prabhakarpandian, M.-C. Shen, J. B. Nichols, I. R. Mills, M. Sidoryk-Wegrzynowicz, M. Aschner, and K. Pant, Sym-bbb: a microfluidic blood brain barrier model. *Lab Chip*, **13**(6), 1093–1101 (2013).
152. Y.-C. Tung, A. Y. Hsiao, S. G. Allen, Y.-S. Torisawa, M. Ho, and S. Takayama, High-throughput 3d spheroid culture and drug testing using a 384 hanging drop array. *Analyst*, **136**(3), 473–478 (2011).
153. L. S. Liu, S. Q. Liu, S. Y. Ng, M. Froix, T. Ohno, and J. Heller, Controlled release of interleukin-2 for tumour immunotherapy using alginate/chitosan porous microspheres. *J. Control. Release*, **43**(1), 65–74 (1997).
154. C. Wang, R. J. Stewart, and J. Kopecek, Hybrid hydrogels assembled from synthetic polymers and coiled-coil protein domains. *Nature*, **397**(6718), 417–420 (1999).
155. X. Qu, A. Wirsen, and A. C. Albertsson, Novel ph-sensitive chitosan hydrogels: swelling behavior and states of water. *Polymer*, **41**(12), 4589–4598 (2000).
156. S. H. Um, J. B. Lee, N. Park, S. Y. Kwon, C. C. Umbach, and D. Luo, Enzyme-catalysed assembly of DNA hydrogel. *Nat. Mater.*, **5**(10), 797–801 (2006).
157. M. T. Frey and Y. Wang, A photo-modulatable material for probing cellular responses to substrate rigidity. *Soft Matter*, **5**(9), 1918–1924 (2009).
158. S. R. Peyton, C. B. Raub, V. P. Keschrumrus, and A. J. Putnam, The use of poly(ethylene glycol) hydrogels to investigate the impact of ecm chemistry and mechanics on smooth muscle cells. *Biomaterials*, **27**(28), 4881–4893 (2006).
159. Y. Chau, Y. Luo, A. C. Y. Cheung, Y. Nagai, S. Zhang, J. B. Kobler, S. M. Zeitels, and R. Langer, Incorporation of a matrix metalloproteinase-sensitive substrate into self-assembling peptides — a model for biofunctional scaffolds. *Biomaterials*, **29**(11), 1713–1719 (2008).
160. H. Vaisocherova, V. M. Faca, A. D. Taylor, S. Hanash, and S. Jiang, Comparative study of spr and elisa methods based on analysis of cd166/alcam levels in cancer and control human sera. *Biosens. Bioelectron.*, **24**(7), 2143–2148 (2009).
161. T. B. Tran, S. Cho, and J. Min, Hydrogel-based diffusion chip with electric cell-substrate impedance sensing (ecis) integration for cell viability assay and drug toxicity screening. *Biosens. Bioelectron.*, **50**, 453–459 (2013).
162. C. K. Choi, C. H. Margraves, S. I. Jun, A. E. English, P. D. Rack, and K. D. Kihm, Opto-electric cellular biosensor using optically transparent indium tin oxide (ito) electrodes. *Sensors*, **8**(5), 3257–3270 (2008).
163. H. Otsuka, M. Nagamura, A. Kaneko, K. Kutsuzawa, and T. Sakata, Label-free and noninvasive monitoring of cell differentiation on spheroid microarray. *Ieice T. Electron.*, **E96C**(3), 353–357 (2013).

Chapter 11

Protein-Immobilized Hydrogel Microstructures for Optical Biosensing

Won-Gun Koh
Department of Chemical and Biomolecular Engineering,
Yonsei University, 50 Yonsei-ro, Seodaemoon-gu, Seoul, South Korea
wongun@yonsei.ac.kr

Hydrogels are attractive support materials for protein immobilization because they provide soft and hydrated environment that is essential for proteins to maintain their structure and activity. Protein-incorporated hydrogels can be easily micropatterned via various microfabrication techniques and have great potential to be used as protein microarrays. Among the various hydrogels, this chapter mainly focuses on poly(ethylene glycol)(PEG)-based hydrogels that are prepared from acrylated PEG macromer with different molecular weights. Physical properties of PEG hydrogel can be tuned by changing the molecular weight (MW) of PEG. Planar or suspension microarrays of PEG hydrogels are fabricated by different lithographic techniques and proteins such as enzymes and antibodies are incorporated into resultant PEG hydrogel microarrays via physical entrapment or covalent surface immobilization. Hydrogel-incorporated proteins can be applied to microarray-based biosensor, where interaction between probe proteins and target molecules are monitored with absorbance or fluorescence. Eventually, hydrogel microstructures would be integrated with microfluidic device to achieve high-throughput and high-content protein-based analysis.

1. Introduction

Following completion of the human genome project, the scientific community has turned its attention toward proteomics, which is the large-scale study of proteins, particularly their structures and functions, to obtain a better understanding of organisms than possible with genomics.[1] Strong demands for achieving high-throughput and high-content protein analysis have brought into focus on the protein microarrays that are able to immobilize a large number of proteins simultaneously in a high-density format to identify protein-protein, protein-DNA, or protein-small molecule interactions.[2–6]

In most of protein microarray-related researches, proteins are generally immobilized on two-dimensional rigid substrates by non-specific adsorption or covalent binding to the monolayer of tethered functional groups on the surfaces.[7,8] However, the amount of protein that can be attached to a fixed area on the planar and rigid substrate is limited, resulting in relatively low assay sensitivity. Furthermore, a rapid evaporation of the aqueous environment eventually causes dehydration and denatures of the immobilized proteins. One solution for these problems is to use a hydrogel, which has a soft and hydrated properties.[9,10] Hydrogels are three-dimensionally (3D) crosslinked polymers that can absorb water or other biological fluids and become swollen. They are attractive support materials for protein immobilization because the soft and hydrated environment of swollen hydrogels can provide proteins with near-physiological conditions that minimize denaturation and allow them to carry out their full biological functions.

Among the various hydrogels, PEG-based hydrogels have been widely used for many years not only in bioassays but also drug delivery devices, tissue engineering, and cell transplantation because of their high water content, hydrophilicity, and biocompatibility.[11–13] PEG molecules of different MW can be easily converted into acrylates such as PEG diacrylate (PEG-DA), and photopolymerization of the acrylated PEGs yields a highly crosslinked hydrogel network.[14] Furthermore, with combination of photopatterning techniques, nano/micro PEG hydrogel patterns can be fabricated.[15–17] The physical properties of PEG-based hydrogels can be easily tuned by varying the MW of the PEG molecules, and the transparent nature of PEG hydrogels also makes them suitable for optical detection schemes in biosensing applications. PEG hydrogels also provide a protective environment for immobilized proteins and inhibit degradation and fouling.

In this chapter, we introduce optical biosensor system based on protein-immobilized PEG hydrogel microstructures. Fabrication of PEG hydrogel micropatterns and immobilization of proteins will be discussed first followed by biosensing application using protein-immobilized PEG hydrogel microstructures.

2. Fabrication of Hydrogel Microstructures

In protein microarrays, a number of proteins can be immobilized either on a planar substrates or on microparticles to realize multiplexed and high-throughput

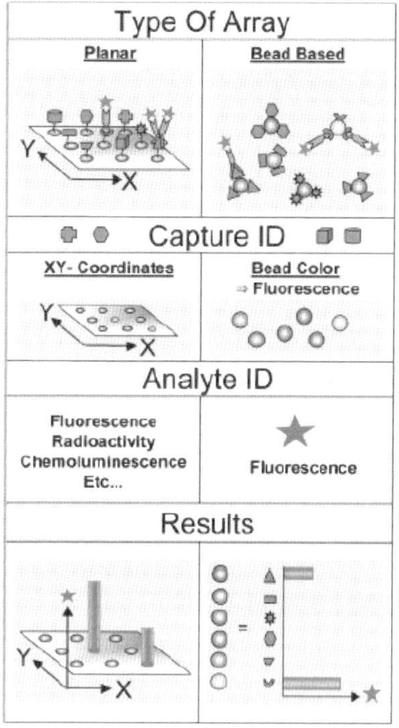

Fig. 1. Types of microarray system. Planar and suspension microarrays represent two different techniques that can be employed for multiplexed protein-based assays [20] (with the permission from Ref. 20, copyright 2003 WILEY-VCH).

protein-based assays,[18,19] which generates two categories of protein microarrays: chip-based planar (positional) arrays and particle-based suspension (solution) arrays as shown in Fig. 1.[20]

Planar arrays have been fabricated on two-dimensional planar substrates such as glass, silicon, and gold via various micropatterning techniques such as photolithography, soft lithography, or pin-spotting. Each protein is identified by x and y coordinates on the microarray. Although a variety of positional microarrays have been developed and widely used for disease diagnosis, prognosis, biochemical analysis, and therapeutic regimes, suspension arrays are becoming popular as an alternative approach for high-throughput assays. Suspension arrays utilize self-encoded particles as array elements instead of positionally-encoded spots on a planar surface. These arrays are believed to display faster reaction kinetics because of the radial diffusion of analytes or probes in solution. Furthermore, because each element in a suspension array is independent, the system has the flexibility to investigate a few or even thousands of proteins, without the need to fabricate a new chip, simply by adding new particles to detect corresponding analytes.

Protein-incorporated PEG hydrogel microstructures can be also prepared as planar array and suspension array formats by using different microfabrication process. This session is focused on the fabrication of different types of hydrogel microstructures which can be potentially applied for protein microarrays.

2.1. *Planar (positional) array type*

Planar array types of PEG hydrogel microstructures were first prepared by simple photolithography techniques.[21] The ability of hydrogel precursor solution containing PEG-DA and photoinitiators to gel upon exposure to ultraviolet (UV) light was used to create negative patterns. Such patterns were created by spin-coating the gel precursor solution onto the glass or silicone substrate and then exposing the coated surface to UV light through photomasks. Light was projected through the photomasks and areas exposed to UV light undergo free radical polymerization and subsequent crosslinking, thus becoming insoluble in water, creating polymerized regions that corresponded to the pattern of the photomasks. Unpolymerized PEG solution was then washed away with water so that only the insoluble, crosslinked patterns remained on the surface. Figure 2(A) describes these procedures to fabricate hydrogel microstructures using photolithography and Fig. 2(B) shows the SEM image of resultant hydrogel microarrays.

When PEG hydrogel micropatterns were fabricated on the bare silicon or glass substrates, surface attachment was weak and hydrogel array elements easily delaminated upon hydration due to swelling of the crosslinked hydrogel matrixes during the developing process. To prevent delamination, a self-assembled monolayer of 3-(trichlorosilyl) propyl methacrylate (TPM) on the silicon/glass substrate was used to create a reactive surface onto which the hydrogel was covalently affixed during

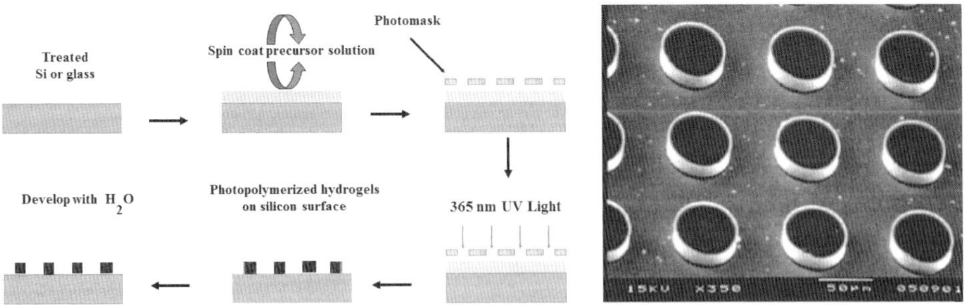

(A) (B)

Fig. 2. Fabrication of planar arrays of PEG hydrogel microstructures using photolithography. (A) Schematic illustration of preparing PEG hydrogel microarray, (B) SEM image of resultant PEG hydrogel microarrays fabricated on the silicon surface[21] (with the permission from Ref. 21, copyright 2001 American Chemical Society).

photopolymerization. These methacrylated surfaces of substrates can take part in the free-radical polymerization reaction during the photopolymerization because it contains unsaturated C=C functionalities. Methacrylate groups on the surfaces of substrates react with radical species present near the surface to effectively anchor the hydrogels to the surface of the silicon/glass substrates. Thus delamination or detachment of PEG hydrogel microstructures from the substrate was significantly reduces, even though extensive swelling did occur in the gels upon exposure to water.

Soft lithography can be also used to fabricate hydrogel microarrays. For example, Koh and Pishko fabricated PEG hydrogel microstructures via photoreaction injection molding using microfluidic networks of PDMS as micromolds.[22] Acrylate or methacrylate modified macromers were injected into the fluidic system and light was used to gel the macromers to form a hydrogel. The PDMS replica was then removed, leaving the gelled microstructures on the substrate surface (Fig. 3(A)). By combining photoreaction injection molding with photolithography, arrays of hydrogel microstructures possessing different chemistries were created as shown in Fig. 3(B). Suh *et al.* fabricated hydrogel micropatterns via capillary force lithography (CFL) using PDMS mold.[23,24]

Besides lithography techniques, hydrogel microstructures can be also fabricated using a pin printer (as is used to create DNA microarrays).[25] Hydrogel precursor solution is deposited using microarray robot in an array format on the

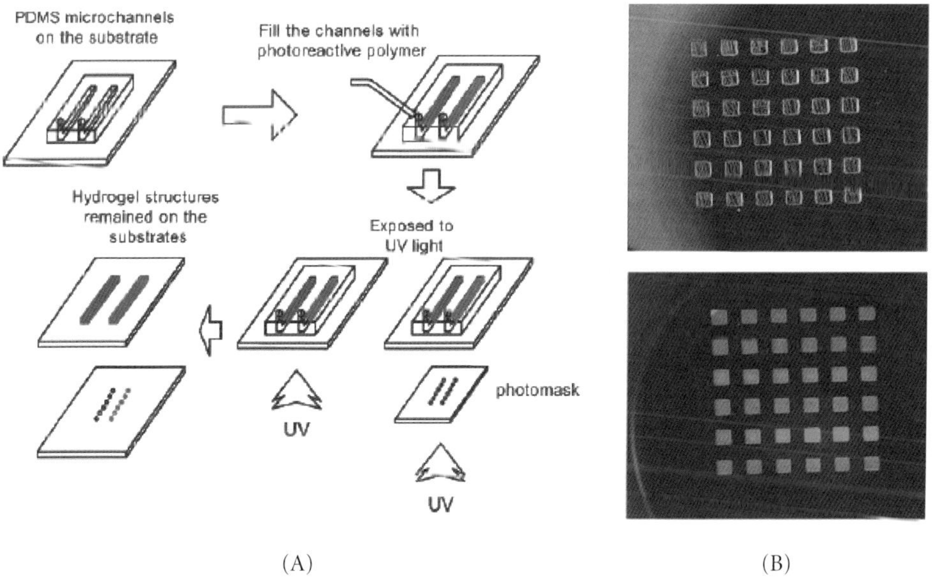

(A) (B)

Fig. 3. Preparation of hydrogel microarrays using photoreaction injection molding.[22] (A) Schematic illustration. (B) An array of hydrogel microstructures containing different components (with the permission from Ref. 22, copyright 2003 American Chemical Society).

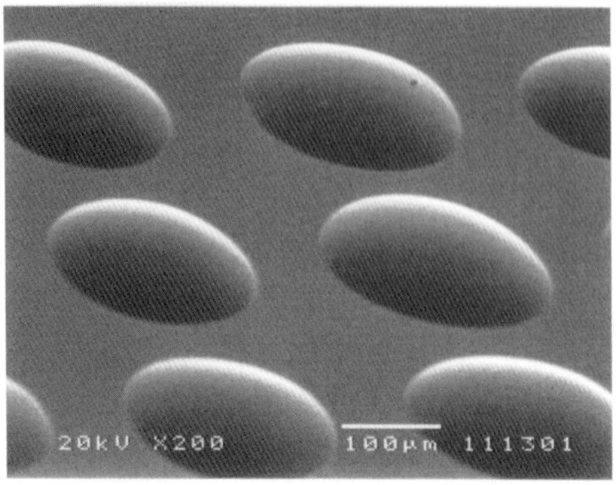

(A)

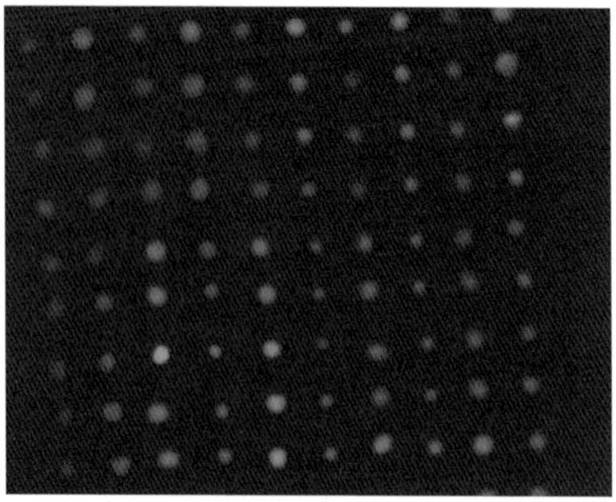

(B)

Fig. 4. (A) SEM image of PEG microstructures prepared using a pin-printing microarrayer robot. (B) Alternating rows containing FITC and TRITC-containing hydrogel microstructures fabricated using a pin printer [25] (with the permission from Ref. 25, copyright 2003 Elsevier).

pretreated silicon or glass surfaces and the polymer spots are photopolymerized, producing PEG hydrogel microarrays as shown in Fig. 4(A). Using pin printer enables easy fabrication of an array of hydrogel microstructures containing different components on the same substrate. Figure 4(B) demonstrates spotting the array of hydrogel microstructures with alternating rows containing Fluorescein isothiocyanate (FITC) and tetramethylrhodamine isothiocyanate (TRITC) dyes in the polymer precursor solution. The arrays were spotted using a single pin with the first precursor solution printed in alternate rows followed by a wash step and printing of the second precursor solution. This was followed by photopolymerization of the pattern by exposure to UV light.

2.2. *Suspension (solution) array type*

Currently, most suspension array systems employ optically-coded spherical microparticles to achieve multiplexed assays.[26,27] For example, fluorescence molecules possessing different fluorescence emission are incorporated into microparticles with various ratios to identify each particle as shown in Fig. 5.[28]

Hydrogel micro/nanoparticles are usually prepared by emulsion techniques, suspension polymerization, or using shear flow in microfluidic channels as spherical shape and can be optically-coded by incorporating fluorescence molecules.[29–33] However, there are several disadvantages to use multiple colors as a means of encoding, including the limited number of color combinations and spectral overlap between an encoding color and the analyte-detection color. Due to above-mentioned disadvantages of optically-coded systems, several groups have proposed using

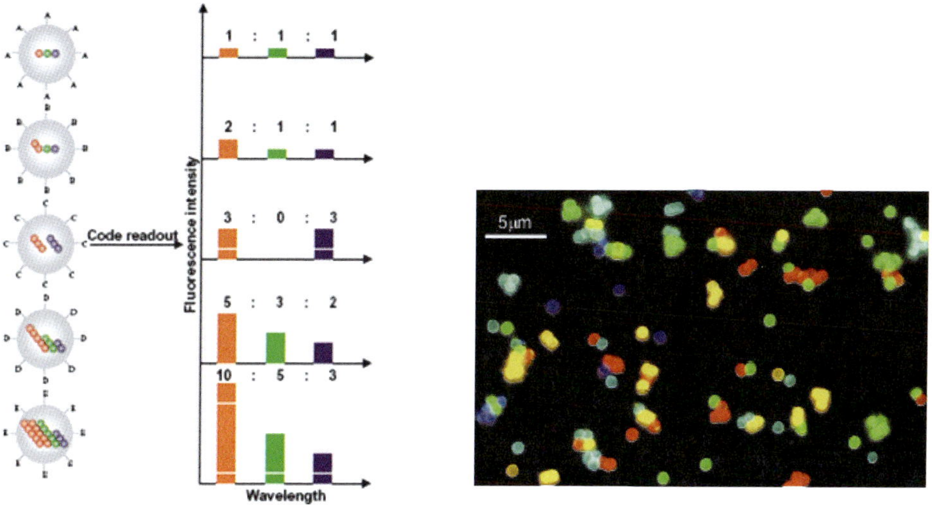

(A) (B)

Fig. 5. (A) Preparation of optically-coded suspension microarrays using quantum dot (QD)-entrapping microparticles. (B) Fluorescence image showing microparticles entrapping red, green, and blue-emitting QDs at different ratio [28] (with the permission from Ref. 28, copyright 2001 Nature Publishing Group).

graphically or shape-coded hydrogel microparticles as a suspension microarray format for multiplexed bioassays.[34,35] Doyle's group developed continuous/stop flow lithography,[36–38] which makes it possible to synthesize various shapes of non-spherical hydrogel microparticles as shown in Fig. 6(A).[37] Furthermore, bifunctional Janus particles were synthesized by exploiting the laminar flow within microchannels (Fig. 6(B)).[37] Based on these techniques, bar-coded PEG hydrogel microparticles consisting of graphical code and detecting region could be fabricated as shown in Fig. 6(C).[39]

Koh *et al.* also fabricated suspension arrays of PEG hydrogel by using photopatterning process. Various shapes of hydrogel microparticles with lateral dimensions ranging from 50 to 300 μm were fabricated using different molecular weights of PEG (700, 3,400, and 8,000 Da), by which the water content and swelling behavior of the hydrogel microparticles could be controlled.[35]

2.3. *Hydrogel microstructures within microfluidic devices*

After the concept of micro total analysis system (μ-TAS) or "labs on a chip" was proposed, microfluidic devices have been developed for a number of protein-based assay. Conventional protein-based assays have been carried out homogeneously within microfluidic devices. However, a lot of recent studies have reported successful implementation of microfluidic-based heterogeneous assays using

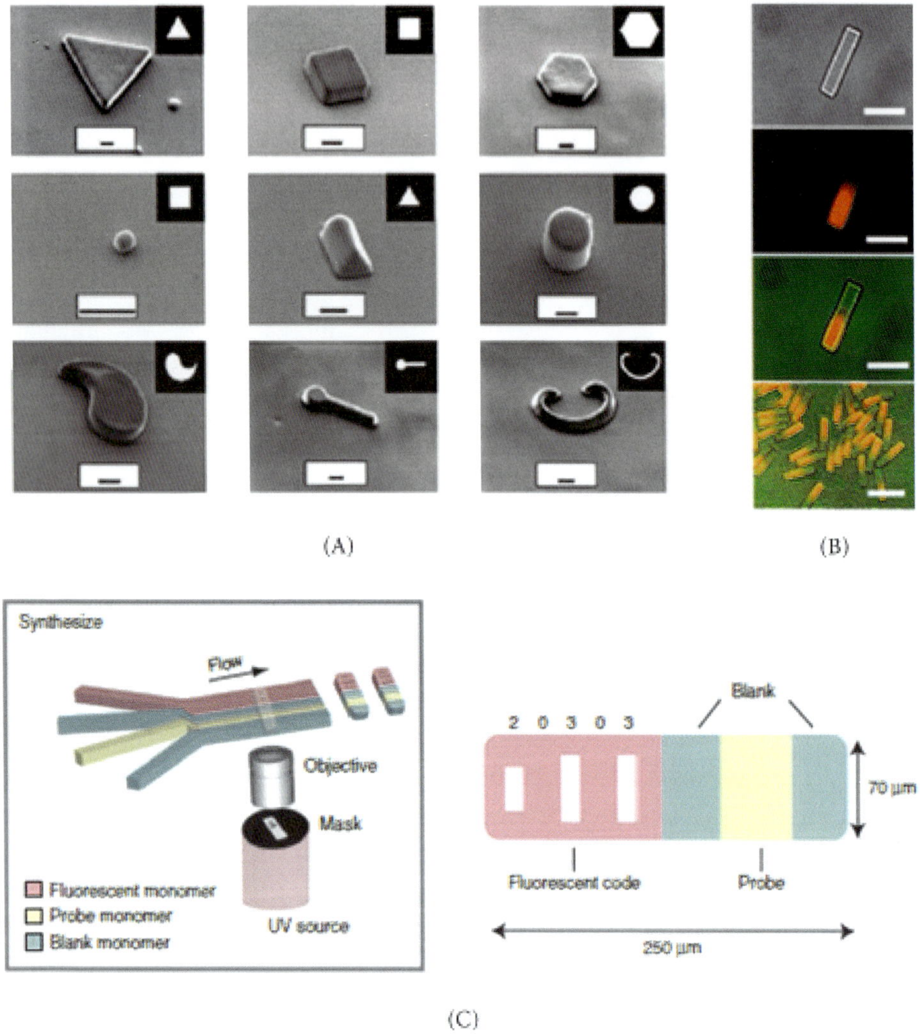

Fig. 6. (A) SEM images of hydrogel microparticles with various shapes prepared via continuous-flow lithography.[37] (B) Optical and fluorescence images of Janus hydrogel microparticles [37] (with the permission from Ref. 37, copyright 2006 Nature Publishing Group) (C) Schematic diagram of the synthesis of bar-coded hydrogel microparticles using stop-flow lithography [39] (with the permission from Ref. 39, copyright 2011 Nature Publishing Group).

proteins immobilized either directly on the walls of microchannels or on a support contained within the channel.[40–44] Since Beebe *et al.* reported the fabrication of 3D hydrogel structures within microfluidic devices to control the flow inside microchannels,[45] planar and suspension array types of hydrogel microstructures were incorporated into microfluidic devices for protein-based assays.

Crooks *et al.* integrated PEG hydrogel micropatches into microchannels. Hydrogel micropatches of different sizes are conveniently prepared by controlling

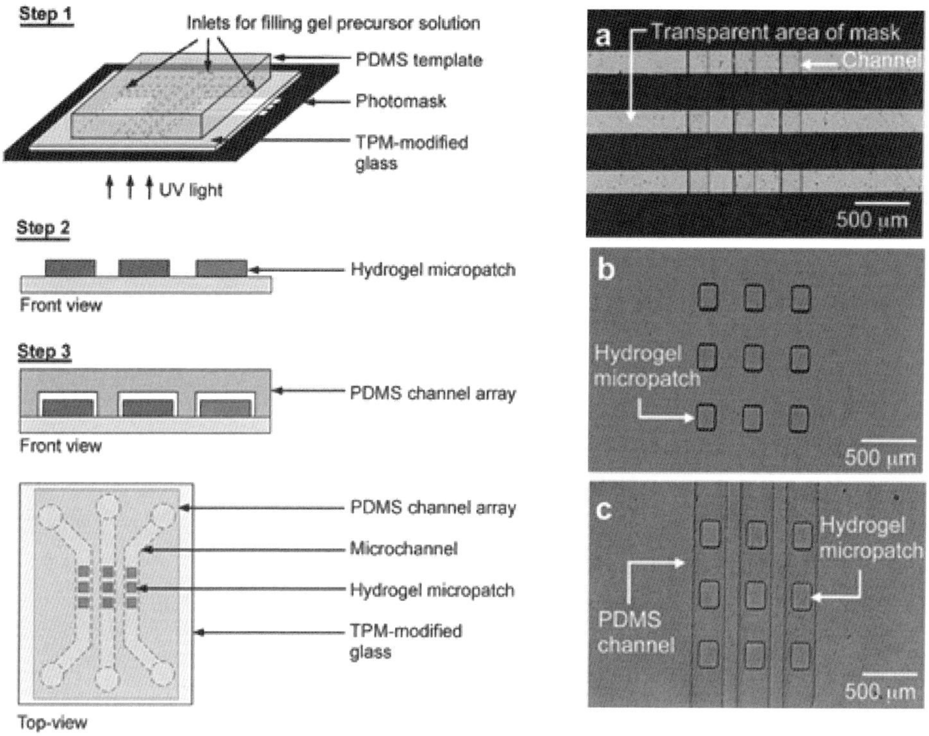

Fig. 7. Schematic illustration of multiple procedure to prepare microfluidic devices incorporating hydrogel micropatch within microchannels (left) and optical images taken at each step (right)[46] (with the permission from Ref. 46, copyright 2005 American Chemical Society).

the dimensions of the microfluidic channels and photomasks.[46,47] The complete device is prepared by replacing microchannels used to fabricate the hydrogel micropatches with a somewhat wider and higher channel than the first one to allow fluid to flow around the micropatches (Fig. 7).[16]

Pishko *et al.* directly fabricated the hydrogel microstructures inside microchannels using photolithography.[48] The microchannels were filled with precursor solutions and then exposed to UV light through a photomask. Finally, by flushing the channel with water, desired hydrogel structures were obtained inside a microchannel as shown in Fig. 8(A). They also fabricated hydrogel microstructures containing different compositions in a single channel by multiple photolithography process. For example, microstructures containing red dye were made inside the channel and then precursor solution containing green dye was polymerized next to red dye-containing hydrogels as shown in Fig. 8(B).

Besides planar array types of hydrogel, suspension array of hydrogel could be incorporated into microfluidic devices. Koh *et al.* constructed a microfluidic device by serially connecting patterning chambers and a microfilter-integrated detection

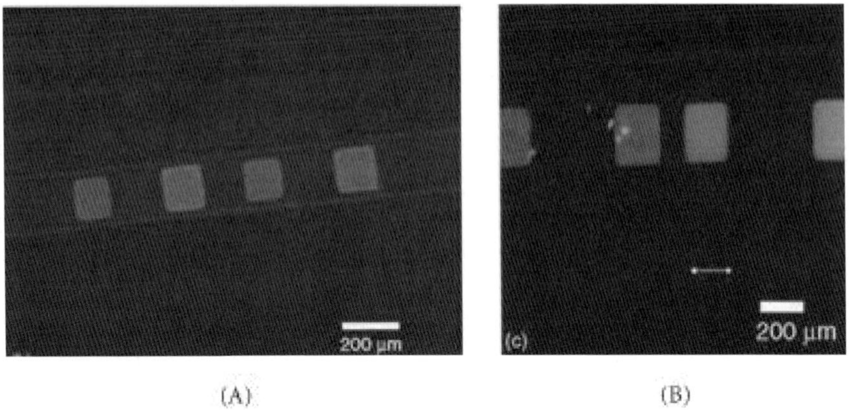

Fig. 8. (A) fluorescence Images of PEG hydrogel microstructures containing fluorophore Microchannels. (B) Microchannels containing four hydrogel microstructures (two labeled with green dye and two labeled with red dye[48] (with the permission from Ref. 48, copyright 2004 Elsevier).

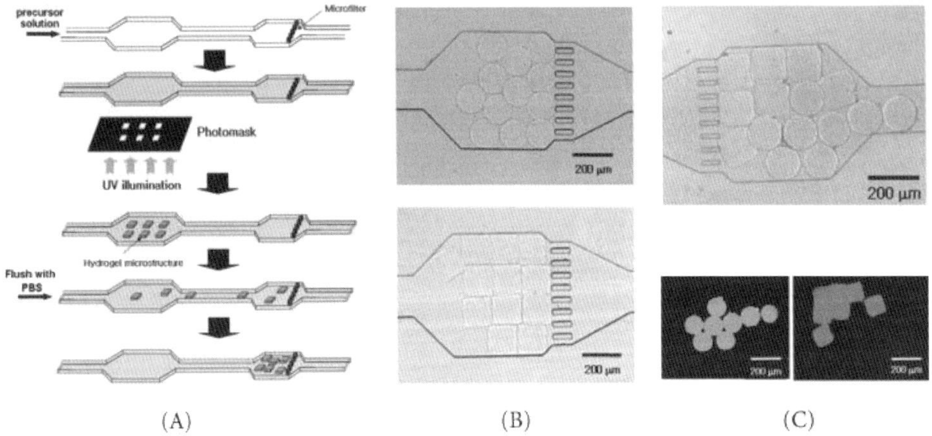

Fig. 9. (A) Schematic illustration of experimental procedure to fabricate and locate hydrogel microparticles inside the microfluidic device using the photopatterning and flushing method. (B) Hydrogel microparticles trapped within the detection chamber. (C) Hydrogel microparticles containing two different molecules produced by multiple photopatterning and flushing process[49] (with the permission from Ref. 49, copyright 2008 Springer-Verlag).

chamber.[49] Various shapes and sizes of hydrogel microparticles were fabricated in the patterning chamber by photopatterning and moved to the detection chamber by pressure-driven flow as described in Fig. 9(A). All of the hydrogel microparticles were retained within the detection chamber due to an in-chamber integrated microfilter with smaller mesh size than hydrogel microparticles (Fig. 9(B)). Furthermore, by using multiple processes, hydrogel microparticles with different shapes or different compositions could be prepared within the reaction chamber (Fig. 9(C)).

3. Incorporation of Proteins into Hydrogel Microstructures

Proteins can be immobilized onto the surface of PEG hydrogel or entrapped within hydrogel matrix.

Highly crosslinked hydrogel networks are capable of physically encapsulating proteins as shown in Fig. 10(A) and hydrogel-entrapped proteins can remain structurally intact maintaining their biological function, owing to the relatively inert aqueous environment within the hydrogel matrix. Various natural and synthetic hydrogels have been used for protein entrapment and it was reported that enzyme immobilization inside hydrogel matrix enhanced the resistance to thermal inactivation, pH, denaturing agents, and organic solvents.[50–53] When proteins are physically entrapped within hydrogel matrix for bioassay, porosity of hydrogel should be carefully controlled so that mesh size of hydrogel is large enough to allow diffusion of target molecules but also small enough to prevent leaching out of proteins. PEG hydrogels is attractive in this aspect compared with other hydrogels because their network structure or microporosity can be easy controlled by simply changing the MW of PEG. Russell *et al.* examined the mass transfer characteristics of hydrogel fabricated from different MW of PEG (MW 575, 1,000, 18,500).[54] In their study, estimated hydrogel mesh sizes ranged from 8.6 to 13.7 Å for hydrogels fabricated using PEG with MWs between 575 and 1,000, to 103 Å for hydrogels based on PEG with a MW of 18,500. Diffusivities of small molecules such as TRITC (MW 443.52) was on the order of 10^{-7}–10^{-9} cm^2/s depending on MW of PEG. Despite many advantages of hydrogel entrapment, activity loss of the entrapped proteins is inevitable because proteins are entrapped within crosslinked hydrogel which hinders diffusion of substrates. Although diffusion of target molecules through hydrogel can be enhanced by increasing water content or porosity of hydrogel, it is restricted because more proteins may leach out from hydrogel.

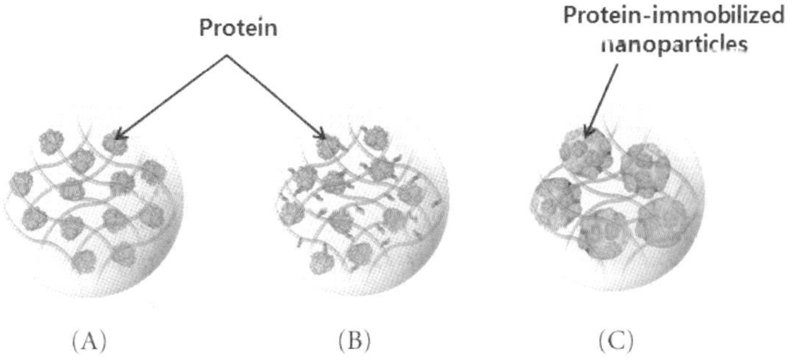

Fig. 10. Different methods to immobilize protein within hydrogel. (A) Physical entrapment of protein within hydrogel matrix. (B) Covalent immobilization of protein within hydrogel matrix via bifunctional linkers. (C) Entrapment of protein-immobilized nanoparticles within hydrogel.

Leaching problem of physically entrapped proteins can be overcome by combining covalent immobilization with physical entrapment. Covalent immobilization can be achieved by with heterobifunctional linkers such as acryloyl-PEG-*N*-hydroxysuccinimide (acryloyl-PEG-NHS).[55,56] Proteins are functionalized with acrylate moiety by conjugation with NHS groups and copolymerization between PEG-DA and acrylated proteins leads to the covalent immobilization of the proteins within the hydrogel structures, where enzymes are covalently tethered to the polymer backbone without leaching from hydrogel matrix as shown in Fig. 10B. Another method that allows the use of hydrogel with large mesh without leaching problem is to utilize nanoparticles. Here, proteins are first covalently pre-immobilized onto nanoparticles such as silica, iron oxide, and gold which are larger than mesh size of hydrogels.[57,58] Resultant nanoparticle-immobilized proteins are subsequently entrapped within hydrogel during crosslinking as shown in Fig. 10C.

When proteins are immobilized to the monolayer of tethered functional groups on the surfaces of hard materials via non-specific adsorption or covalent binding, the immobilized proteins closely contact with solid substrates and eventually lose their native structure and function. A solution for the minimization of protein denatures would be use of hydrogel-coated solid substrate or hydrogel itself as supports for protein immobilization, because the soft hydrogels can provide "cushioning effect" as well as large hydration volume of surface.[59,60] Although PEG hydrogels possess a lot of advantages to be used as biomaterial for various applications, the dynamic swelling, soft elasticity, unavailable functional groups, and protein-repelling behavior of PEG hydrogels limit the immobilization of proteins on their surfaces.

Frank *et al.* photochemically modified PEG-based hydrogel to enable covalent linkages with proteins and to facilitate the adhesion of cells.[61] To couple collagen type I to the PEG hydrogel surfaces, the heterobifunctional crosslinker, 5-azidonitrobenzoyloxy N-hydroxysuccinimide (Azide-NHS) was used, which has a phenyl azide group on one end and a protein-binding NHS group on the other (Fig. 11A). Substituted phenyl azide have been shown to react with light to generate aromatic nitrenes, which insert into a variety of covalent bonds. Attachment of the linker to the hydrogel via the phenyl azide group then allows the NHS groups to react with free amines on proteins, and in turn, tether them to the hydrogel surface. This surface modification method can be applied to generate micropatterns on hydrogel surfaces using photomask due to the nature of the photochemically driven reaction.[62] UV exposure of bifunctional linker-coated hydrogel substrates through a photomask can create distinct regions of unexposed PEG hydrogel that are protein repellent and regions of NHS-terminated PEG hydrogel that can covalently immobilize protein. When proteins are incubated with micropatterned PEG hydrogel substrates, proteins are immobilized only on the modified regions, enabling the spatial control of protein immobilization to a micrometer scale on PEG hydrogel substrates as shown in Fig. 11(B).

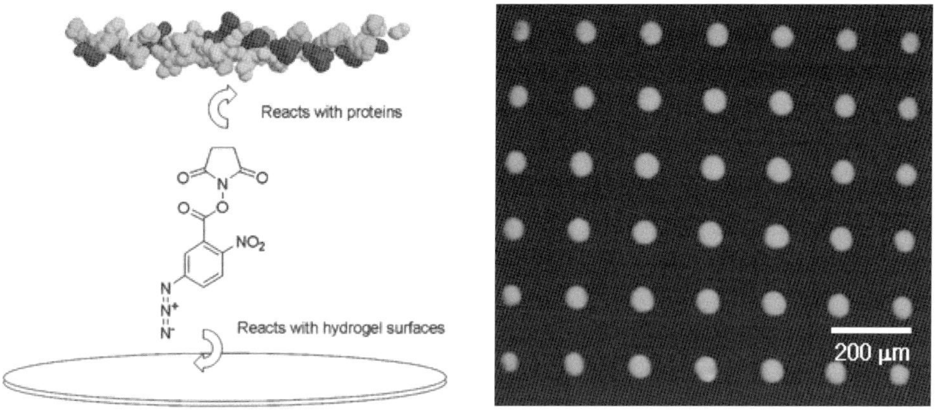

Fig. 11. Generation of protein micropatterns on the surface of PEG hydrogel. (A) Schematic illustration of covalent immobilization of proteins onto PEG hydrogel via heterobifunctional linker[61] (with the permission from Ref. 61, copyright 2007 Springer-Verlag). (B) Fluorescence image of protein micropatterns on PEG hydrogels fabricated by photochemical modification and photopatterning[62] (with the permission from Ref. 62, copyright 2008 Society of Chemical Industry).

PEG hydrogel surfaces can be also modified by grafting poly(acrylic acid) (PAA) via photoinduced surface-initiated polymerization using acrylic acid as a monomer and benzophenone as a surface initiator. Subsequently, the carboxyl groups in PAA are activated using ethyl(dimethylaminopropyl) carbodiimide (EDC) and NHS to covalently immobilize the proteins.[63]

4. Biosensing Using Protein-Immobilized Hydrogel Microstructures

Biosensors are usually classified into metabolism sensors and affinity sensors based on their sensing mechanism. Metabolism sensors are based on analyte recognition plus conversion reaction (substrate consumption and product formation), and affinity sensors use binding between the analytes and immobilized biomolecules without analytes conversion. Although various detection methods have been used for different protein-based assay, this chapter mainly focuses on the absorbance or fluorescence-based optical detection.

4.1. *Metabolism-based bioassay*

Most of enzyme-based biosensors belong to metabolism sensor because the concentration of substrate can be detected by monitoring the product formation through enzyme-catalyzed reactions. Glucose sensor based on specific reaction between glucose and glucose oxidase (GOX) has been widely studied and many researches have been carried out using hydrogel-entrapped GOX. Reaction between glucose and hydrogel-entrapped GOX can be monitored by a colorimetric determination of the amount of the hydrogen peroxide formed, which is based on the oxidation of *o*-dianisidine through a peroxidase (POD)-coupled system as shown in Eq. (1).[56] Therefore, peroxidase (POD) as well as GO_X was immobilized inside hydrogels and solution containing *o*-dianisidine and different concentration of glucose are allowed to react with hydrogel-entrapped enzymes.

Koh *et al.* showed the time dependent absorbance profile and color change of enzyme-containing hydrogels induced by two consecutive enzyme reactions. Due to the transparent nature of PEG hydrogel, color change by enzyme-catalyzed reaction inside hydrogel was optically observed and intensity of absorbance increased with reaction time by the formation of more hydrogen peroxide.[64] If reaction time is fixed, absorbance can be correlated with glucose concentration.

$$\beta\text{-D-glucose} + O_2 + H_2O \xrightarrow{\text{GOX}} \text{D-gluconic acid} + O_2 + H_2O_2,$$
$$H_2O_2 + o\text{-dianisidine (reduced)} \xrightarrow{\text{Peroxidase}} o\text{-dianisidine (oxidized)} + H_2O\cdot \quad (1)$$

Similarly, the reaction between glucose and GO_X can be analyzed via fluorescence detection method using Amplex Red which reacts with hydrogen peroxide in the presence of POD inside the hydrogel microparticles to produce fluorescent resorufin as described in Fig. 12(A). Figure 12(B) shows the suspension array of hydrogel microparticles emitting a strong red color resulting from the sequential bienzymatic reaction.[35] This assay can be also performed within microfluidic devices.[49]

Although it was reported that physical entrapment of enzyme inside the hydrogel network based on a low MW of PEG (for example, MW 575) was sufficient to prevent enzyme leaching and allow diffusion of small substrates such as glucose, using hydrogel synthesized with higher MW of PEG is essential to improve assay performance. That is because larger mesh size of hydrogel prepared from higher MW of PEG allows fast diffusion of substrate into hydrogel matrix, i.e., higher accessibility of the substrate to the active site of the immobilized enzyme, resulting in the lower value of apparent K_m and higher relative activity of entrapped enzyme. However, use of less-crosslinked hydrogels is restricted by the increased leaching of enzymes from the hydrogel. Enzyme leaching problems associated with using hydrogel made from large MW of PEG can be overcome by either covalently

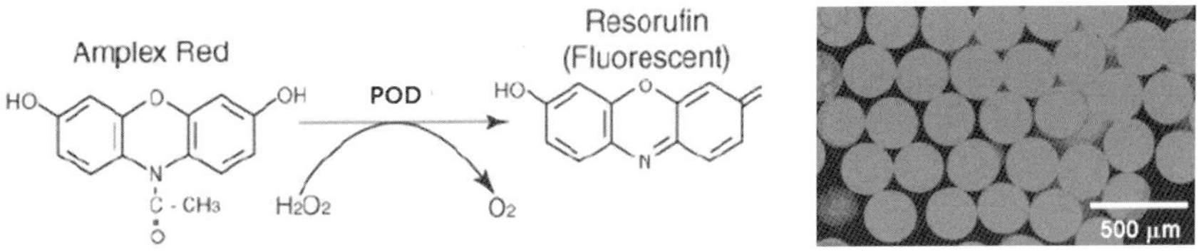

(A) (B)

Fig. 12. (A) Conversion of non-fluorescence Amplex Red into fluorescence resorufin by the reaction between POD and H_2O_2. (B) Fluorescence image of enzyme-entrapped hydrogel microparticles reacted with glucose and Amplex Red [35] (with the permission from Ref. 35 copyright 2008 Springer Science+Business Media).

immobilizing enzyme into hydrogel matrix using bifunctional linkers such as acryloyl-PEG-NHS or by entrapping enzyme that was covalently-immobilized onto nanoparticles that is larger than mesh size of hydrogels as explained earlier. For example, Koh *et al.* prepared hydrogel microstructures entrapping enzyme-linked silica or magnetic nanoparticles.[57,58] By using nanoparticle-immobilized enzymes, it was possible to use hydrogel based on the higher MW of PEG without enzyme leaching, which could improve the reaction rate of entrapped enzymes due to reduced mass transport resistance.

The detection approaches based on *o*-dianisidine and Amplex Red can be extended to other biosensors based on not only GO_X but also on any H_2O_2-generating oxidase such as galactose oxidase, cholesterol oxidase, and alcohol oxidase. Jang *et al.* demonstrated that glucose and alcohol could be successfully detected with enzyme-entrapping hydrogel microparticles using Amplex Red.[65] Furthermore, they showed the simultaneous detection of glucose and alcohol in a single microchannel using shape coded hydrogel microparticles entrapping GO_X/POD (circle) and AOX/POD (square) as shown in Fig. 13(A). When a solution containing either glucose or ethanol was injected into a microchannel, only the hydrogel microparticles entrapping a corresponding enzyme emitted red fluorescence as shown in Fig. 13(B). On the other hand, introduction of a solution containing both glucose and ethanol resulted in fluorescence from both hydrogel microparticle shapes (Fig. 13(B)).

Monitoring pH is another detection method for enzyme-based assays, because most of enzyme-catalyzed reactions accompany pH change. Pishko *et al.* utilized carboxyl seminaphthofluorescein (SNAFL), a ratiometric pH-dependent fluorophore, for the analysis of different enzyme-catalyzed reaction.[48] In their study, SNAFL-conjugated alkaline phosphatase (AP) was incorporated in hydrogel

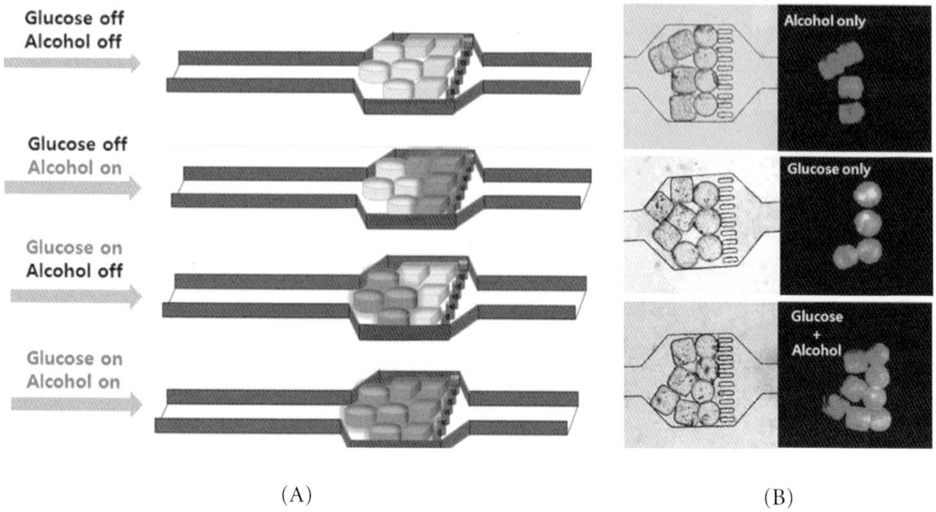

Fig. 13. Analysis of multiple enzyme-based reaction using shape-coded hydrogel microparticles within microfluidic device.[65] (A) Assay logic to detect glucose and alcohol simultaneously (B) Optical and fluorescence image of enzyme-entrapped hydrogel microparticles that were reacted with different sample composition (with the permission from Ref. 65, copyright 2009 Elsevier).

microstructures inside microfluidic channels and pNPP as a substrate was introduced to microchannels. The reaction between AP and pNPP decreased the microenvironment pH in the hydrogel through the production of phosphoric acid as a result of the enzyme-catalyzed reaction. On the other hands, the reaction of hydrogel-entrapped urease upon exposure to urea caused the increase of pH due to the formation NH_4^+. These pH changes resulted from two enzyme-catalyzed reactions caused the change of fluorescence intensity of SNAFL, which could be correlated with concentration of substrates.

Semiconductor quantum dots (QDs) are widely used in the field of optical biosensors as alternatives to organic fluorescence dyes due to their superior optical properties such as high quantum yield, long fluorescence lifetime, large extinction coefficient, and stability against photobleaching.[66–68] Most of QD-based protein assays still rely on the fluorescence resonance energy transfer (FRET) processes,[69–71] but electron transfer quenching has been recently proposed as a new detection method for enzyme-based metabolism sensor since Willner *et al.* reported the possibility of monitoring tyrosinase activity via the quinone-induced electron transfer quenching of QDs.[72,73] Different optical biosensors have been developed using different H_2O_2-generating oxidases based on the fact that H_2O_2, a product of numerous oxidases, quenches the photoluminescence of QDs by the electron

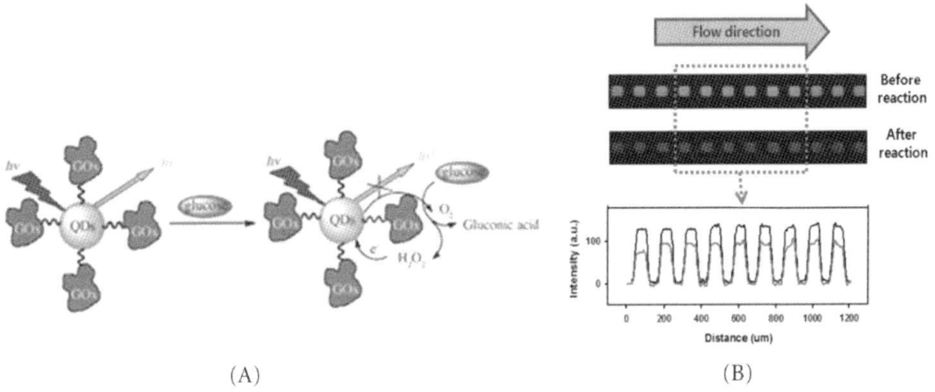

Fig. 14. (A) Mechanism that H_2O_2 quenches the photoluminescence of QD by the electron transfer reaction[76] (with the permission from ref 74, copyright 2008 WILEY-VCH) (B) Change of fluorescence intensity by the reaction between hydrogel-entrapped QD-GOX conjugates and glucose within microchannels[79] (with the permission from Ref. 79, copyright 2011 Elsevier).

transfer reaction as described in Fig. 14(A).[74–78] Enzymes are usually covalently conjugated to QDs and resultant QD-enzyme hybrid systems are directly used for sensing assays in solution state or immobilized in appropriate solid support for more useful sensing approaches. The QD-enzyme conjugates can be also entrapped within hydrogel microstructures that were fabricated within the microchannels by a photopatterning process.[79] For example, hydrogel-entrapped GO_X are able to perform enzyme-catalyzed oxidation of glucose to produce H_2O_2, which subsequently quenched the fluorescence of the conjugated QDs as shown in Fig. 14(B). The fluorescence intensity of the hydrogel microstructures decreases as the glucose concentrations increased.

4.2. *Affinity-based bioassay*

Biotin-streptavidin assay, antigen-antibody assay, and competitive binding assay are examples of affinity sensor that can be used for protein-based assay.

The strong streptavidin-biotin binding can be used to attach various proteins to one another or onto a solid support for the bioassay purpose. Molecular recognition-mediated specific binding between biotin and streptavidin was successfully assayed using microarrays on PEG hydrogels that were photochemically modified with heterobifunctional crosslinker or grafted with PAA by photoinduced surface-initiated polymerization.[62,63]

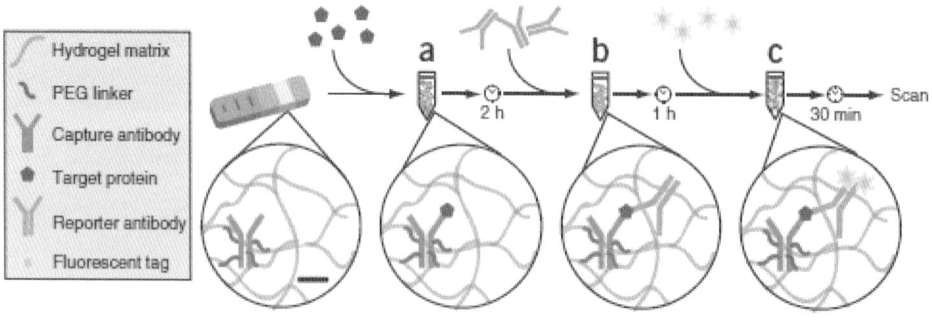

Fig. 15. Schematic illustration of the sandwich immunoassay using graphically-coded PEG hydrogel microparticles [39] (with the permission from Ref. 39, copyright 2011 Nature Publishing Group).

Hydrogel-entrapped antibodies can be also used for various bioassays via specific binding with target antigens. Doyle *et al.* demonstrated multiplexed immunoassay with graphically encoded PEG hydrogel microparticles as illustrated in Fig. 15.[39] In their system, capture antibodies were covalently incorporated within PEG hydrogel matrix via heterobifunctional PEG linker. The assay involves assembling an antibody sandwich around the target protein by exposing particles to the sample, adding a biotinylated reporter antibody and labeling the bound reporters with a streptavidin-phycoerythrin (SAPE) complex.[80]

Unlike enzyme-based assay where target molecules are usually small enough to diffuse into hydrogel, target molecules of most immunoassay are large proteins that do not penetrate PEG hydrogel network easily. The diffusion of proteins into the interior of PEG hydrogel can be enhanced by the use of porogens. Palecek *et al.* developed macroporous PEG hydrogel by the use of porogen.[81] They demonstrated that proteins immobilized in the interior of these macroporous PEG hydrogels retained their structure and were accessible for interaction with antibodies, in contrast with proteins immobilized in non-porous PEG hydrogels formed without porogens, which remained inaccessible to large proteins.

Most well-known competitive binding assay is fluorescence-based glucose detection by exploiting the concanavalin A (Con A) affinity for polysaccharides. Russell *et al.* developed fluorescence-based glucose biosensor based on a PEG hydrogel incorporating fluorescein isothiocyanate dextran (FITC-dextran) and tetramethylrhodamine isothiocyanate-Con A (TRITC-Con A) chemically conjugated into the hydrogel network via bifunctional PEG linkers. In the absence of glucose, TRITC-Con A binds with FITC-dextran, and the FITC fluorescence is quenched through FRET. Competitive glucose binding to TRITC-Con A liberates FITC-dextran, resulting in increased FITC fluorescence proportional to the glucose concentration.[82]

5. Conclusions and Outlook

Protein-based biosensors constitute a promising field that has numerous applications ranging from pharmaceutical screening to detection of pathogen and toxicant. The trends toward miniaturization of protein-based biosensor continue to spur development of protein microarray integrated into microfluidic devices. In spite of recent advance in microarray technologies, the formation of protein microarrays capable of fully understand protein function has so far been a tremendous challenge due to the fact that proteins need to be in a wet environment in order to remain structurally intact and maintain their activity. Immobilization of proteins within hydrogel materials would be the best solution to overcome this barrier because the environment that surrounds proteins can be nearly water. Among various hydrogels, this chapter mainly focused on the protein-incorporated PEG hydrogel microarrays and their application to fluorescence based- assays. Future efforts should be made on the development of well-designed hydrogel materials and detection strategies to provide better environment to proteins and to achieve improved assay performance such as sensitivity and detection limit.

Acknowledgments

This work was supported by the National Research Foundation (NRF) grant (NRF-2007-0056566 "Active Polymer Center for Pattern Integration at Yonsei University" and NRF-2011-0022709).

References

1. M. Tyers and M. Mann, *Nature*, **422**, 193 (2003).
2. P. Arenkov, A. Kukhtin, A. Gemmell, S. Voloshchuk, V. Chupeeva, and A. Mirzabekov, *Anal. Biochem.*, **278**, 123 (2000).
3. M. Hartmann, J. Roeraade, D. Stoll, M. Templin, and T. Joos, *Anal. Bioanal. Chem.*, **393**, 1407 (2009).
4. Y. Hu, M. Uttamchandani, and S. Q. Yao., *High Throughput Screen.*, **9**, 203 (2006).
5. G. MacBeath and S. L. Schreiber, *Science*, **289**, 1760 (2000).
6. P. Mitchell, *Nat. Biotechnol.*, **20**, 225 (2002).
7. J. D. Hoheisel, *Nat. Rev. Genet.*, **7**, 200 (2006).
8. D. S. Wilson and S. Nock, *Angew. Chem. Int. Ed.*, **42**, 494 (2003).
9. S. Kiyonaka, K. Sada, I. Yoshimura, S. Shinkai, N. Kato, and I. Hamachi, *Nat. Mater.*, **3**, 58 (2004).
10. S. Zhang, *Nat. Mater.*, **3**, 7 (2004).
11. V. A. Li and S. N. Bhatia, *Biomed. Microdevices*, **4**, 257 (2002).
12. C. P. Pathak, A. S. Sawhney, and J. A. Hubbell, *J. Am. Chem. Soc.*, **114**, 8311 (1992).

13. R. J. Russell, M. V. Pishko, A. L. Simonian, and J. R. Wild, *Anal. Chem.*, **71**, 4909 (1999).
14. G. M. Cruise, D. S. Scharp, and J. A. Hubbell, *Biomaterials*, **19**, 1287 (1998).
15. M. S. Hahn, L. J. Taite, J. J. Moon, M. C. Rowland, K. A. Ruffino, and J. L. West, *Biomaterials*, **27**, 2519 (2006).
16. S. H. Lee, J. J. Moon, and J. L. West, *Biomaterials*, **29**, 2962 (2008).
17. Y. Hong, P. Krsko, and M. Libera, *Langmuir*, **20**, 11123 (2004).
18. T. O. Joos, D. Stoll, and M. F. Templin, *Curr. Opin. Chem. Biol.*, **6**, 76 (2001).
19. H. Zhou, S. Roy, H. Schulman, and M. J. Natan, *Trends Biotechnol.*, **19**, S34 (2001).
20. M. F. Templin, D. Stoll, J. M. Schwenk, O. Potz, S. Kramer, and T. O. Joos, *Proteomics*, **3**, 2155 (2003).
21. A. Revzin, R. J. Russell, V. K. Yadavalli, W. Koh, C. Deister, D. D. Hile, M. B. Mellott, and M. V. Pishko, *Langmuir*, **17**, 5440 (2001).
22. W. G. Koh and M. Pishko, *Langmuir*, **19**, 10310 (2003).
23. K. Y. Suh, Y. S. Kim, and H. H. Lee, *Adv. Mater.*, **13**, 1386 (2001).
24. K. Y. Suh, M. C. Park, and P. Kim, *Adv. Funct. Mater.*, **19**, 2699 (2009).
25. V. K. Yadavalli, W. G. Koh, G. J. Lazur, and M. V. Pishko, *Sensor. Actuat. B*, **97**, 290 (2004).
26. J. P. Nolan and F. F. Mandy, *Cell. Mol. Biol.*, **47**, 1241 (2001).
27. J. P. Nolan and L. A. Sklar, *Trends Biotechnol.*, **20**, 9 (2002).
28. M. Y. Han, X. H. Gao, J. Z. Su, and S. Nie, *Nat. Biotechnol.*, **19**, 631 (2001).
29. W. J. Jeong, J. Y. Kim, J. Choo, E. K. Lee, C. S. Han, D. J. Beebe, G. H. Seong, and S. H. Lee, *Langmuir*, **21**, 3738 (2005).
30. H. Kawaguchi, *Prog. Polym. Sci.*, **25**, 1171 (2000).
31. K. Kesenci and E. Piskin, *Macromol. Chem. Phys.*, **199**, 385 (1998).
32. B. Kim and N. A. Peppas, *Biomed. Microdevices.*, **5**, 333 (2003).
33. N. A. Peppas and M. Torres-Lugo, *J. Nanopart. Res.*, **4**, 73 (2002).
34. J. E. Meiring, M. J. Schmid, S. M. Grayson, B. M. Rathsack, D. M. Johnson, R. Kirby, R. Kannappan, K. Manthiram, B. Hsia, Z. L. Hogan, A. D. Ellington, M. V. Pishko, and C. G. Willson, *Chem. Mater.*, **16**, 5574 (2004).
35. W. Lee, D. Choi, J. H. Kim, and W. G. Koh, *Biomed. Microdevices*, **10**, 813 (2008).
36. D. Dendukuri, S. S. Gu, D. C. Pregibon, T. A. Hatton, and P. S. Doyle, *Lab Chip*, **7**, 818 (2007).
37. D. Dendukuri, D. C. Pregibon, J. Collins, T. A. Hatton, and P. S. Doyle, *Nat. Mater.*, **5**, 365 (2006).
38. D. Dendukuri and P. S. Doyle, *Adv. Mater.*, **21**, 4071 (2009).
39. D. C. Appleyard, S. C. Chapin, R. L. Srinivas and P. S. Doyle, *Nat. Protoc.*, **6**, 1761 (2011).
40. Y. L. Bai, C. G. Koh, M. Boreman, Y. J. Juang, I. C. Tang, L. J. Lee, and S. T. Yang, *Langmuir*, **22**, 9458 (2006).
41. A. Dodge, K. Fluri, E. Verpoorte, and N. F. de Rooij, *Anal. Chem.*, **73**, 3400 (2001).
42. E. Eteshola and D. Leckband, *Sensor Actuat. B*, **72**, 129 (2001).
43. H. Mao, T. Yang, and P. S. Cremer, *Anal. Chem.*, **74**, 379 (2002).

44. J. Yakovleva, R. Davidsson, A. Lobanova, M. Bengtsson, S. Eremin, T. Laurell, and J. Emneus, *Anal. Chem.*, **74**, 2994 (2002).
45. D. J. Beebe, J. S. Moore, J. M. Bauer, Q. Yu, R. H. Liu, C. Devadoss, and B. H. Jo, *Nature*, **404**, 588 (2000).
46. J. Heo and R. M. Crooks, *Anal. Chem.*, **77**, 6843 (2005).
47. W. Zhan, G. H. Seong, and R. M. Crooks, *Anal. Chem.*, **74**, 4647 (2002).
48. W. G. Koh and M. Pishko, *Sensor. Actuat. B*, **106**, 335 (2005).
49. D. Choi, E. Jang, J. Park, and W. G. Koh, *Microfluid. Nanofluid.*, **5**, 703 (2008).
50. S. S. Betigeri and S. H. Neau, *Biomaterials*, **23**, 3627 (2002).
51. E. Kalogeris, Y. Sanakis, D. Mamma, P. Christakopoulos, D. Kekos, and H. Stamatis, *Enzyme Microb. Technol.*, **39**, 1113 (2006).
52. S. S. Kanwar, S. Pathak, H. K. Verma, S. Kumar, R. Gupta, S. S. Chimni, and G. S. Chauhan, *J. Appl. Polym. Sci.*, **100**, 4636 (2006).
53. N. Pekel, B. Salih, and O. Guven, *J. Biomater. Sci.-Polym. Ed.*, **16**, 253 (2005).
54. R. J. Russell, A. C. Axel, K. L. Shields, and M. V. Pishko, *Polymer.*, **42**, 4893 (2001).
55. S. A. DeLong, J. J. Moon, and J. L. West, *Biomaterials*, **26**, 3227 (2005).
56. D. Choi, W. Lee, Y. Lee, D. N. Kim, J. Park, and W. G. Koh, *J. Chem. Technol. Biotechnol.*, **83**, 252 (2008).
57. E. Jang, S. Park, S. Park, Y. Lee, D. N. Kim, B. Kim, and W. G. Koh, *Polym. Adv. Technol.*, **21**, 476 (2010).
58. S. Park, Y. Lee, D. N. Kim, S. Park, E. Jang, and W. G. Koh, *React. Funct. Polym.*, **69**, 293 (2009).
59. M. R. Burnham, J. N. Turner, D. Szarowski, and D. L. Martin, *Biomaterials*, **27**, 5883 (2006).
60. M. Zourob, J. E. Gough, and R. V. Ulijn, *Adv. Mater.*, **18**, 655 (2006).
61. D. Myung, W. Koh, A. Bakri, F. Zhang, A. Marshall, J. M. Ko, J. Noolandi, M. Carrasco, J. R. Cochran, C. W. Frank, and C. N. Ta, *Biomed. Microdevices.*, **9**, 911 (2007).
62. D. N. Kim, W. Lee, and W. G. Koh, *J. Chem. Technol. Biotechnol.*, **84**, 279 (2009).
63. W. J. Lee, D. K. Choi, Y. Lee, D. N. Kim, J. W. Park, and W. G. Koh, *Sensor. Actuat. B*, **129**, 841 (2008).
64. D. Choi, W. Lee, J. Park, and W. Koh, *Bio-Med. Mater. Eng.*, **18**, 345 (2008).
65. E. Jang and W. G. Koh, *Sensor. Actuat. B*, **143**, 681 (2010).
66. J. M. Costa-Fernández, R. Pereiro, and A. Sanz-Medel, *TrAC Trends Anal. Chem.*, **25**, 207 (2006).
67. M. Frasco and N. Chaniotakis, *Anal. Bioanal. Chem.*, **396**, 229 (2010).
68. C. Han and H. Li, *Anal. Bioanal. Chem.*, **397**, 1437 (2010).
69. H. Mattoussi, A. R. Clapp, I. L. Medintz, H. T. Uyeda, B. R. Fisher, E. R. Goldman, and M. G. Bawendi, *J. Am. Chem. Soc.*, **127**, 18212 (2005).
70. J. M. Mauro, I. L. Medintz, A. R. Clapp, H. Mattoussi, E. R. Goldman, and B. Fisher, *Nat. Mater.*, **2**, 630 (2003).
71. Z. Rosenzweig, L. F. Shi, V. De Paoli, and N. Rosenzweig, *J. Am. Chem. Soc.*, **128**, 10378 (2006).

72. R. Gill, R. Freeman, J.-P. Xu, I. Willner, S. Winograd, I. Shweky, and U. Banin, *J. Am. Chem. Soc.*, **128**, 15376 (2006).
73. M. G. Sandros, D. Gao, and D. E. Benson, *J. Am. Chem. Soc.*, **127**, 12198 (2005).
74. L. Cao, J. Ye, L. Tong, and B. Tang, *Chem. Euro. J.*, **14**, 9633 (2008).
75. H. D. Duong and J. I. Rhee, *Talanta*, **73**, 899 (2007).
76. R. Gill, L. Bahshi, R. Freeman, and I. Willner, *Angew. Chem. Int. Ed.*, **120**, 1700 (2008).
77. M. Hu, J. Tian, H. T. Lu, L. X. Weng, and L. H. Wang, *Talanta*, **82**, 997 (2010).
78. C.-P. Huang, S.-W. Liu, T.-M. Chen, and Y.-K. Li, *Sensors Actuat. B*, **130**, 338 (2008).
79. E. Jang, S. Kim, and W. G. Koh, *Biosens. Bioelectron.*, **31**, 529 (2012).
80. D. C. Appleyard, S. C. Chapin, and P. S. Doyle, *Anal. Chem.*, **83**, 193 (2011).
81. A. G. Lee, C. P. Arena, D. J. Beebe, and S. P. Palecek, *Biomacromolecules*, **11**, 3316 (2010).
82. R. J. Russell, M. V. Pishko, C. C. Gefrides, M. J. McShane, and G. L. Cote, *Anal. Chem.*, **71**, 3126 (1999).

Chapter 12

Cell-Encapsulating Hydrogels for Biosensing

Pu Chen[1,*], Shuqi Wang[1,2,3,4,*], Fatih Inci[1], Sinan Güven[1,#], Savas Tasoglu[1,†], and Utkan Demirci[1]

[1]*Bio-Acoustic MEMS in Medicine (BAMM) Lab, Department of Radiology, School of Medicine, Canary Center for Early Cancer Detection, Stanford University, CA, 94304*

[2]*State Key Laboratory for Diagnosis and Treatment of Infectious Diseases, First Affiliated Hospital, College of Medicine, Zhejiang University, Hangzhou, Zhejiang, 310003, China*

[3]*Collaborative Innovation Center for Diagnosis and Treatment of Infectious Diseases, Hangzhou, Zhejiang, 310003, China*

[4]*Institute for Translational Medicine, Zhejiang University, Hangzhou, Zhejiang, 310029, China*

** These authors contribute equally to this work.*

[#]Present Address: Izmir Biomedicine and Genome Center, Dokuz Eylul University, 35340, Izmir — Turkey

[†]Present Address: Departments of Mechanical and Biomedical Engineering, University of Connecticut, Storrs, 06269 CT

Cell-based biosensors (CBBs) are emerging as a sensing platform in which live cells are utilized to sense external stimuli including physical, chemical, and biological changes. Till now, CBBs have demonstrated a broad range of applications including diagnostics, drug screening, environment monitoring, and biosafety monitoring. Although promising, current CBBs are normally based on cells cultured in two-dimensional (2D) surfaces, which brings challenges such as ease of contamination and limited capability for long-time preservation. In addition, cells grown in 2D culture environments cannot fully represent microenvironment in three-dimensional (3D) native tissues and may result in analytical variations such as in drug screening. One potential strategy to overcome these challenges is to incorporate cells in 3D hydrogels, which provide cytocompatible microenvironments for prolonged cell preservation and improved phenotypic similarity with cells in native tissues. In this chapter, we present an overview of cell-encapsulating hydrogel based biosensors (CHBBs) and highlight the unique features of CHBBs as opposed to traditional CBBs.

1. Introduction

Cell-based biosensors (CBBs), also named as whole-cell biosensors, are analytical devices incorporating whole cells (e.g., native cells, genetically modified cells, synthetic cells, etc.) with a signal transducer (e.g., electronics, optics).[1–7] In CBBs, cells function as sensing units that interact with analytes and respond in a cytophysiological manner. The transducer then converts cellular responses into readable signals. By correlating these signal readouts with the original stimulus, characterization of biological samples such as sample type and target concentration can be quantitatively or qualitatively obtained (Fig. 1).

In most of CBBs, cells are patterned/cultured on two-dimensional (2D) surfaces. Although surface-based cell culture brings advantages such as ease of implementation, and compatibility for optical/electric detection, they are associated with issues such as ease of contamination for field applications, and incompatibility for long-time preservation of mammalian cells or for immobilization of bacteria. Moreover, when cells are cultured in a monolayer, they exhibit significant differences in phenotypes compared to cells in native tissues. Hence, cells on 2D surfaces do not effectively represent cells in complex three-dimensional (3D) tissue environments.[8–10]

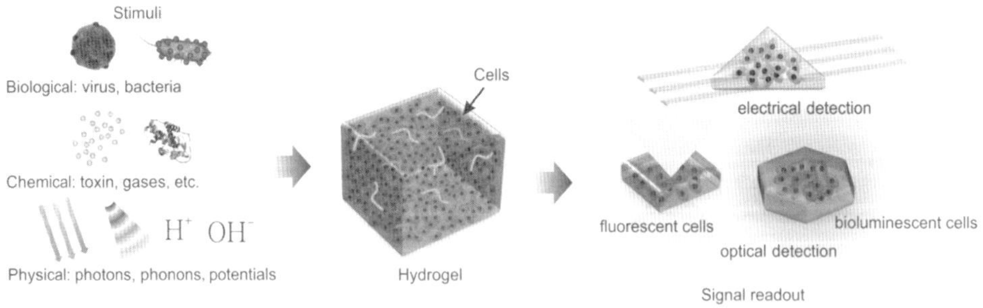

Fig. 1. Schematic of a cell-encapsulating hydrogel-based biosensor. CHBBs are composed of three basic components, namely, cell, hydrogel and signal transducer. Cells function as a sensing unit in response to physical, chemical and biological stimuli in the surrounding microenvironment. Hydrogels provide a 3D ECM. Cellular responses are converted to readable signals by optical/electronic transducers.

Hydrogels (such as collagen, fibrin, alginate, and polyethylene glycol (PEG)) are extensively used as biomaterials to mimic extracellular matrix (ECM) due to their cytocompatibility, moldability, tunability in mechanical properties, high-water content, and porosity.[10–12] The strategy of encapsulating cells within 3D hydrogel constructs has been successfully used in multiple applications such as creating 3D tissue structures for bottom-up tissue engineering, minimizing contamination, and reducing immune rejection for cell delivery *in vivo*.[10,13–16] By bringing together hydrogels, cells, and signal transducers, cell-encapsulating hydrogel-based biosensors (CHBBs) present improved functionality similar to cells in native tissues and are promising to become a new paradigm for CBBs. In this chapter, we first discuss three basic components of CHBBs, namely, hydrogels, cells and signal readout methods, and we then present various applications of CHBBs. Emphasis is placed on the unique features of CHBBs compared to traditional CBBs. Current challenges and their potential solutions of CHBBs are also discussed.

2. Components of Cell-Encapsulating Hydrogels

2.1. *Hydrogels*

2.1.1. *Hydrogel materials*

Hydrogels have attracted tremendous research interest over years in both scientific and industrial applications because of high-water content and biocompatibility. Cell-encapsulating hydrogels provide mimicries for 3D ECM, which improves cell viability, proliferation and maintains metabolic activities such as secretion of cell specific cytokines. Hydrogels can be categorized into natural and synthetic hydrogels.[16,17] Natural hydrogels are extracted from animal tissues such as collagen, gelatin, fibrin, hyaluronic acid, chitosan, as well as plants such as seaweeds. Compared to natural hydrogels, synthetic hydrogels provide a better control over reproducibility with minimal batch-to-batch variations, which is critical in biosensing applications. PEG, poly(vinyl alcohol) and poly(hydroxyethyl methacrylate) are the most widely used synthetic polymers for cell encapsulation.[18] Here, we discuss hydrogel materials that have been demonstrated for 3D cell encapsulation for biosensing applications.

Collagen — Collagen is the most abundant protein in ECM, comprising approximately 25% to 35% of the total protein mass in animal body.[19] Collagen is among the most popular natural hydrogels for cell encapsulation due to its biocompatibility and availability. Cell homing and bio-mimicry of the collagen hydrogels provide highly sensitive cellular responses in biosensor studies. Collagen type-I is the most widely used among all collagen types. Polymeric structure of the collagen has peptide domains with amino acid sequences such as Asp–Gly–Glu–Ala (DGEA) and Gly–Phe–Gln–Glu–Arg (GFQGER) that regulate cell adhesion and phenotypic activities. However, collagen hydrogels have poor mechanical properties. They are soft with elastic moduli, ~5 kPa. Mixing collagen with other hydrogels is a common practice to enhance the mechanical rigidity as well as specializing hydrogels for culturing sensitive cell types such as neurons.[20] Due to its biocompatibility in biosensing applications, collagen has been used to encapsulate a wide variety of cells (e.g., stem cells and cancer cells). In addition, collagen is used to coat the surface of biosensor rods to increase biocompatibility and to prevent immune rejection for *in vivo* transplantation.[22]

PEG — PEG is a synthetic hydrophilic polymer with good biocompatibility, non-immunogenicity, and high resistance to protein adsorption. These properties make PEG one of the most popular synthetic polymers in medicine and biomedical applications. PEG can be synthesized in different chain lengths for specific purposes. PEG can be chemically modified with methacrylate groups, which becomes photo-crosslinkable, flexible to form the desired shape and spatial organization. Further, PEG hydrogels can be modified with various biological agents such as short peptides (i.e., Arg–Gly–Asp, RGD), enzymes (i.e., matrix metallopeptidase 13, MMP-13), growth factors (i.e., vascular endothelial growth factor, VEGF) or polymers to induce degradation (i.e., disulfide group, polyester, and fumarate). The mechanical properties of PEG hydrogels vary by molecular weight and crosslinking efficiency. The synthesis and modifications of PEG hydrogels can be standardized to enhance reproducibility. In this regard, biologically modified PEG hydrogels are promising materials for encapsulating cells for biosensing applications.[23] Cell-encapsulating PEG hydrogels have been developed as oxygen sensors utilizing fluorescent particles.[24] Moreover, PEG hydrogels can be utilized to respond to changes in pH[24] or cytokines[26] as well as to study the binding kinetics of antibodies.[27,28]

Alginate — Alginate, also known as alginic acid, is derived from brown algae. It is an anionic polysaccharide with high-water absorption capacity. Alginate is biocompatible and is widely used in food industry, drug delivery, and wound dressing applications. In long-term cell encapsulation studies, biological properties of alginate might be tuned with addition of peptides and/or growth factors to generate more favorable microenvironments for cells. RGD peptides are widely used to promote cell adhesion and to functionalize encapsulated cells. Due to its easy handling and crosslinking properties, alginate also finds applications in cell-encapsulating hydrogels for biosensing. Due to its anionic nature, alginate is sensitive to electric changes, which makes it a good candidate for electric sensing. In industrial applications, fungal cells have been encapsulated in alginate to detect changes in pH,[29] and fibroblast encapsulating hydrogels have been used for detection of specific chemical agents.[30]

Agarose — Agarose is a polysaccharide derived from seaweed and is one of the main components of agar. The stiffness and porosity of agarose gels can be tuned by varying densities and concentrations. It is mostly used in gel electrophoresis to separate proteins and nucleic acids. Cell-encapsulating agarose can be used for migration assays and growing bacterial cells. Agarose hydrogels create favorable microenvironments for bacterial cells to detect chemical contaminations in drinking water.[31]

2.1.2. *Design and fabrication of hydrogels for biosensing applications*

The design of a cell-encapsulating hydrogel should take into account that the hydrogel should be biocompatible, porous and provide an essential for the cell survival and functionality. Besides, the hydrogel should not interfere with the read out signal, introducing background signal noise or luminescence shielding of the cell response. In addition, hydrogels should be highly standardized, and the batch-to-batch variability should be kept at a minimum range. Commercialization efforts for such systems require the accessibility, simplicity in use, cost efficiency and high throughput in mass production as well as long shelf life. Merging more than one biosensing features to biosensor design can significantly promote the wide applicability of the biosensing devices.

Assembly approaches — Fabrication techniques for cell-encapsulating hydrogels use a similar approach to have cells suspended in hydrogel prepolymer solution followed by crosslinking. The general cell-encapsulating approach is to mix the cells with the liquid form of polymer, and crosslink by temperature (e.g., collagen, agarose), chemicals (e.g., crosslinking alginate in calcium chloride solution) or with ultraviolet (UV) light (e.g., methacrylated PEG). There is a growing interest in engineering 3D living constructs from microscale cell-encapsulating hydrogel units[16] using different assembly principles such as magnetic,[32–35] acoustic,[36] surface tension,[37] liquid-based template,[38] ratchet-based,[39] microfluidic,[40–42] and robotic principles.[43] Advantages and disadvantages of these microscale assembly approaches were discussed in previous chapters.[44–48]

Bioprinting — Bioprinting emerged as a powerful method to pattern cell-encapsulating droplets on a receiving surface for biological applications, including tissue-like structures[49] and biosensors,[50–52] or cell-free polymer blocks for numerous applications such as 3D printed micro-battery[53] or bionic ear.[54] Here, we focus on the use of bioprinting technologies in developing cell-based biosensors, and illustrate underlying mechanisms of bioprinting to deposit cell-encapsulating droplets. Bioprinting offers high throughput, computerized xyz controlled deposition of multiple bioactive factors and cells to a receiving substrate. Bioprinting has been demonstrated in several applications, including tissue engineering, microphysiological system engineering, as well as biosensor fabrication.[55] With these capabilities, bioprinting can reduce the cost of fabricating living biosensors. A biosensor platform was presented with a lensless charge-coupled device (CCD) and bioprinted smooth muscle cells on a microfluidic chip.[50,51] The CCD was utilized to evaluate cellular changes in morphology and viability to an external stimulus. Cell alignment and orientation were quantified in seconds without labeling. Recently, bioprinting has been also used to pattern mouse myotubes onto micronized cantilevers and ultimately to fabricate muscle-powered biosensors[52] (Fig. 2A). Results showed that mature myotubes were formed only after 4 days in the bioprinted samples, whereas the same process took >14 days in the control group of randomly seeded cells. Synchronous responses of cells to electrical and chemical stimuli were also investigated. These results are promising to demonstrate fabrication of biosensors at high throughput via bioprinting.

Computational simulations can help to understand how parameter, prior to actual experimentation, affect cell encapsulation or post-printing cell proliferation and viability.[56–58] Encapsulation of a cell or multiple cells into pico/nano-liter droplets is a probabilistic phenomenon[59,60] (Fig. 2B). Statistical models were presented to provide an understanding of cell-encapsulation process.[59] Results were presented as probability distributions, P(Xt) as a function of different target cell densities and types of cell loading (Fig. 2C). These statistical functions evaluated the probability of a single target cell suspended in a heterogeneous cell mixture to be encapsulated in an ejected droplet. While the ratio of target cells and homogeneity decreased in cell suspension, each probability function followed a Poisson distribution (Fig. 2C).

During the landing of ejected cell-laden droplets, mechanical aspects such as hydrodynamic pressure, capillary forces, and shear stress, might cause deformation of droplet and cell surface[61,62] (Figs. 2D and 2E). Eventually, this process can lead to programmed cell death (apoptosis). However, these experimental dynamics can be adjusted by decreasing/increasing ejection velocity or by replacing encapsulating fluids with better combination of material properties. Cell fate may depend on the hydrophobicity/hydrophilicity of the receiving surface, which is highly correlated with contact angle between droplet medium and surface. Prediction of cell fate using simulations can provide more parametric control over biosensor designs as well as complex viable tissue surrogate designs.[32–36,44,63] A finite-difference/front-tracking simulation model was reported for deposition of viscous compound droplets onto a receiving surface as a model for cell printing process.[64–66] By simulation, physical parameters were optimized for minimum cell deformation. Analyses were carried out for a set of dimensionless parameters, i.e., Weber

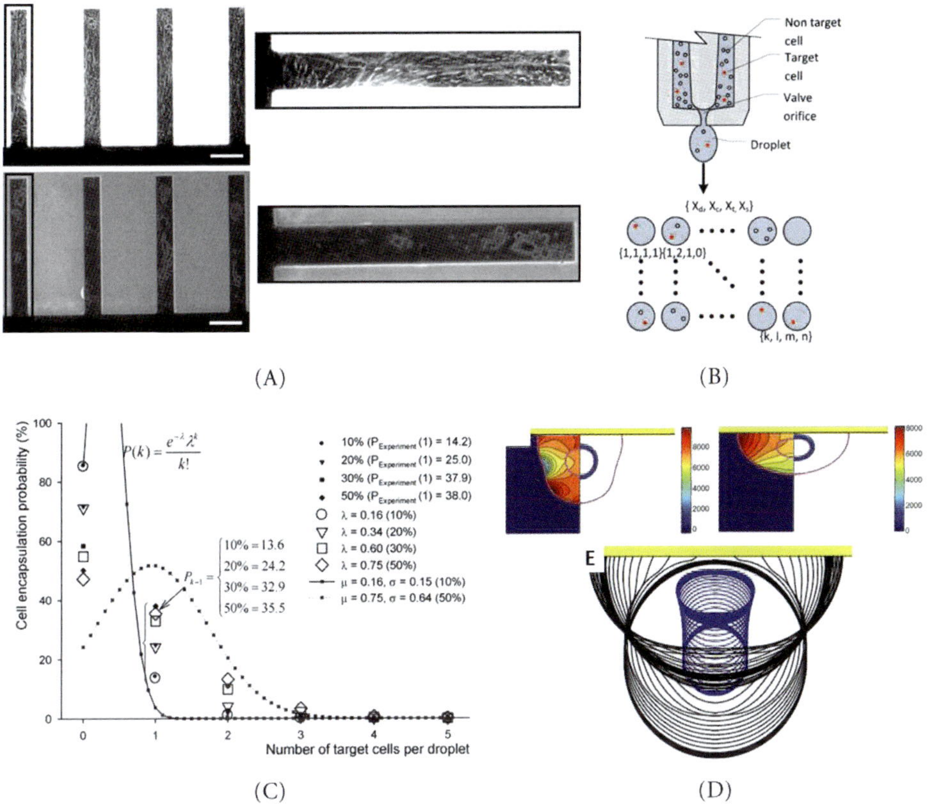

Fig. 2. (A) Biosensor fabrication via bioprinting cells onto cantilevers. Mouse muscle myoblast cells bioprinted on cantilevers. (top) Bioprinted cells formed myofibers on all cantilevers after 4 days in culture. (bottom) Non-printed cells randomly distributed on cantilevers and no myofibers were observed after 7 days. Scale bars 200 μm. (B–E) Statistical and computational modeling of cell encapsulation and printing process. (B) A droplet ejector was filled with heterogeneous mixture, including target and non-target cells for random cell encapsulation process. (X_d) the number of droplets that contain cells, (X_c) number of cells per droplet, (X_t) number of target cells, and (X_s) droplets encapsulating a single target cell, were mapped onto a matrix of cell-encapsulating droplets. (C) Cell encapsulation probability, P(X_t), as a function of number of target cells per droplet for cell concentration = 1.5 × 10^5 cells ml^{-1}. (D) Simulation of cell printing process. Pressure contours and pressure distribution on the cell were plotted at the left half and the right half, respectively. Governing non-dimensional numbers are: We = 0.5, Re = 30, $d_o/d_i = 2.85$, $\sigma_o/\sigma_i = 2541$, $\mu_c/\mu_d = 10$. (E) Sequential impact images of cell-encapsulating droplet. *(A) is reproduced with permission,*[52] *(B) with permission,*[59] *(C) with permission,*[66] *and (D, E) with permission.*

number (We), diameter ratio (d_o/d_i), viscosity ratio (μ_c/μ_d), Reynolds number (Re), surface tension ratio (σ_o/σ_i), and equilibrium contact angle (θ_e). We and Re are widely-used non-dimensional numbers in fluid dynamics[67] to evaluate the influence of inertial forces compared to surface tension and viscous forces, respectively.[68–70] The computational results demonstrated that the geometrical deformation of cell monotonically increased: (i) as d_o/d_i decreased; (ii) as q_e decreased (iii) as μ_o/μ_1 increased; (iv) as Re increased; or (v) as μ_c/μ_d decreased. On the other hand, a local minimum, at least, of maximum geometrical deformation was obtained at We = 2. Results demonstrated that θ_e and μ_c/μ_d were strongly correlated with cell fate.

Such computational models can accelerate the development of more precise and reliable biosensors via bioprinting. Next-generation models should incorporate non-Newtonian features of fluid flows,[71,72] smaller contact angles, microstructured models of cells, and multiple patterning of droplets.

2.2. *Cell-based sensing units*

2.2.1. *Introduction of cell-based sensing units*

Cell is the basic unit of structure and function in most organisms, spontaneously perceives physical, chemical, and biological changes in its surrounding microenvironment, and responds in a cytophysiologically relevant manner.[2] Cell membrane is a selectively permeable interface that consists of a phospholipid bilayer with embedded proteins (e.g., receptors, channel protein), permitting intracellular–extracellular mass exchange. Of these proteins, receptors serve as microsensors and monitor environmental changes. Activation of the receptors by an external stimulus gives rise to a series of intracellular signaling cascade, which may result in complex cell physiological changes at a wide spatiotemporal scale (from the molecular level to the cellular level ranging from nanoseconds to hours). For example, folding of β-hairpins occurs in microseconds[73]; activation of G protein-coupled receptors (GPCRs),[74] inositol triphosphate receptor (IPR) mediated calcium release,[75] cytoplasmic protein and membrane protein translocation[76] ion transmembrane flux through ligand-gated ion channel[77] can be completed in milliseconds; chemotaxis induced neutrophil migration to an inflammation site takes seconds[78]; activation of protein- and lipid-mediated kinase cascades can be completed in minutes; and phenotypic changes generally take hours and even days. By correlating cell physiological changes with external stimuli, cells can be explored as a sensing unit to detect the existence of pathogens and toxins in food, clinical and environmental samples. In addition, these cytophysiological changes can also be used to deduce functional information (e.g., toxicity and efficacy of pharmacological agent) of a known stimulus, which can be useful for drug screening.

Biological macromolecules such as antibodies, enzymes, and nucleic acids have been widely used as sensing units in broad applications owing to their specificity and sensitivity. Although CBBs explore the same molecular recognition mechanism with molecule-based biosensors (MBBs) via interactions between cellular membrane proteins and external stimuli, CBBs have their own unique features. First, some cell types are more robust than biomolecules under harsh conditions (e.g., thermophiles at high temperatures and acidophilus under high acidic conditions), whereas biomolecules, including proteins and deoxyribonucleic acid (DNAs) may lose their functions as a result of denaturation or degradation when exposed to the similar conditions. Second, CBBs enable spatiotemporal amplification of stimulation signals via intracellular signaling networks, which make it possible to detect subtle interactions between cells and stimuli. On the other hand, MBBs may require specific labeling using materials such as gold nanoparticles and quantum dots for signal amplification. Third, CBBs, especially bacteria based biosensors, are more cost-effective than MBBs. Bacteria can rapidly proliferate, which reduces the assay time and cost. Fourth, CBBs are capable of responding to a variety of biohazard substances (e.g., toxins, pathogens) in a cytophysiologically relevant manner akin to human beings and animals, thus offering an alternative platform for drug screening.[2] Fifth, CBBs allow more sophisticated applications than MBBs by bioengineering cells with functional exogenous genes or synthesizing artificial cells.[79–81] Comparison of MBBs and CBBs is given in Table 1.

Table 1. Comparison of cell-based and molecule-based biosensors. Mammalian-cell based biosensors and bacteria-based biosensors are listed separately due to their significant difference.

	Cell-based biosensors		
	Eukaryotic cells (e.g., Mammalian cells)	**Prokaryotic cells (e.g., bacteria)**	**Molecule-based biosensors**
Pros	1. Highly specific receptors 2. Spontaneous signal amplification 3. Provide functional information	1. Long-term stability and shelf life 2. Inexpensive to expand bacterial population 3. Spontaneous signal amplification	1. High specificity 2. Easy to immobilize 3. Linear response and suitable for quantitative analysis.
Cons	1. Expensive to expand mammalian cell population 2. Vulnerable to environmental changes 3. Difficult to deduce original stimulus in case of downstream cytophysiological events	1. Complex signaling transduction pathway 2. Difficult to deduce original stimulus	1. Expensive to manufacture 2. Easy to be degraded and denatured at the field

2.2.2. *Strategies for cell-based biosensing*

Bacteria and mammalian cells have been widely used as a sensing unit in CBBs. Bacteria are robust to harsh microenvironments, which make them suitable for applications such as deployment on the field. In addition, it is not expensive to maintain and expand a bacterial population. Compared to CBBs that immobilize bacteria on a 2D surface, bacterium-encapsulating hydrogels have advantages, including improved immobilization/packaging, minimization of contamination from other microorganisms and ease of deployment. However, wild-type bacteria usually develop comprehensive panels of membrane receptors with complex signaling cascade networks to survive in nature. Therefore, one downstream cytophysiological change in bacteria can be initiated by multiple upstream cytophysiological events, which brings difficulty in analyzing the presence of an unknown stimulus based on only downstream cytophysiological change. Adult mammalian cells are usually highly differentiated and responsible for specific tasks in a multicellular organism. Some specialized mammalian cells such as taste-receptor cells, photoreceptor cells and gustatory-receptor cells can be used for highly sensitive detection of a specific external stimulus. Despite their great promise, mammalian cells are fragile to environment changes and require well-controlled culture environment (e.g., proper temperature, humidity, carbon dioxide concentration, and culture media) to maintain their normal cytophysiology, and thus it is a challenge to deploy mammalian cells in harsh environments. In addition, maintenance of mammalian cells is labor-intensive, which raises the cost of CBBs. Here, we review cell-based sensing units reported in CHBBs, from the perspective of sensing mechanisms. We discuss four types of sensing mechanisms (i.e., voltage gated channels, GPCR receptors, nuclear receptors and immunoglobulin receptors) that are widely used in CBBs (Fig. 3). More detailed information regarding natural receptors in sensors can be found in literature.[31,82,83]

Voltage gated ion channel — Voltage-gated calcium channel (VGCC) is one of the trans-membrane ion channels with permeability to calcium ions 1000 times greater than sodium ions. VGCCs are widely found in the membrane of excitable cells (e.g., neurons, cardiac cells, muscle cells, and secretory cells) and explored as a mechanism for biosensing ions and small molecules. Under normal cytophysiological conditions, VGCCs are closed, and the concentration of intracellular Ca^{2+} is several thousand times higher than extracellular Ca^{2+}. Exposed to a high concentration of extracellular Ca^{2+} ions, VDCCs are activated due to membrane depolarization. Extracellular Ca^{2+} rush into cytoplasm, resulting in series of cell physiological changes ranging from muscular contraction, excitation of neurons, gene expression, to exocytosis of hormones or neurotransmitters, which can be utilized for biosensing.[84,85] Detection of the heavy-metal ions was demonstrated using a cardiac

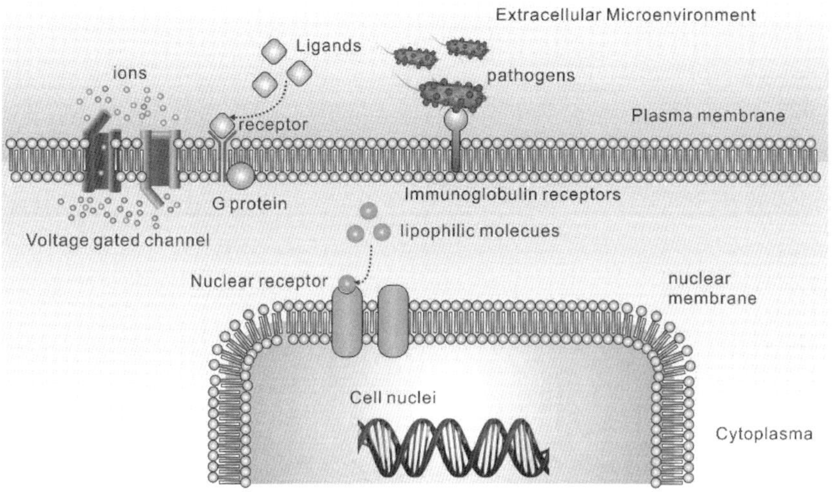

Fig. 3. Schematic of cell sensing mechanisms. Four types of typical sensing mechanisms are schematically illustrated, including voltage gated channels, GPCR receptors, nuclear receptors, and immunoglobulin receptors.

cell-based biosensor.[86] Beating cardiomyocytes generated periodic extracellular potential that was detected by a light-addressable potentiometric sensor. Exposure to heavy-metal ions, cardiomyocytes experienced characteristic changes in their beating parameters (i.e., frequency, duration, and amplitude) based on the physiological effects of metal ions. Fe^{3+} and Cu^{2+} induced very small changes in cardiomyocytes beating, while Hg^{2+} and Cd^{2+} resulted in significant changes in the beating pattern. By exploring the frequency-duration shifting in phase diagrams, different ions can be distinguished. The device provided an alternative way for heavy-metal monitoring and toxicity screening. Human neuroblastoma cells were incorporated into a 3D collagen type-I based biosensor.[87,88] Activation of VGCCs was demonstrated by supplying high concentration of K^+ ions (50 mM) that led to depolarization. Intracellular Ca^{2+} dynamics were transformed into fluorescence signals by calcium green-1, a calcium fluorescence indicator, and finally quantified using a fluorescence microscope. In addition, the membrane potential dynamics were fluorescently quantified using a potentiometric fluorescent dye.

G protein coupled receptor — G protein coupled receptors (GPCRs) are a super family of membrane receptors that enable a variety of cytophysiologically relevant processes, including visual sensing, gustatory sensing, olfactory sensing, and cell density sensing. GPCRs are activated by external signals such as electromagnetic radiation and exogenous molecules, including odors, pheromones, hormones, and neurotransmitters. The activation of GPCRs results in signal transduction and ultimate cell physiological changes at cellular levels (e.g., cell growth and metastasis) and subcellular levels (e.g., increase in cytosolic calcium ion).[23,89] Thus, interaction between GPCRs and ligands or other signal mediators can be used as a sensing mechanism for CBBs. Kang and coworkers reported neuronal micro-circuit arrays for drug screening.[90] Primary hippocampal neurons were maintained in an agarose microwell with a microelectrode immobilized at the bottom to record drug's cytophysiological effect on neuronal circuit activity. Several drugs (AP, bicuculline, and NMDA) were tested using neurotransmitter γ-aminobutyric acid and glutamate as agonist or inhibitory receptors. This cell-based biosensor can potentially be a high-throughput drug-screening platform for excitable cells. Sun and coworkers developed a microfluidic cell-based biosensor that enables detection of transient chemical stimuli using GPCRs with a temporal resolution of milliseconds.[91,92] Spatiotemporally-controllable chemical stimulation of cells was achieved by hydrodynamic-gated sample injection.[93,94] Endogenous P2Y receptors on the membrane of NIH 3T3 fibroblast cells were activated using Adenosine Triphosphate, resulting in a series of subcellular signaling cascades, including activation of phospholipase C, hydrolysis of phosphatidylinositol bisphosphate into inositol triphosphate (IP3) and diacylglycerol by phospholipase C, binding between IP3 and IP3 receptors, and Ca^{2+} release from the internal Ca^{2+} stores to the cytoplasm. Intracellular Ca^2+ dynamics was indicated using Fluo-3, an intracellular calcium indicator and recorded by a fluorescence microscope. Feng *et al.*, bioengineered endogenous P2Y receptors (ATP–GPCRs) into a bullfrog fibroblast cell line and coupled these cells with microelectrode arrays for chemical sensing. Bullfrog fibroblast cells demonstrated advantages over mammalian cells for cell-based sensing, as they can be deployed under ambient atmospheric conditions and are suitable for long-term storage.[95]

Nuclear receptor — Nuclear receptors are a class of endogenous proteins found in metazoan cells. Nuclear receptors can sense lipophilic substances such as steroid, thyroid hormones and vitamins A and D and regulate the expression of specific genes.[73,96,97] Genetic engineered *Saccharomyces cerevisiae* cells with nuclear receptors were demonstrated for detection of estrogen.[98–100] *Saccharomyces cerevisiae* strains were genetically engineered with a human androgen receptor, estrogen receptor alpha, or estrogen receptor beta together with firefly luciferase initiated by a respective hormone responsive promoter. These cells are immobilized in 3D polyvinyl acetate and alginate hydrogel matrix. Detection of 17-β-estradiol, an estrogen, was demonstrated with a detection limit of 0.08 μg L^{-1} and a half maximal effective concentration of 0.64 μg L^{-1}. *Saccharomyces cerevisiae* encapsulated in alginate microbeads showed a high viability for luminescence measurements even after 1 month of storage at −80°C. The assay reproducibility for each test was illustrated by coefficient of variation ranging from 4.35 to 18.47%.

Immunoglobulin superfamily receptors — Immunoglobulin superfamily receptors are a class of soluble membrane proteins featured with immunoglobulin-like structures. Immunoglobulin receptors are found in immune cells (e.g., T and B lymphocytes), and they are involved in the recognition and binding of antigen. For example, B lymphocytes express B-cell receptors (an immunoglobulin receptor) that allow highly specific sensing of pathogens or toxin via antibody-antigen interaction.[5] Kim and coworkers developed a strategy to increase the detection sensitivity of lymphocyte based biosensor by exploring B cell as antigen presenting cells for T cell.[101] A confluent layer of B-cells was layered on T cells that were immobilized on hydrogel microwells. B cells captured and internalized exogenous proteins, proteolyzed these proteins into short peptides, and further presented these short peptides to the contacted T cells. Activation of T-cell-receptors by the presented peptides resulted in an increase in the cytosolic calcium level in T cells, which was monitored by fluorescence imaging of calcium via calcium sensitive fluorescence dye Fura-2. In addition, collagen-encapsulated Ped-2E9 cells (a B-cell hybridoma) were demonstrated for CBBs.[102] Rapid cytotoxicity assays of pathogens or their toxins were achieved by quantitatively measuring alkaline phosphatase (ALP) released from infected Ped-2E9 cells.

2.3 *Signal readout systems*

Basically, CHBBs have a signal readout system, which interfaces with cells and transduces cellular responses into quantitative or qualitative signals for detection of analyte. When cells are cultured on a 2D/3D surface, signal readout from individual cells can be obtained using multiple approaches, including electrochemical, mechanical, and optical platforms.[103–114] However, when cells are encapsulated in 3D hydrogels,[115,116] signal readout is geometrically limited, which makes it difficult for some sensing technologies (e.g., electrochemical, mechanical transduction) for single-cell analysis. Optical detection becomes a feasible way for single-cell analysis in CHBBs due to its advantages such as contactless detection, high spatial resolution and readability for naked eyes. Especially, bioluminescence provides an alternative strategy to deliver observable signals without transducers. In the future, microfluidic and lab-chip technologies will have considerable impact on CHBBs by monitoring cellular density profiles without the need for labelling. In this section, we focus on various sensing platforms from a CHBB perspective.

2.3.1. *Optical-based signal readout systems*

Optical detection is a widely-used approach to monitor absorbance, luminescence, or fluorescence signals that are functions of cytophysiological changes and cellular metabolite production.[7]

Fluorescence detection — In CHBBs, fluorescence dyes are employed convert physicochemical or biochemical events into fluorescence signals. Alternatively, fluorescence genes are fused into cells as a reporter to indicate expression of target genes. Diverse fluorescence detection technologies have proven to be invaluable gadgets for monitoring molecular changes in cellular mechanisms such as protein and gene expression.[86,117–119] Fluorescent-labeled gene and protein probes have also been employed to monitor cellular function in genetically engineered organisms for decades.[120] Recently, a fluorescence-based sensor has been developed to monitor continuous blood-glucose levels *in vivo* using glucose-responsive fluorescent hydrogel fibers.[121] The implanted fluorescent PEG-bonded polyacrylamide (PAM) hydrogel fibers transmitted fluorescent signals transdermally of the blood-glucose concentration.[121] In this platform, the fibrous structure of hydrogels further allowed this sensor to remain at the implantation site for a long time period (up to 140 days). Thus, a minimally invasive and transdermal hydrogel-based glucose sensor has been developed to increase the quality of life of diabetic patients.[121] On the other hand, the fluorometric imaging plate reader (FLIPR™) and highly sensitive fluorometric assays have been developed to assess membrane potential, intracellular calcium mobilization and cellular signaling processes.[82,122,123] In contrast to other labeling assays such as radiolabeling, fluorescence-based methods are safe-to-use, and do not cause any mutations and DNA damages as observed in the prolonged exposure of radioactive dyes.[124–126] However, auto-fluorescence of molecules can potentially interfere with fluorescent assays and cause false-positive results.[119]

Bioluminescence — Bioluminescence describes production and emission of light by a living organism. Bioluminescence naturally exists in some microorganisms, fishes, and fungi. Bioluminescent microorganisms include *Vibrio fischeri* and *Pyrodimium bahamense*. For example, *Vibrio fischeri* is utilized to detect mercury and selenium with a detection of limit of 1 ppb.[37] Compared to fluorescence, bioluminescence has advantages, including no need for light excitation source and long emission time. When transfected into cells, bioluminescence genes can be used to indicate the expression of a target gene in response to environmental stimuli. For example, Michelini and co-workers reported a yeast-based biosensor for bioluminescent detection of androgen-like compounds. Recombinant *Saccharomyces cerevisiae* cells were genetically modified with a human androgen receptor and a reporter gene *YIpLuc* to indicate activation of the human androgen receptor.[99]

Plasmonic detection — Plasmonic-based readout systems are one type of optical signal transduction methods that monitor minute changes at the close vicinity of a metallic substrate.[131–135] Surface plasmon-based approaches report vital information in terms of wavelength, extinction intensity and the angle of reflected light, which are particularly sensitive to binding and capture events as well as the dielectric properties of medium on the metal substrate.[127–131,136] Surface plasmon-based detection strategies also provide a great opportunity to tune the sensitivity and specificity by altering the type of nanostructures at the sensor substrate.[131] Surface plasmon resonance (SPR) approach allows implementing hydrogels onto plasmonic substrates, and monitors physiological behavior of cell and cellular substances encapsulated in hydrogels. In these platforms, the presence of analyte causes hydrogels to expand in volume, and thus, the refractive index differs from that of the original state.[113] By monitoring local refractive index changes, SPR imaging technologies are employed to examine the hydrogel network and assist in improving the optimal conditions for large-scale production of biosensor arrays.[113,137] By changing coupling method (e.g., waveguide coupling) in SPR techniques, the other characteristics of hydrogel film such as thickness can also be examined.[138] Another plasmonic-based platform, localized surface plasmon resonance (LSPR), utilizes strong electromagnetic response of metal nanoparticles and assesses collective oscillation of nanoparticles.[139–141] LSPR biosensing platforms have been broadly used to detect nucleic acids, proteins, toxins, and viruses.[133,141–145] Hydrogels have recently been applied to LSPR-based platforms to monitor the physiological behavior of cells and to enhance the biosensing capacity of stimuli-responsive hydrogel-metal nanoparticles for medical applications.[146,147] For instance, a glucose oxidase was immobilized into the stimuli-responsive hydrogels to evaluate the effect of glucose binding.[146] The interparticle distance of silver nanoparticles in the hydrogel increased, when glucose was applied to the hydrogels. As a result of these interactions, the absorbance intensity of the LSPR peak wavelength decreased, and limit of detection was observed to be 10 pM for glucose sensing.[146]

Holographic sensing. By immobilizing nanoparticles within a hydrogel matrix, holographic sensors have been developed by incorporating ionizable monomers into hydrogel films.[148] In contrast to other optical pH sensors, the shrinkage and swelling behavior of hydrogels, were evaluated by monitoring changes in the holographic diffraction wavelength of holograms.[148] This holographic biosensor presented a sensitivity down to mili-pH changes, and this platform was also validated by quantifying the changes in proton concentrations in milk samples undergoing homolactic fermentation in the presence of *Lactobacillus casei*.[148] The major obstacle in this method is the fabrication of nanoparticle arrays and the need for a frequency doubled Nd-YAG laser.[113,148]

2.3.2. *Electrochemical-based signal readout systems*

Electrochemical sensing finds a variety of applications ranging from pH to conductivity sensing systems.[149,150] These systems can also be combined with other biosensing platforms, including optical and mechanical sensors.[151] Hydrogels are used as a supporting material, where biomolecules such as DNA, peptides, enzymes are immobilized for biomolecular detection. Hydrogels are also utilized as a stabilizing material for the encapsulated bioactive component.[152–155] Further, hydrogels provide a large surface area, thus increasing the loading capacity and improving the detection sensitivity.[113,156] Conductometric measurements were used to quantify electrochemical signals from hydrogels.[157,158] These platforms utilized interdigitated electrodes and measured the electrical conductivity of hydrogels.[137] Recently, these signal transduction systems have been used to monitor cellular mechanisms such as viability, apoptosis and proliferation, and specific intracellular and extracellular cellular reactions for drug screening.[151,160–162]

3. Applications

CHBBs have recently gained a great deal of interest in drug screening, pathogen detection, and environmental monitoring. As a sensing unit, cells can quickly respond to delicate or drastic stimuli from their microenvironment, and the response can be transformed to qualitative and/or quantitative signals for biosensing. For example, cell morphological changes, altered protein expression profiles, and cell viability can be used to directly indicate the impact of environmental changes. Often, bioengineered cells are preferably used to detect specific proteins and emit fluorescence signals. Here, we summarize the applications of bioengineered cells for rapid pathogen detection and drug screening without reference to traditional methods such as enzyme-linked immunosorbent assay (ELISA) or polymerase chain reaction (PCR).

Rider *et al.* genetically engineered B lymphocytes for specific sensing a variety of pathogens at low concentrations.[83] First, B cell lines were engineered to stably express cytosolic aequorin, as a bioluminescent reporter. Second, these cell lines were transfected with plasmids that express antibodies specific for a pathogen of interest on the cell surface. The crosslinking of antibodies triggered by their corresponding antigens from a specific pathogen can elevate the level of intracellular calcium and thus lead to a fluorescence emission by aequorin. As reported, a number of B cell lines have been developed for a variety of bacteria and viruses including *Yersinia pestis*, orthopoxviruses, *Venezuelan equine encephalitis* (VEE) virus, *Escherichia coli* strain O157:H7, *B. anthracis* and foot-and-mouth virus. Most importantly, detection can be completed within 3–5 min, and no signal or nucleic acid amplification is required. Although the preparation of these cell lines seems to be complicated, it takes less than 1h hands-on time over two days, indicating that scaling-up of this platform technology can be made economically. Thus, this B cell sensing platform can be widely used for pathogen detection in clinical diagnosis, bio-defense, and food safety. Similarly, another study demonstrated that fluorescent dye loaded mast cells detected *Escherichia coli* and *Listeria monocytogenes* at a picomolar level via a chimeric protein (fusion of Fc region of the IgE antibody and CD14).[163]

Recently, 3D culture has been developed on-chip to sense pathogens[101] or assess the toxicity of drug candidates.[164] For example, Kim *et al.* developed a T-cell/B-cell based microchip biosensor to detect intact pathogens.[101] In this biosensor, hydrogel microwells were first fabricated. The surface of microwells was then functionalized with antibodies for immobilizing T cells in an array format. On the top of the T cell layer, antigen-presenting B cells were seeded. When B cells presented specific antigens to T cells via T cell receptors, the level of intracellular calcium accordingly increases proportionally to the concentration antigenic peptides. The concentration of intracellular calcium can be measured by measuring the release of calcium-sensitive fluorescent dyes, which were preloaded. As demonstrated, this live cell array quantified a model peptide analyte ranging range 0.05 to 5 μM within minutes.[101]

Banerjee *et al.* reported a portable CHBBs platform using collagen-encapsulated B lymphocytes that can detect bacterial contamination and their toxin derivatives.[102] In this study, they fabricated a Type-I collagen gel matrix, in which lymphocyte origin murine Ped-2E9 cells were seeded at a density of 2.5×10^7 cells mL^{-1} in a 48 well plate. The cells were continuously cultured at 37°C for 24–96 h prior to addition of bacteria or toxin. It took 3–6 h for bacteria and toxin to diffuse through the collagen gel to Ped 2E9 cells and for the cells to secrete ALP. The supernatant was then transferred to another 48 well plate, which was preloaded with an ALP liquid substrate, leading to a color development. The color intensity was measured using a handheld spectrophotometer at a wavelength of 405/595 nm. Alternatively, the ALP liquid substrate can be added directly to the culture plate, and the color development can be measured using a standard spectrophotometer. Using this CHBBs platform, the authors achieved rapid detection of pathogenic *Listeria* and *Bacillus* species and the toxins from these organisms. Potentially, this portable platform can be used for improving food safety and environmental monitoring.

Currently, drug discovery, especially at the early stage, relies on cell-based assays to screen for therapeutic candidates and to assess toxicities. However, cells grown in a 2D format, either on a substrate or in a suspension, may respond differently to drug candidates compared to their 3D counterparts *in vivo*. Accumulating evidence has shown that 3D culture, in contrast to 2D planar culture on-dish, can better mimic the mechanical, chemical, and biological features of ECM.[165–167] Thus, 3D cell culture arrays have also been used to screen for drug candidates and to assess toxicity in a high-throughput manner. For example, Lee *et al.* fabricated a miniaturized 3D cell-culture array system, which consisted of one microchip, named as DataChip, with 1,080 individual cell cultures in collagen or alginate gels spotted on a functionalized glass slide, and one microchip, termed as MetaChip, contained their cytochrome P450 — generated metabolites P450 in a complementary format.[164] When the DataChip was in contact with the MetaChip by stamping, the interaction between drug candidates and mammalian cells was initiated. The MetaChip was further rinsed, cultured, stained, and subsequently scanned using a fluorescence microarray scanner. The analysis of IC50 was used to assess the toxicity of drug candidates and their P450-generated metabolites. The IC50 values obtained from the DataChip were similar to that obtained on 2D and 3D microtiter plates despite nearly 2,000-fold miniaturization. However, due to the nature of cytotoxicity analysis, the assay time was fairly long (more than 3 days). Nevertheless, the DataChip allowed the high-throughput screening of drug candidates given the number of cell spots arrayed on the DataChip.

In another study, a 3D hydrogel-based microchip was developed for anticancer drug analysis, in which controlled morphology of microstructure was created without using photomasks.[168] For fabrication, cell suspension in hydrogel precursors was injected into a microchannel and local UV photopolymerization was applied to the photosensitive precursors with the aid a fluorescence microscope to form cell encapsulated hydrogel structures in the microchannel. The unreacted hydrogel precursors were then washed away using **Phosphate-buffered saline** and replaced with anti-cancer drug candidates. Via this system, the authors examined the apoptosis in HepG2 and A549 cells in the presence of two intracellular redox parameters such as glutathione (GHS) and reactive oxygen species (ROS). The cell viability and the levels of these intracellular redox parameters were monitored using a fluorescence microscope. Compared to the assessment on a 96-well plate, the 3D culture on-chip demonstrated similar levels of intracellular GSH and ROS contents. This method allows selective encapsulation of cells in hydrogels for 3D drug screening.

4. Conclusions and Perspectives

Encapsulating cells into hydrogels presents an emerging and promising strategy to develop cell-based biosensors. Hydrogels provide a cytocompatible 3D microenvironment for cell immobilization, maintenance and packaging. In addition, cell-encapsulating hydrogels mimic native tissues and provide an attractive drug screening platform. However, CHBBs are still limited by multiple challenges for real-world practice. First, new hydrogels need to be developed to enable a long-term cell preservation with high cell viability, especially for mammalian cells. Second, cells are randomly encapsulated in hydrogels, which brings a challenge in standardizing CHBBs. Thus, fine cell manipulation technologies are needed to pattern individual cells in 3D hydrogels down to a single-cell resolution. Third, CHBBs inherit common issues of CBBs such as lack of selectivity and inability for quantitative analysis. To address these challenges, development of synthetic biology and genetic technologies is highly needed. Despite these challenges, enabling technologies such as 3D bioprinting, microfluidics, microscale assembly and microscale opto-electromechanical systems provide significant opportunities to advance CBBs toward practical applications. For example, microfluidic technologies can isolate rare cells from cell mixtures and manipulate individual cells to physiochemical stimuli for biosensing.[91–94,169–171] Microscale assembly of hydrogels significantly facilitates design and fabrication of CHBBs for both basic and clinical research.[32–36,44] With integration of miniaturized electronic/optical signal read out systems,[172] CHBBs can provide a powerful analytical toolbox for broad applications in environmental monitoring, drug screening, and clinical diagnostics in the near future.

Acknowledgments

We would like to acknowledge NIH R21-HL112114 and 1R01EB015776-01A1. This work is partially supported by the National Science Foundation under NSF CAREER Award Number 1150733. Any opinions, findings, and conclusions or recommendations expressed in this work are those of the author(s) and do not necessarily reflect the views of the National Science Foundation. UD is a founder of, and has an equity interest in: (i) DxNow Inc., a company that is developing microfluidic and imaging technologies for point-of-care diagnostic solutions, and (ii) Koek Biotech, a company that is developing microfluidic IVF technologies for clinical solutions. UD's interests were reviewed and are managed by the Brigham and Women's Hospital and Partners HealthCare in accordance with their conflict of interest policies.

References

1. P. Wang, G. Xu, L. Qin, Y. Xu, Y. Li, and R. Li, Cell-based biosensors and its application in biomedicine. *Sensor. Actuat B-Chem.*, **108**(1–2), 576–584 (2005).
2. P. Banerjee and A. K. Bhunia, Mammalian cell-based biosensors for pathogens and toxins. *Trend. Biotechnol.*, **27**(3), 179–188 (2009).
3. Y. Lei, W. Chen and A. Mulchandani, Microbial biosensors. *Anal. Chim. Acta*, **568** (1–2), 200–210 (2006).
4. S. Belkin, Microbial whole-cell sensing systems of environmental pollutants. *Curr. Opin. Microbiol.*, **6**(3), 206–212 (2003).

5. P. Banerjee, B. Franz, and A. Bhunia, Mammalian Cell-Based Sensor System. In *Whole Cell Sensing Systems I, Advances in Biochemical Engineering/Biotechnology*, (Eds.) S. Belkin and M. B. Gu, Berlin Heidelberg: Springer, **117**, 21–55 (2010).
6. T. Elad, J. Lee, M. Gu, and S. Belkin, Microbial cell arrays. In *Whole Cell Sensing Systems I, Advances in Biochemical Engineering/Biotechnology*, Eds. S. Belkin and M. B. Gu, Berlin Heidelberg: Springer, pp. 85–108 (2010).
7. L. Zhou, G. Huang, S. Wang, J. Wu, W. Lee, Y. Chen, F. Xu, and T. Lu, Advances in cell-based biosensors using three-dimensional cell-encapsulating hydrogels. *Biotechnol. J.*, **6**(12),1466–1476 (2011).
8. A. Birgersdotter, R. Sandberg, and I. Ernberg, Gene expression perturbation *in vitro* — A growing case for three-dimensional (3D) culture systems. *Semin. Cancer Biol.*, **15**(5), 405–412 (2005).
9. M. W. Tibbitt and K. S. Anseth, Hydrogels as extracellular matrix mimics for 3D cell culture. *Biotechnol Bioeng*, **103**(4), 655–663 (2009).
10. H. Geckil, F. Xu, X. Zhang, S. Moon, and U. Demirci, Engineering hydrogels as extracellular matrix mimics. *Nanomedicine*, **5**(3), 469–484 (2010).
11. M. C. Cushing and K. S. Anseth, Hydrogel cell cultures. *Science*, **316**(5828), 1133–1134 (2007).
12. M. W. Tibbitt and K. S. Anseth, Hydrogels as extracellular matrix mimics for 3D cell culture. *Biotechnol. Bioeng.*, **103**(4), 655–663 (2009).
13. Y. Ling, J. Rubin, Y. Deng, C. Huang, U. Demirci, J. M. Karp, and A. Khademhosseini, A cell-laden microfluidic hydrogel. *Lab Chip*, **7**(6), 756–762 (2007).
14. J. Yeh, Y. Ling, J. Karp, J. Gantz, A. Chandawarkar, G. Eng, J. Blumling, R. Langer, and A. Khademhosseini, Micromolding of shape-controlled, harvestable cell-laden hydrogels. *Biomaterials*, **27**, 5391–5398 (2006).
15. Y. S. Song, R. L. Lin, G. Montesano, N. G. Durmus, G. Lee, S. S. Yoo, E. Kayaalp, E. Haeggstrom, A. Khademhosseini, and U. Demirci, Engineered 3D tissue models for cell-laden microfluidic channels. *Anal. Bioanal. Chem.*, **395**(1), 185–193 (2009).
16. S. Guven, P. Chen, F. Inci, S. Tasoglu, B. Erkmen, and U. Demirci, Multiscale assembly for tissue engineering and regenerative medicine. *Trends in Biotechnology*, **33**(5), 269–279 (2015).
17. K. Y. Lee and D. J. Mooney, Hydrogels for tissue engineering. *Chem. Rev.*, **101**(7), 1869–1880 (2001).
18. J. L. Drury and D. J. Mooney, Hydrogels for tissue engineering: Scaffold design variables and applications. *Biomaterials*, **24**(24), 4337–4351 (2003).
19. M. Guvendiren and J. A. Burdick, Engineering synthetic hydrogel microenvironments to instruct stem cells. *Curr. Opin. Biotech.*, **24**(5), 841–846 (2013).
20. R. DeVolder and H. J. Kong, Hydrogels for *in vivo*-like three-dimensional cellular studies. *Wires Syst. Biol. Med.*, **4**(4), 351–365 (2012).
21. T. Valero, G. Moschopoulou, S. Kintzios, P. Hauptmann, M. Naumann, and T. Jacobs, Studies on neuronal differentiation and signalling processes with a novel impedimetric biosensor. *Biosens. Bioelectron.*, **26**(4), 1407–1413 (2010).

22. Y. M. Ju, B. Yu, L. West, Y. Moussy, and F. Moussy, A novel porous collagen scaffold around an implantable biosensor for improving biocompatibility. II. Long-term *in vitro/in vivo* sensitivity characteristics of sensors with NDGA- or GA-crosslinked collagen scaffolds. *J. Biomed. Mater. Res A*, **92**(2), 650–658 (2010).
23. J. Zhu, Bioactive modification of poly(ethylene glycol) hydrogels for tissue engineering. *Biomaterials*, **31**(17), 4639–4656 (2010).
24. M. A. Acosta, P. Ymele-Leki, K. V. Kostov, and J. B. Leach, Fluorescent microparticles for sensing cell microenvironment oxygen levels within 3D scaffolds. *Biomaterials*, **30**(17), 3068–3074 (2009).
25. R. J. Russell, M. V. Pishko, A. L. Simonian, and J. R. Wild, Poly(ethylene glycol) hydrogel-encapsulated fluorophore-enzyme conjugates for direct detection of organophosphorus neurotoxins. *Anal. Chem.*, **71**(21), 4909–4912 (1999).
26. D. C. Appleyard, S. C. Chapin, and P. S. Doyle, Multiplexed protein quantification with barcoded hydrogel microparticles. *Anal. Chem.*, **83**(1), 193–199 (2011).
27. J. Moorthy, R. Burgess, A. Yethiraj, and D. Beebe, Microfluidic based platform for characterization of protein interactions in hydrogel nanoenvironments. *Anal. Chem.*, **79**(14), 5322–5327 (2007).
28. D. A. Zubtsov, E. N. Savvateeva, A. Y. Rubina, S. V. Pan'kov, E. V. Konovalova, O. V. Moiseeva, V. R. Chechetkin, and A. S. Zasedatelev, Comparison of surface and hydrogel-based protein microchips. *Anal. Biochem.*, **368**(2), 205–213 (2007).
29. M. N. Arip, L. Y. Heng, M. Ahmad, and S. Ujang, A cell-based potentiometric biosensor using the fungus Lentinus sajor-caju for permethrin determination in treated wood. *Talanta*, **116**, 776–781 (2013).
30. V. Varelas, N. Sanvicens, M. P. Marco, and S. Kintzios, Development of a cellular biosensor for the detection of 2,4,6-trichloroanisole (TCA). *Talanta*, **84**(3), 936–940 (2011).
31. N. Buffi, D. Merulla, J. Beutier, F. Barbaud, S. Beggah, H. van Lintel, P. Renaud, and J. R. van der Meer, Development of a microfluidics biosensor for agarose-bead immobilized Escherichia coli bioreporter cells for arsenite detection in aqueous samples. *Lab Chip*, **11**(14), 2369–2377 (2011).
32. F. Xu, C. A. M. Wu, V. Rengarajan, T. D. Finley, H. O. Keles, Y. R. Sung, B. Q. Li, U. A. Gurkan, and U. Demirci, Three-dimensional magnetic assembly of microscale hydrogels. *Adv. Mater.*, **23**(37), 4254–4260 (2011).
33. S. Tasoglu, D. Kavaz, U. A. Gurkan, S. Guven, P. Chen, R. Zheng, and U. Demirci, Paramagnetic levitational assembly of hydrogels. *Adv. Mater.*, **25**(8), 1137–1143 (2013).
34. S. Tasoglu, C.H. Yu, H.I. Gungordu, S. Guven, T. Vural and U. Demirci, Guided and magnetic self-assembly of magnetoceptive gels. *Nature Communications*, **5**, 4702 (2014).
35. S. Tasoglu, C.H. Yu, V. Liaudanskaya, S. Guven, C. Migliaresi and U. Demirci, Magnetic levitational assembly for living material fabrication. *Advanced Healthcare Materials*, doi:10.1002/adhm. 201500092, in press (2015).

36. F. Xu, T. D. Finley, M. Turkaydin, Y. Sung, U. A. Gurkan, A. S. Yavuz, R. O. Guldiken, and U. Demirci, The assembly of cell-encapsulating microscale hydrogels using acoustic waves. *Biomaterials*, **32**(31), 7847–7855 (2011).
37. Y. Du, E. Lo, S. Ali, and A. Khademhosseini, Directed assembly of cell-laden microgels for fabrication of 3D tissue constructs. *Proc. Natl. Acad. Sci. USA*, **105**(28), 9522–9527 (2008).
38. P. Chen, Z. Luo, S. Güven, S. Tasoglu, A. V. Ganesan, A. Weng, and U. Demirci, Microscale assembly directed by liquid-based template. *Adv. Mater.*, 26(34), 5936–5941 (2014).
39. K. Sekeroglu, U. A. Gurkan, U. Demirci, and M. C. Demirel, Transport of a soft cargo on a nanoscale ratchet. *Appl. Phys. Lett.*, **99**(6), 63703–637033 (2011).
40. S. E. Chung, W. Park, S. Shin, S. A. Lee, and S. Kwon, Guided and fluidic self-assembly of microstructures using railed microfluidic channels. *Nat. Mater.*, **7**(7), 581–587 (2008).
41. W. Park, H. Lee, H. Park, and S. Kwon, Sorting directionally oriented microstructures using railed microfluidics. *Lab Chip*, **9**(15), 2169–2175 (2009).
42. S. E. Chung, Y. Jung, and S. Kwon, Three-dimensional fluidic self-assembly by axis translation of two-dimensionally fabricated microcomponents in railed microfluidics. *Small*, **7**(6), 796–803 (2011).
43. S. Tasoglu, E. Diller, S. Guven, M. Sitti, and U. Demirci, Untethered micro-robotic coding of three-dimensional material composition. *Nat. Commun.*, **5**, 3124 (2014).
44. U. A. Gurkan, S. Tasoglu, D. Kavaz, M. C. Demirel, and U. Demirci, Emerging technologies for assembly of microscale hydrogels. *Adv. Healthc. Mater.*, **1**(2), 149–158 (2012).
45. K. A. Athanasiou, R. Eswaramoorthy, P. Hadidi, and J. C. Hu, Self-organization and the self-assembling process in tissue engineering. *Annu. Rev. Biomed. Eng.*, (2012).
46. N. N. Kachouie, Y. A. Du, H. Bae, M. Khabiry, A. F. Ahari, B. Zamanian, J. Fukuda, and A. Khademhosseini, Directed assembly of cell-laden hydrogels for engineering functional tissues. *Organogenesis*, **6**(4), 234–244 (2010).
47. P. Gong, W. Zheng, D. Xiao, and X. Jiang, Microscale methods to assemble mammalian cells into tissue-like structures. *Sci. China Life Sci.*, **55**(10), 862–871 (2012).
48. J. S. Liu and Z.J. Gartner, Directing the assembly of spatially organized multicomponent tissues from the bottom up. *Trends Cell Biol.*, **22**(12), 683–691 (2012).
49. G. Villar, A. D. Graham, and H. Bayley, A tissue-like printed material. *Science*, **340**(6128), 48–52 (2013).
50. F. Xu, S. Moon, A. E. Emre, C. Lien, E. S. Turali, and U. Demirci, Cell bioprinting as a potential high-throughput method for fabricating cell-based biosensors (CBBs). *IEEE Sensors*, 387–391 (2009).
51. F. Xu, S. Moon, E. Hefner, T. Beyazoglu, A. E. Emre, T. Manzur, and U. Demirci, A high-throughput label-free cell-based biosensor (CBB) system, Proc. of SPIE, 7693, 76931B-7 (2010).
52. X. Cui, G. Gao, and Y. Qiu, Accelerated myotube formation using bioprinting technology for biosensor applications. *Biotechnol. Lett.*, **35**(3), 315–321 (2013).

53. K. Sun, T.-S. Wei, B. Y. Ahn, J. Y. Seo, S. J. Dillon, and J. A. Lewis, 3D printing of interdigitated Li-Ion microbattery architectures. *Adv. Mater.*, 25(33), 4539–4543 (2013).
54. M. S. Mannoor, Z. Jiang, T. James, Y. L. Kong, K. A. Malatesta, W. O. Soboyejo, N. Verma, D. H. Gracias, and M. C. McAlpine, 3D Printed Bionic Ears. *Nano Lett.*, **13**(6), 2634–2639 (2013).
55. N. G. Durmus, S. Tasoglu, and U. Demirci, Bioprinting: functional droplet networks. *Nat. Mater.*, **12**(6), 478–479 (2013).
56. U. Demirci, A. Khademhosseini, R. Langer, and J. Blander, *Microfluidic Technologies for Human Health*, Singapore: 241–292 World Scientific Publishing Company (2012).
57. S. Tasoglu, U. A. Gurkan, S. Q. Wang, and U. Demirci, Manipulating biological agents and cells in micro-scale volumes for applications in medicine. *Chem. Soc. Rev.*, **42**(13), 5788–5808 (2013).
58. S. Tasoglu and U. Demirci, Bioprinting for stem cell research. *Trends Biotechnol.*, **31**(1), 10–19 (2013).
59. S. Moon, E. Ceyhan, U. A. Gurkan, and U. Demirci, Statistical modeling of single target cell encapsulation. *PLoS One*, **6**(7), e21580 (2011).
60. E. Ceyhan, F. Xu, U. A. Gurkan, A. E. Emre, E. S. Turali, R. El Assal, A. Acikgenc, C. M. Wu, and U. Demirci, Prediction and control of number of cells in microdroplets by stochastic modeling. *Lab Chip* 12(22), 4884–4893 (2012).
61. U. Demirci U and G. Montesano, Cell encapsulating droplet vitrification. *Lab Chip*, **7**(11), 1428–1433 (2007).
62. U. Demirci and G. Montesano, Single Cell Epitaxy by Acoustic Picoliter Droplets. *Lab Chip*, **7**, 1139–1145 (2007).
63. S. Tasoglu, D. Kavaz, U. A. Gurkan, and U. Demirci, Magnetic 3D assembly of microgels for tissue engineering and regenerative medicine. *J. Tissue Eng. Regen. Med.*, **6(Suppl. 1)**, 224 (2012).
64. S. F. Kistler (ed.), *Wettability*, New York: Dekker (1993).
65. M. Muradoglu and S. Tasoglu, A front-tracking method for computational modeling of impact and spreading of viscous droplets on solid walls. *Comput. Fluids*, **39**(4), 615–625 (2010).
66. S. Tasoglu, G. Kaynak, A. J. Szeri, U. Demirci, and M. Muradoglu, Impact of a compound droplet on a flat surface: a model for single cell epitaxy. *Phys. Fluids*, **22**(8), 082103 (2010).
67. S. Tasoglu, L. C. Rohan, D. F. Katz, and A. J. Szeri, Transient swelling, spreading and drug delivery by a dissolved anti-HIV microbicide-bearing film. *Phys. Fluids*, **25**(3), 031901 (2013).
68. A. J. Szeri, S. C. Park, S. Tasoglu, S. Verguet, A. Gorham, Y. Gao, and D. F. Katz, Epithelial coating mechanisms by semi-solid materials: application to microbicide gels. *Biophys. J.*, **98**(3), 604 (2010).
69. S. Tasoglu, J. J. Peters, S. C. Park, S. Verguet, D. F. Katz, and A. J. Szeri, The effects of inhomogeneous boundary dilution on the coating flow of an anti-HIV microbicide vehicle. *Phys. Fluids*, **23**(9), 093101 (2011).

70. S. Tasoglu, U. Demirci, and M. Muradoglu, The effect of soluble surfactant on the transient motion of a buoyancy-driven bubble. *Phys. Fluids*, **20**(4), 040805 (2008).
71. S. Tasoglu, S. C. Park, J. J. Peters, D. F. Katz, and A. J. Szeri, The consequences of yield stress on deployment of a non-Newtonian anti-HIV microbicide gel. *J. Nonnewton. Fluid Mech.*, **166**(19–20), 1116–1122 (2011).
72. S. Tasoglu, D. F. Katz, and A. J. Szeri, Transient spreading and swelling behavior of a gel deploying an anti-HIV microbicide. *J. Nonnewton. Fluid Mech.*, **187**, 36–42 (2012).
73. V. Munoz, P. A. Thompson, J. Hofrichter, and W. A. Eaton, Folding dynamics and mechanism of (beta)-hairpin formation. *Nature*, **390**(6656), 196–199 (1997).
74. J. P. Vilardaga, M. Bunemann, C. Krasel, M. Castro, and M. J. Lohse, Measurement of the millisecond activation switch of G protein-coupled receptors in living cells. *Nat. Biotechnol.*, **21**(7), 807–812 (2003).
75. J. Sneyd, J. Keizer, and M. J. Sanderson, Mechanisms of calcium oscillations and waves: a quantitative analysis. *FASEB J.*, **9**(14), 1463–1472 (1995).
76. M. Schaefer, N. Albrecht, T. Hofmann, T. Gudermann, and G. Schultz, Diffusion-limited translocation mechanism of protein kinase C isotypes. *FASEB J.*, **15**(9), 1634–1636 (2001).
77. A. Surprenant, G. Buell, and R. Alan North, P2x receptors bring new structure to ligand-gated ion channels. *Trends in Neurosciences*, **18**(5), 224–229 (1995).
78. C. A. Parent and P. N. Devreotes, A cell's sense of direction. *Science*, **284**(5415), 765–770 (1999).
79. C. Chiarabelli and P. L. Luisi, Chemical synthetic biology. *Sci. Progr.*, **97**(1), 48–61 (2014).
80. D. A. Hammer and N. P. Kamat, Towards an artificial cell. *FEBS Lett.*, **586**(18), 2882–2890 (2012).
81. N. Dalchau, M. J. Smith, S. Martin, J. R. Brown, S. Emmott, and A. Phillips, Towards the rational design of synthetic cells with prescribed population dynamics. *J. R. Soc. Interface*, **9**(76), 2883–2898 (2012).
82. I. B. Marshall, D. Owen, and S. McNulty, Measuring Ca2+ changes in multiwell format using the fluorometric imaging plate reader (FLIPR®). In *Calcium Signaling Protocols, Methods in Molecular Biology*, Eds. D. G. Lambert and R. D. Rainbow, New York: Humana Press, pp. 103–109 (2013).
83. T. H. Rider, M. S. Petrovick, F. E. Nargi, J. D. Harper, E. D. Schwoebel, R. H. Mathews, D. J. Blanchard, L. T. Bortolin, A. M. Young, J. Chen, and M. A. Hollis, A B cell-based sensor for rapid identification of pathogens. *Science*, **301**(5630), 213–215 (2003).
84. M. C. R. Shastry and H. Roder, Evidence for barrier-limited protein folding kinetics on the microsecond time scale. *Nat. Struct. Biol.*, **5**(5), 385–392 (1998).
85. A. W. Catterall, Structure and function of voltage-gated ion Channels. *Annu. Rev. Biochem.*, **64**(1), 493–531 (1995).
86. R. J. Aguilera, J. Montoya, T. P. Primm, and A. Varela-Ramirez, Green fluorescent protein as a biosensor for toxic compounds. 463–467, Springer, US (2006).

87. A. Desai, W. S. Kisaalita, C. Keith, and Z. Z. Wu, Human neuroblastoma (SH-SY5Y) cell culture and differentiation in 3-D collagen hydrogels for cell-based biosensing. *Biosens. Bioelectron*, **21**(8), 1483–1492 (2006).
88. C. Mao and W. S. Kisaalita, Characterization of 3-D collagen hydrogels for functional cell-based biosensing. *Biosens. Bioelectron.*, **19**(9), 1075–1088 (2004).
89. N. C. Hunt and L. M. Grover, Cell encapsulation using biopolymer gels for regenerative medicine. *Biotechnol. Lett.*, **32**(6), 733–742 (2010).
90. G. Kang, J. H. Lee, C. S. Lee, and Y. Nam, Agarose microwell based neuronal micro-circuit arrays on microelectrode arrays for high throughput drug testing. *Lab Chip*, **9**(22), 3236–3242 (2009).
91. J. Sun, P. Chen, X. J. Feng, W. Du, and B. F. Liu, Development of a microfluidic cell-based biosensor integrating a millisecond chemical pulse generator. *Biosens. Bioelectron.*, **26**(8), 3413–3419 (2011).
92. J. Sun, J. J. Wang, P. Chen, X. J. Feng, W. Du, and B. F. Liu, A chemical signal generator for resolving temporal dynamics of single cells. *Anal. Bioanal. Chem.*, **400**(9), 2973–2981 (2011).
93. P. Chen, X. J. Feng, J. Sun, Y. Wang, W. Du, and B. F. Liu, Hydrodynamic gating for sample introduction on a microfluidic chip. *Lab Chip*, **10**(11), 1472–1475 (2010).
94. P. Chen, P. Chen, X. Feng, W. Du, and B.-F. Liu, Analysis of intercellular communication by flexible hydrodynamic gating on a microfluidic chip. *Anal. Bioanal. Chem.*, **405**(1), 307–314 (2013).
95. X. Feng, J. Castracane, N. Tokranova, A. Gracias, G. Lnenicka, and B. G. Szaro, A living cell-based biosensor utilizing G-protein coupled receptors: Principles and detection methods. *Biosens. Bioelectron.*, **22**(12), 3230–3237 (2007).
96. T. G. Golos, L. M. Pollastrini, and B. Gerami-Naini, Human embryonic stem cells as a model for trophoblast differentiation. *Semin. Reprod. Med.*, **24**(5), 314–321 (2006).
97. M. Shimomura-Shimizu and I. Karube, Applications of Microbial Cell Sensors. In *Whole Cell Sensing System II*, Advances in Biochemical Engineering/Biotechnology, Eds. S. Belkin and M. B. Gu, Berlin Heidelberg: Springer, pp. 1–30 (2010).
98. P. Leskinen, E. Michelini, D. Picard, M. Karp, and M. Virta, Bioluminescent yeast assays for detecting estrogenic and androgenic activity in different matrices. *Chemosphere*, **61**(2), 259–266 (2005).
99. E. Michelini, P. Leskinen, M. Virta, M. Karp, and A. Roda, A new recombinant cell-based bioluminescent assay for sensitive androgen-like compound detection. *Biosens, Bioelectron.*, **20**(11), 2261–2267 (2005).
100. T. Fine, P. Leskinen, T. Isobe, H. Shiraishi, M. Morita, R. S. Marks, and M. Virta, Luminescent yeast cells entrapped in hydrogels for estrogenic endocrine disrupting chemical biodetection. *Biosens. Bioelectron.*, **21**(12), 2263–2269 (2006).
101. H. Kim, R. E. Cohen, P. T. Hammond, and D. J. Irvine, Livel lymphocyte arrays for biosensing. *Adv. Funct. Mater.*, **16**(10), 1313–1323 (2006).

102. P. Banerjee, D. Lenz, J. P. Robinson, J. L. Rickus, and A. K. Bhunia, A novel and simple cell-based detection system with a collagen-encapsulated B-lymphocyte cell line as a biosensor for rapid detection of pathogens and toxins. *Lab. Invest.*, **88**(2), 196–206 (2008).
103. S.Q. Wang, F. Inci, T. L. Chaunzwa, A. Ramanujam, A. Vasudevan, S. Subramanian, A.C.F. Ip, B. Sridharan, U. A. Gurkan, and U. Demirci, Portable microfluidic chip for detection of *Escherichia coli* in produce and blood. *International Journal of Nanomedicine*, **7**, 2591 (2012).
104. S.Q. Wang, F. Inci, G. De Libero, A. Singhal, and U. Demirci, Point-of-care assays for tuberculosis: role of nanotechnology/microfluidics. *Biotechnology Advances*, **31**(4), 438–449 (2013).
105. V. Mani, S.Q. Wang, F. Inci, G. De Libero, A. Singhal, and U. Demirci, Emerging technologies for monitoring drug-resistant tuberculosis at the point-of-care. *Advanced Drug Delivery Reviews*, **78**, 105–117 (2014).
106. C. Lissandrello, F. Inci, M. Francom, M. R. Paul, U. Demirci, and K. L. Ekinci, Nanomechanical motion of Escherichia coli adhered to a surface. *Applied Physics Letters*, **105**(11), 113701 (2014).
107. H. Shafiee, S.Q. Wang, F. Inci, M. Toy, T. J. Henrich, D.R. Kuritzkes, and U. Demirci, Emerging technologies for point-of-care management of HIV infection. *Annual Review of Medicine*, **66**, 387–405 (2015).
108. U.H. Yildiz, F. Inci, S.Q. Wang, M. Toy, H. C. Tekin, A. Javaid, D. T.-Y. Lau, and U. Demirci, Recent advances in micro/nanotechnologies for global control of hepatitis B infection. *Biotechnology Advances*, **33**(1), 178–190 (2015).
109. A.C. Sobieranski, F. Inci, H. C. Tekin, E. Comunello, A. Von Wangenheim, and U. Demirci, Portable digital in-line holography platform for sperm cell visualization and quantification. In *Graphics, Patterns and Images (SIBGRAPI), 2014 27th SIBGRAPI Conference on*, pp. 274–281. IEEE (2014).
110. N. G. Durmus, E. N. Taylor, F. Inci, K.M. Kummer, K. M. Tarquinio, and T. J. Webster, Fructose-enhanced reduction of bacterial growth on nanorough surfaces, *International Journal of Nanomedicine*, **7**, 537–545 (2012).
111. S. Tasoglu, H. C. Tekin, F. Inci, S. Knowlton, S. Q. Wang, F. Wang-Johanning, G. Johanning, D. Colevas, and U. Demirci, Advances in nanotechnology and microfluidics for human papillomavirus diagnostics. *Proceedings of the IEEE*, **103**(2), 161–178 (2015).
112. U. Demirci, S.Q. Wang, and F. Inci, Editorial for advanced health care technologies. *Advanced Health Care Technologies*, **1**, 1–2 (2015).
113. K. Gawel, D. Barriet, M. Sletmoen, and B. Stokke, Responsive hydrogels for label-free signal transduction within biosensors. *Sensors*, **10**(5), 4381–4409 (2010).
114. M. Guenther and G. Gerlach, Hydrogels for Chemical Sensors. *Hydrogel Sensors and Actuators, Springer Series on Chemical Sensors and Biosensors*, Eds. G. Gerlach and K.-F. Arndt, pp. 165–195 (2010).
115. F. Xu, F. Inci, O. Mullick, U. A. Gurkan, Y. Sung, D. Kavaz, B. Li, E. B. Denkbas, and U. Demirci, Release of magnetic nanoparticles from cell-encapsulating biodegradable nanobiomaterials. *ACS Nano*, 6(8), 6640–6649 (2012).

116. S. Guven, P. Chen, F. Inci, S. Tasoglu, B. Erkmen, and U. Demirci, Multiscale assembly for tissue engineering and regenerative medicine. *Trends in Biotechnology*, **33**(5), 269–279 (2015).
117. K. Dunn, S. Mayor, J. Myers, and F. Maxfield, Applications of ratio fluorescence microscopy in the study of cell physiology. *FASEB J.*, **8**(9), 573–582 (1994).
118. K. Giuliano, P. Post, K. Hahn, and D. Taylor, Fluorescent protein biosensors: measurement of molecular dynamics in living cells. *Annu. Rev. Biophy.Biomol. Struct.*, **24**, 405–434 (1995).
119. J. Pancrazio, J. Whelan, D. Borkholder, W. Ma, and D. Stenger, Development and application of cell-based biosensors. *Ann. Biomed. Eng.*, **27**(6), 697–711 (1999).
120. R. Day and M. Davidson, The fluorescent protein palette: tools for cellular imaging. *Chem. Soc. Rev.*, **38**(10), 2887–2921 (2009).
121. Y. Heo, H. Shibata, T. Okitsu, T. Kawanishi, and S. Takeuchi, Long-term *in vivo* glucose monitoring using fluorescent hydrogel fibers. *Proceedings of the National Academy of Sciences of the United States of America*, **108**(33), 13399–13403 (2011).
122. A. K. Theresa, P. A. Stephen, and L. D.-R. Diana, Rapid assessment of ligand actions with nicotinic acetylcholine receptors using calcium dynamics and FLIPR. *Drug Develop. Res.*, **44**, 14–20 (1998).
123. K. Whiteaker, J. Sullivan, and M. Gopalakrishnan, Cell-based assays using the fluorometric imaging plate reader (FLIPR). 9.2.1-9.2.23, John Wiley & Sons, Inc. (2001). In *Current Protocols in Pharmacology*, Ed. S. J. Enna *et al.* (2001).
124. C. Cistulli, T. Sorger, J. Marsella, C. Vaslet, and A. Kane, Spontaneous p53 mutation in murine mesothelial cells: increased sensitivity to DNA damage induced by asbestos and ionizing radiation. *Toxicol. Appl. Pharm.*, **141**(1), 264–271 (1996).
125. Z. Kote-Jarai, R. Williams, N. Cattini, M. Copeland, I. Giddings, R. Wooster, R. tePoele, P. Workman, B. Gusterson, J. Peacock, G. Gui, C. Campbell, and R. Eeles, Gene expression profiling after radiation-induced DNA damage is strongly predictive of BRCA1 mutation carrier status. *Clin. Cancer Res.*, **10**(3), 958–963 (2004).
126. R. Ballal, A. Cheema, W. Ahmad, E. Rosen, and T. Saha, Fluorescent oligonucleotides can serve as suitable alternatives to radiolabeled oligonucleotides. *JBT*, **20**(4), 190–194 (2009).
127. S. Sevimli, F. Inci, H. M. Zareie, and V. Bulmus, Well-defined cholesterol polymers with pH-controlled membrane switching activity. *Biomacromolecules*, **13**(10), 3064–3075 (2012).
128. F. Inci, U. Celik, B. Turken, H. Ö. Özer, and F. N. Kok, Construction of P-glycoprotein incorporated tethered lipid bilayer membranes. *Biochemistry and Biophysics Reports*, **2**, 115–122 (2015).
129. O. Tokel, U. H. Yildiz, F. Inci, N. G. Durmus, O. O. Ekiz, B. Turker, C. Cetin *et al.*, Portable microfluidic integrated plasmonic platform for pathogen detection. *Scientific Reports*, **5**, (2015).
130. H. Shafiee, W. Asghar, F. Inci, M. Yuksekkaya, M. Jahangir, M. H. Zhang, N. G. Durmus, U. A. Gurkan, D. R. Kuritzkes, and U. Demirci, Paper and flexible substrates

as materials for biosensing platforms to detect multiple biotargets. *Scientific Reports*, **5**, (2015).
131. M. Stewart, C. Anderton, L. Thompson, J. Maria, S. Gray, J. Rogers, and R. Nuzzo, Nanostructured plasmonic sensors. *Chem. Rev.*, **108**(2), 494–521 (2008).
132. F. Inci, O. Tokel, S. Wang, U. Gurkan, D. Kuritzkes, and U. Demirci, Nanoplasmonic biosensing platform for multiple pathogen detection. In *Solid-State Sensors, Actuators and Microsystems (TRANSDUCERS & EUROSENSORS XXVII), 2013 Transducers & Eurosensors XXVII: The 17th International Conference on*, (IEEE), pp. 2431–2434 (2013).
133. F. Inci, O. Tokel, S. Wang, U. Gurkan, S. Tasoglu, D. Kuritzkes, and U. Demirci, Nanoplasmonic quantitative detection of intact viruses from unprocessed whole blood. *Acs Nano*, **7**(6), 4733–4745 (2013).
134. H. Shafiee, E. A. Lidstone, M. Jahangir, F. Inci, E. Hanhauser, T. J. Henrich, D. R. Kuritzkes, B. T. Cunningham, and U. Demirci, Nanostructured optical photonic crystal biosensor for HIV viral load measurement. *Scientific Reports*, **4**, (2014).
135. O. Tokel, F. Inci, and U. Demirci, Advances in plasmonic technologies for point of care applications. *Chem. Rev.* **114**(11), 5728–5752 (2014).
136. J. Anker, W. Hall, O. Lyandres, N. Shah, J. Zhao, and R. Van Duyne, Biosensing with plasmonic nanosensors. *Nat. Mater.*, **7**(6), 442–453 (2008).
137. O. Andersson, A. Larsson, T. Ekblad, and B. Liedberg, Gradient hydrogel matrix for microarray and biosensor applications: an imaging SPR study. *Biomacromolecules*, **10**(1),142–148 (2009).
138. A. Aulasevich, R. Roskamp, U. Jonas, B. Menges, J. Dostálek, and W. Knoll, Optical waveguide spectroscopy for the investigation of protein-functionalized hydrogel films. *Macromol. Rapid Comm.*, **30**(9–10), 872–877 (2009).
139. U. Demirci, and F. Inci, Detection, capture and quantification of biological moieties from unprocessed bodily fluids using nanoplasmonic platform. U.S. Patent 20,150,160,218, issued June 11, 2015.
140. F. Inci, C. Filippini, M. Baday, M. O. Ozen, S. Calamak, N. G. Durmus, S.Q. Wang, E. Hanhauser, K.S. Hobbs, F. Juillard, P. P. Kuang, M. L. Vetter, M. Caroccik, H. S. Yamamoto, Y. Takagi, U. H. Yildiz, D. Akin, D. R. Wesemann, A. Singhal, P. L. Yang, M. L. Nibert, R. N. Fichorova, D.T.-Y. Lau, T.J. Henrich, K. M. Kaye, S.C. Schachter, D. R. Kuritzkes, L.M. Steinmetz, S. S. Gambhir, R.W. Davis, and U. Demirci, Multitarget, quantitative nanoplasmonic electrical field-enhanced resonating device for diagnostics. *Proceedings of the National Academy of Sciences*, (2015).
141. K. A. Willets and R. P. Van Duyne, Localized surface plasmon resonance spectroscopy and sensing. *Annu. Rev. Phys. Chem.*, **58**(1), 267–297 (2007).
142. L.-K. Chau, Y.-F. Lin, S.-F. Cheng, and T.-J. Lin, Fiber-optic chemical and biochemical probes based on localized surface plasmon resonance. *Sensor. Actuat. B-Chem.*, **113**(1), 100–105 (2006).
143. W.-C. Law, K.-T. Yong, A. Baev, and P.N. Prasad, Sensitivity improved surface plasmon resonance biosensor for cancer biomarker detection based on plasmonic enhancement. *Acs Nano*, **5**(6), 4858–4864 (2011).

144. T. Endo, K. Kerman, N. Nagatani, Y. Takamura, and E. Tamiya, Label-free detection of peptide nucleic acid–DNA hybridization using localized surface plasmon resonance based optical biosensor. *Anal. Chem.*, **77**(21), 6976–6984 (2005).
145. T. Nagatsuka, H. Uzawa, K. Sato, S. Kondo, M. Izumi, K. Yokoyama, I. Ohsawa, Y. Seto, P. Neri, H. Mori, Y. Nishida, M. Saito, and E. Tamiya, Localized surface plasmon resonance detection of biological toxins using cell surface oligosaccharides on glyco chips. *ACS Appl. Mater. Inter.*, **5**(10), 4173–4180 (2013).
146. T. Endo, R. Ikeda, Y. Yanagida, and T. Hatsuzawa, Stimuli-responsive hydrogel-silver nanoparticles composite for development of localized surface plasmon resonance-based optical biosensor. *Anal. Chim. Acta*, **611**(2), 205–211 (2008).
147. I. Tokarev, I. Tokareva, V. Gopishetty, E. Katz, and S. Minko, Specific biochemical-to-optical signal transduction by responsive thin hydrogel films loaded with noble metal nanoparticles. *Adv. Mater.*, **22**(12),1412–1416 (2010).
148. A. Marshall, J. Blyth, C. Davidson, and C. Lowe, pH-sensitive holographic sensors. *Anal. Chem.*, **75**(17), 4423–4431 (2003).
149. S. Viswanathan, T. N. Narayanan, K. Aran, K. D. Fink, J. Paredes, P. M. Ajayan, S. Filipek *et al.*, Graphene–protein field effect biosensors: Glucose sensing. *Materials Today* (2015).
150. H. Shafiee, M. Jahangir, F. Inci, S. Wang, R. Willenbrecht, F. F. Giguel, A. Tsibris, D. R. Kuritzkes, and U. Demirci, Acute on — chip hiv detection through label — Free electrical sensing of viral nano — Lysate. *Small*, **9**(15), 2553–2563 (2013).
151. J. Wang, C. Wu, N. Hu, J. Zhou, L. Du, and P. Wang, Microfabricated electrochemical cell-based biosensors for analysis of living cells *in vitro*. *Biosensors*, **2**(2), 127–170 (2012).
152. H. Qi, M. Ghodousi, Y. Du, C. Grun, H. Bae, P. Yin, and A. Khademhosseini, DNA-directed self-assembly of shape-controlled hydrogels. *Nat. Comm.*, **4** (2013).
153. J. H. Kim, S. Y. Lim, D. H. Nam, J. Ryu, S. H. Ku, and C. B. Park, Self-assembled, photoluminescent peptide hydrogel as a versatile platform for enzyme-based optical biosensors. *Biosens. Bioelectron.*, **26**(5), 1860–1865 (2011).
154. H.-M. Chen, T.-H. Huang, and R.-M. Tsai, A biotin–hydrogel-coated quartz crystal microbalance biosensor and applications in immunoassay and peptide-displaying cell detection. *Anal. Biochem.*, **392**(1), 1–7 (2009).
155. B. Kim, Y. Lee, K. Lee, and W.-G. Koh, Immobilization of enzymes within hydrogel microparticles to create optical biosensors for the detection of organophosphorus compounds. *Curr. Appl. Phys.*, **9**(4, Supplement), e225–e228 (2009).
157. B.-F. Ye, Y.-J. Zhao, Y. Cheng, T.-T. Li, Z.-Y. Xie, X.-W. Zhao, and Z.-Z. Gu, Colorimetric photonic hydrogel aptasensor for the screening of heavy metal ions. *Nanoscale*, **4**(19), 5998–6003 (2012).
158. J. Yan, V. A. Pedrosa, A. L. Simonian, and A. Revzin, Immobilizing enzymes onto electrode arrays by hydrogel photolithography to fabricate multi-analyte electrochemical biosensors. *ACS Appl. Mater. Inter.*, **2**(3), 748–755 (2010).
159. Jr. N. F. Sheppard, M. J. Lesho, P. McNally, and A. Shaun Francomacaro, Microfabricated conductimetric pH sensor. *Sensor. Actuat. B-Chem.*, **28**(2), 95–102 (1995).

160. D. Du, J. Cai, H. Ju, F. Yan, J. Chen, X. Jiang, and H. Chen, Construction of a biomimetic zwitterionic interface for monitoring cell proliferation and apoptosis. *Langmuir*, **21**(18), 8394–8399 (2005).
161. L. Ding, D. Du, J. Wu, and H. Ju, A disposable impedance sensor for electrochemical study and monitoring of adhesion and proliferation of K562 leukaemia cells. *Electrochem. Comm.*, **9**(5), 953–958 (2007).
162. Q. Liu and P. Wang, *Cell-Based Biosensors: Principles and Applications*, London: Artech House (2009).
163. T. Curtis, R. M. Z. G. Naal, C. Batt, J. Tabb, and D. Holowka, Development of a mast cell-based biosensor. *Biosens. Bioelectron.*, **23**(7), 1024–1031 (2008).
164. M.-Y. Lee, R. A. Kumar, S. M. Sukumaran, M. G. Hogg, D. S. Clark, and J. S. Dordick, Three-dimensional cellular microarray for high-throughput toxicology assays. *Proceedings of the National Academy of Sciences*, **105**(1), 59–63 (2008).
165. L. Ding, T. L. Saunders, G. Enikolopov, and S. J. Morrison, Endothelial and perivascular cells maintain haematopoietic stem cells. *Nature*, **481**(7382), 457–462 (2012).
166. J. V. Chakkalakal, K. M. Jones, M. A. Basson, and A. S. Brack, The aged niche disrupts muscle stem cell quiescence. *Nature*, **490**(7420), 355–360 (2012).
167. J. Wagers Amy, The stem cell niche in regenerative medicine. *Cell Stem Cell*, **10**(4), 362–369 (2012).
168. D. Gao, J. J. Liu, H. B. Wei, H. F. Li, G. S. Guo, and J. M. Lin, A microfluidic approach for anticancer drug analysis based on hydrogel encapsulated tumor cells. *Anal. Chim. Acta*, **665**(1), 7–14 (2010).
169. L. Wu, P. Chen, Y. Dong, X. Feng, and B. -F. Liu, Encapsulation of single cells on a microfluidic device integrating droplet generation with fluorescence-activated droplet sorting. *Biomed. Microdevices*, **15**(3), 1–8 (2013).
170. P. Chen, X. J. Feng, W. Du, and B. F. Liu, Microfluidic chips for cell sorting. *Front. Biosci.-Landmrk*, **13**, 2464–2483 (2008).
171. P. Chen, X. J. Feng, R. Hu, J. Sun, W. Du, and B. F. Liu, Hydrodynamic gating valve for microfluidic fluorescence-activated cell sorting. *Anal. Chim. Acta*, **663**(1),1–6 (2010).
172. G. Xu, Y. Wu, R. Li, P. Wang, W. Yan, and X. Zheng, Cell-based biosensors: towards the development of cellular monitoring. *Chin. Sci. Bull.* **47**(22), 1849–1856 (2002).

Index